I0065814

Electric Power Engineering: Design, Development and Applications

Electric Power Engineering: Design, Development and Applications

Editor: Helena Walker

NY RESEARCH
P R E S S

New York

Published by NY Research Press
118-35 Queens Blvd., Suite 400,
Forest Hills, NY 11375, USA
www.nyresearchpress.com

Electric Power Engineering: Design, Development and Applications
Edited by Helena Walker

© 2017 NY Research Press

International Standard Book Number: 978-1-63238-536-9 (Hardback)

This book contains information obtained from authentic and highly regarded sources. Copyright for all individual chapters remain with the respective authors as indicated. All chapters are published with permission under the Creative Commons Attribution License or equivalent. A wide variety of references are listed. Permission and sources are indicated; for detailed attributions, please refer to the permissions page and list of contributors. Reasonable efforts have been made to publish reliable data and information, but the authors, editors and publisher cannot assume any responsibility for the validity of all materials or the consequences of their use.

The publisher's policy is to use permanent paper from mills that operate a sustainable forestry policy. Furthermore, the publisher ensures that the text paper and cover boards used have met acceptable environmental accreditation standards.

Trademark Notice: Registered trademark of products or corporate names are used only for explanation and identification without intent to infringe.

Cataloging-in-Publication Data

Electric power engineering : design, development and applications / edited by Helena Walker.
    p. cm.
Includes bibliographical references and index.
ISBN 978-1-63238-536-9
1. Electric power. 2. Electrical engineering. 3. Electric power systems. 4. Electric power production.
5. Renewable energy sources. I. Walker, Helena.
TK1001 .E44 2017
621.042--dc23

Printed in the United States of America.

Contents

Preface

Electrical power engineering is defined as the design and manufacture of systems that aid in electric power distribution and transmission. This book on electric power engineering deals with topics related to energy efficient technologies and the manufacture of energy grids as well as storage technologies. This book discusses innovative designs, development techniques and the diverse applications of electric power engineering. The various sub-fields of electric power engineering along with technological progress that have future implications are glanced at. This text is a compilation of chapters that discuss the most vital concepts and emerging trends in this field. With state-of-the-art inputs by acclaimed experts of this field, this book targets students and professionals. A number of latest researches have been included to keep the readers up-to-date with the global concepts in this area of study.

The information shared in this book is based on empirical researches made by veterans in this field of study. The elaborative information provided in this book will help the readers further their scope of knowledge leading to advancements in this field.

Finally, I would like to thank my fellow researchers who gave constructive feedback and my family members who supported me at every step of my research.

Editor

Rotating Machine Based DG Islanding Detection Analysis Using Wavelet Transform

Lucas Ongondo Mogaka[1], D. K. Murage[2], Michael Juma Saulo[1]

[1]Electrical and Electronics Department, Technical University of Mombasa, Mombasa, Kenya
[2]Electrical and Electronics Department, JKUAT, Nairobi, Kenya

Email address:
mogaka.Lucas@gmail.com (L. O. Mogaka), dkmurage25@yahoo.com (D. K. Murage), michaelsaulo@yahoo.com (M. J. Saulo)

Abstract: The increased use of distributed generation in the power system due to increased load demand has brought about many benefits to the power grids. This is due to the concerns about whether the technology in use currently in power generation and distribution, is sufficient to cover the future increasing demand with the limited supply. In response to this problem of increased load demand, efforts have been made to decentralize this infrastructure through the use of distributed generators. The benefits of using distributed generation include; improved reliability and increased efficiency in power supply, avoidance of transmission and distribution capacity upgrades, improved power quality and reduced line losses, minimize peak load demand, reduce voltage flicker, eliminate the need of having high spinning reserve among others. Despite these advantages, un-intentional islanding remains a big challenge and has to be addressed in integration of Distributed Generation to the power system. Unlike inverter based distributed generators, rotating machine based generators with fast response governors and AVRs are highly capable of sustaining an island. Therefore, anti-islanding protection for these generators is a more challenging problem in comparison with the inverter-based DG. This paper analyses the use of wavelet transform in islanding detection for rotating based distributed generators.

Keywords: Distributed Generation, Islanding Detection, Rotating Machines

1. Introduction

Electric rotating machinery can be defined as any form of apparatus which has a rotating member and generates, converts, transforms, or modifies electric power, such as a motor, generator, or synchronous generator. Although there are many variations, the two basic rotating machine types are synchronous and induction machines.

The current trend in the increasing use of Distributed Generation (DG) is due to energy exhaustion, efforts to improve power quality, reduce the need of having high spinning reserves in the system, reduction of the voltage flickers in the system and recent environmental issues. This practice enables the collection of electrical energy from a variety of sources and leads to decreased environmental impacts and improved security of supply. They are typically in the range of 1 kW to 10,000 kW and include wind farms, micro hydro turbines, photovoltaic (PV) system and other small generators which are supplied with biomass or geothermal fuel [1].

Among the many advantages of DG integration include: improved system reliability in the power supply, reduction of system peak loads, spinning reserve size reduction, increased efficiency, avoidance of transmission capacity upgrades, improved power quality, security and reduced transmission line losses and environmental benefits (excluding diesel reciprocating engines often used as back-up distributed generators which tend to be the worst performers in terms of greenhouse gas emissions [2]).

Despite the above mentioned merits of incorporating DGs in the distribution system, it has major drawback of unintentional islanding. Islanding condition occurs when the DG continues to power a part of the grid system even after the connection to the rest of the system has been lost, either intentionally or unintentionally. The unintentional islanding mode of operation is not desirable because of a number of reasons. For instance; it poses a threat to the line workers' safety, the islanded system may not be properly grounded resulting in high voltage in the other phases when an earth fault occurs, possibility of creating an ungrounded system depending on the transformer connections and most importantly, the distributed generators may not be able to maintain the voltage and frequency within desired limits in the distribution system when it is islanded.

The rest of the paper is organized as follows; section II discusses the general islanding detection methods, then the various recent islanding detection methods are covered in section III, the islanding detection methods assessment tools for rotating machine based generators are explained in section IV, Methodology for this study in section V, section VI discusses the results and analysis and finally the conclusion in section VII.

2. General Islanding Detection Methods

As per IEEE standard 1547-2003, the distributed generators must sense the unplanned power grid and trip it within two seconds, failure to which may lead to several problems in terms of power quality, safety and operational problems [3].

Most of the commonly used islanding detection methods are suitable for all generators and hence the synchronous and induction machines. The general islanding detection techniques can be categorized as shown in the following diagram.

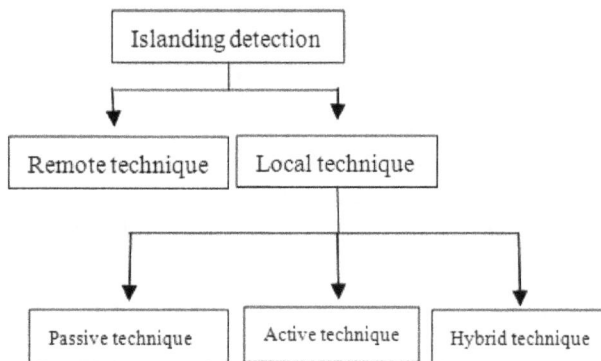

Figure 1. Islanding detection techniques [4].

2.1. Remote Methods

This is a method of islanding detection for the DG through communication by the transmitters and receivers located in the power utility and the DG sides respectively. This is achieved though continuous signaling between the two ends. In case there is disruption in the grid that hinders this communication, then this leads to the conclusion that islanding has occurred. Examples of remote islanding detection methods are the power line communication (PLC), supervisory control and data acquisition (SCADA), transfer-trip among others. These methods do not have a non-detection zone (NDZ). In terms of reliability, remote methods are far much better compared with local methods only that it is expensive and thus it is uneconomical to implement especially in small networks.

2.2. Passive Methods

Passive islanding detection techniques are preferred in island detection especially when the mismatch between the

generated power and the size of the load is very large. However, when the mismatch is very small, it is difficult to detect the islanding state because the variations in voltage or frequency at the point of common coupling (PCC) are also very small [5]. The weaknesses of some of the passive islanding detection methods are highlighted in the table 1.

Table 1. Passive islanding detection [6].

Method	Implementation speed	Weakness
UFP/OFP UVP/OVP	Easy but reaction time unpredictable and variable	Large non-detected zones (NDZs)
Phase jump detection (PJD)	Difficult in implementation and hard to choose threshold	Fails to detect islanding when DG power generation matches the power demand of local load
Total harmonic distortion (THD)	Easy but hard to choose threshold	Fails to detect island in case of low distortion of voltage and current output of inverter or high quality load
Voltage Unbalance		Not applicable to single phase system

2.3. Active Methods

On the other hand, as compared to passive methods, the active methods have smaller Non-Detection Zones (NDZ). However they compromise the power quality of the system by injecting small signals at certain frequencies to the system. Some of these methods, their properties and drawbacks are shown in table 2.

Some of the most common active methods used for islanding detection for synchronous DGs include; reactive power compensation, load fluctuation, impedance measurement, reactive power fluctuation, and QC-mode frequency shift method.

Table 2. Active islanding detection methods [6].

Method	Implementation and speed	Weakness
Impedance measurement	Easy and fast	
Slip-mode frequency shift (SMS)	Medium and slow	Ineffective under certain load eg RLC resonant load
Active frequency Drift (AFD)	Easy and medium	
Sandia frequency shift (SFS)	Difficult and relatively fast	Problem in power quality, system stability
Sandia voltage shift(SVS)	Medium and fast	Increase harmonic distortion

2.4. Hybrid Methods

In short, active and passive techniques have their strong and weak points. Thus these methods are at times merged to benefit from their strong points. This technique is called Hybrid islanding detection [7]. Again, there is no islanding detection scheme currently that can serve all situations in distributed system. Therefore, the method is normally selected according to the nature of the distributed generator [8].

5. Recent Islanding Detection Methods

Since the islanding condition should be detectected as fast as possible as it is stipulated in the set international standards, researchers are continuously looking for better methods of correctly detecting this condition. To achieve this objective, signal processing tools come in handy in extracting the features from measured signals. Then artificial intelligent tools are used to classify the extracted signals to either islanded or non-islanded condition. The general steps usually followed in determining islanding state classification are shown in figure 2 below.

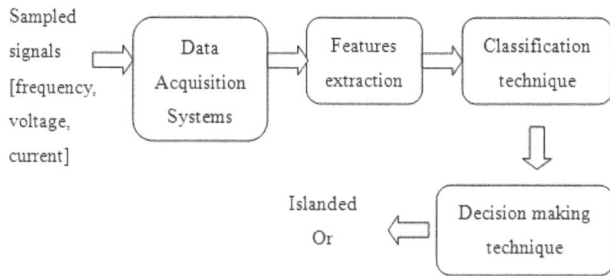

Figure 2. General islanding detection steps.

5.1. The Wavelet Transform

A wavelet can be simply defined as a small wave. A wavelet transform (WT) convert a signal into a series of wavelets that are used for analyzing waveforms that are bound in both frequency and time. Alternatively, the WT can be defined as a collection of functions that are used for analyzing non-stationary signals in both frequency and time domains in MATLAB platform. It actually decomposes signals being analyzed layer by layer into different frequency bands for analysis purposes.

It is actually a new mathematical tool developed for analyzing non-stationary and fast changing wide-band signals for instance in islanding detection in synchronous generators. The wavelet transform tool has a number of advantages. One of the main advantage of wavelet transform is that it needs not to assume the stationery or periodicity of a signal as it is able to simultaneously distinguish both time and frequency signal information due to its multi-resolution characteristic. Hence, it becomes useful in analyzing discontinuous and time varying signals especially in islanding detection.

It has three useful properties which make it applicable in engineering applications and most importantly in islanding detection. First, it has ability to reconstruct back the signal from its wavelet transform. This is achieved due to WT ability of the resolution of identity, the ability to conserve energy in the time-scale space and the wavelet admissible condition. Secondly the WT is a local operator in both time and frequency domains. Hence, the regularity condition is usually imposed on the wavelets. Lastly, the WT has a property related to a multi-resolution signal analysis. It has the ability to analyze both high and low frequency signals. The high frequency signal analysis is done using narrow windows and the low frequency analysis is done using wide windows [9].

Wavelet analysis can be categorized into two main techniques. That is; Continuous wavelet transform (CWT) and discrete wavelet transform (DWT).

5.2. Continuous Wavelet Transform

The CWT is defined as the sum over all time of the signal multiplied by scaled, shifted versions of the wavelet function. The time-scale information provided by wavelet makes it easy to extract signal features that change with time. Mathematically CWT can be expressed as follows;

$$CWT_X^\varphi(a,b) = \frac{1}{\sqrt{a}} \int_{-\infty}^{\infty} x(t).\varphi * \frac{t-b}{a} d \qquad (1)$$

Where: a is the scale parameter of the wavelet, b is the translation or position parameter of the wavelet, x (t) is the analyzed signal, and φ is the mother wavelet and is defined by:

$$\varphi_{a,b}(t) = \frac{1}{\sqrt{a}} \varphi(\frac{t-b}{a}) \qquad (2)$$

The CWT is actually continuous in terms of the shift b during calculation and its operation; the wavelet that is used for analyzing the signal is shifted smoothly over the entire domain of the analyzed function. It is actually the measurement of the similarity of the wavelet to the original signal through calculating the coefficient. Thus practically continuous wavelet transform may give redundant information especially when the measured similarity coefficient is large as the original signal and the wavelet will be similar. Hence for the sake of computation, there is need to discretize the signal.

5.3. Discrete Wavelet Transform

The discrete wavelet transform (DWT) is found by filtering the low pass and high pass signals successfully. The voltage and current transients of a power system do have unique characteristics that signify the cause of transient occurrence. So there should be a process to extract these features to speed up response in classifying. To this end, wavelet transform seems to be suitable [10].

The discrete wavelet transform is actually the continuous wavelets with the discrete scale and translation factors. Here the wavelet transforms are evaluated at discrete scales and translations. This means that when time localization of the signal is required, the discrete wavelet transform is the one appropriate. This is especially in islanding detection. The DWT function can be defined mathematically as shown in the equation 3 below:

$$\varphi_{a,b}(t) = |a|^{-\frac{1}{2}}\varphi\left(\frac{t-b}{a}\right) \quad a,b \in R, a \neq 0 \qquad (3)$$

If (a, b) take discrete value in R^2, we get DWT. A popular approach to select (a, b) is

$$a = \frac{1}{a_0^m}, a_0 = 2,$$

$$a = a_0^{-m} = < 1, \frac{1}{2}, \frac{1}{4}, \frac{1}{8}, \dots >, m: integer$$

$$b = \frac{nb_o}{a_0^m}, a_0 = 2, b_o = 1$$

$$b = \frac{n}{2^m}, \text{ n, m: integer}$$

Then

$$\varphi_{a,b}(t) = |a|^{-\frac{1}{2}}\varphi\left(\frac{t-b}{a}\right) = 2^{\frac{m}{2}}\left(\frac{t-\frac{n}{2^m}}{\frac{1}{2^m}}\right) = 2^{\frac{m}{2}}\varphi(2^m t - n) \quad (4)$$

Generally, it is a requirement that the response time of the islanding detection method should be shorter hence a lower decomposition level should be selected.

The following are some of the merits that make DWT applicable in islanding detection in synchronous and induction generators.

a) It gives enough information for analysis purposes
b) It sufficiently lowers the computation time
c) Its implementation is easy
d) Analyzes signals in different resolutions at different frequency bands
e) Decompose the signal into a coarse approximation and detail information

The ability of DWT to approximate the signals in various scales corresponding to different resolutions. This is known as multi resolution decomposition. The process can be repeated iteratively by breaking signals into many components of lower resolutions. This produces what is called wavelet decomposition tree. The figure 3 below shows an example of a wavelet decomposition tree.[11]

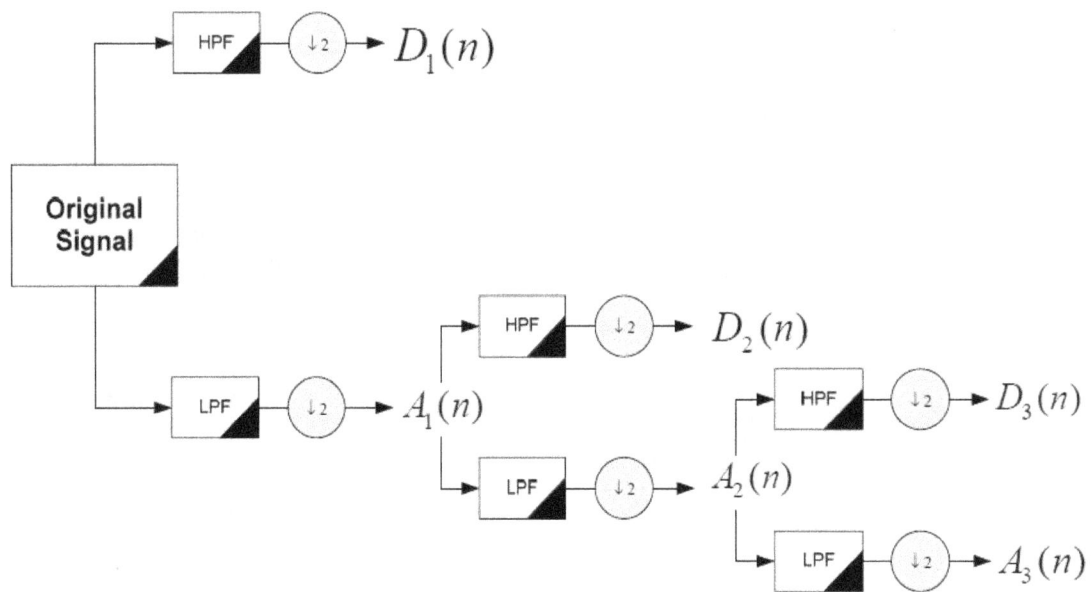

Figure 3. *Wavelet decomposition tree.*

5.4. Fuzzy Logic Controller

A fuzzy logic controller is a control algorithm based on several linguistic control rules and it is used to analyze continuous signals. The fuzzy rule base has the ability to handle more uncertainties in the signal being analyzed that fall along the slope of the fuzzy trapezoidal membership function unlike the crisp classifiers like decision tree which have sharp boundaries, and large data base. Thus, the superior approximation capabilities of the fuzzy systems over crisp classifiers help to develop algorithms that meet the real time application with wide range of uncertainties. Hence the fuzzy logic controller can easily and accurately be used in islanding detection for synchronous and induction generators.

Some of the recent applications of fuzzy logic in islanding detection are as follows; in [12], FL was introduced from the transformation of DT, where the combination of fuzzy membership functions (MFs) and the rule base were used to develop the fuzzy rule base. This technique was easy to implement for online islanding detection and could handle uncertainties such as noise. In [13], however, the band pass

filter was used to replace the function of DWT and still worked out pretty well.

6. Islanding Detection Method Assessment Tools for Rotating Machine Based Generators

6.1. Non-Detection Zones

The effectiveness of any islanding detection method is usually based on the evaluation of the Non-Detection Zone (NDZ) index and the time for islanding detection. Non-detection zones are defined as a loading condition for which an islanding detection method would fail to operate in a timely manner. It is normally evaluated by the use of active and reactive power mismatch space. The non-detected zones for synchronous distributed generators (SDGs) are affected by the following factors among others:

a. Load type
b. generator inertia

c. generator excitation control mode and

d. relay settings

The following equations (5) and (6) are used to illustrate the definition of the NDZ [14].

$$(\frac{v}{v_{max}})^2 - 1 \leq \frac{\Delta p}{p} \leq (\frac{v}{v_{min}})^2 - 1 \qquad (5)$$

$$Q_f[1 - (\frac{f}{f_{min}})^2] \leq \frac{\Delta Q}{p} \leq Q_f[1 - (\frac{f}{f_{max}})^2] \qquad (6)$$

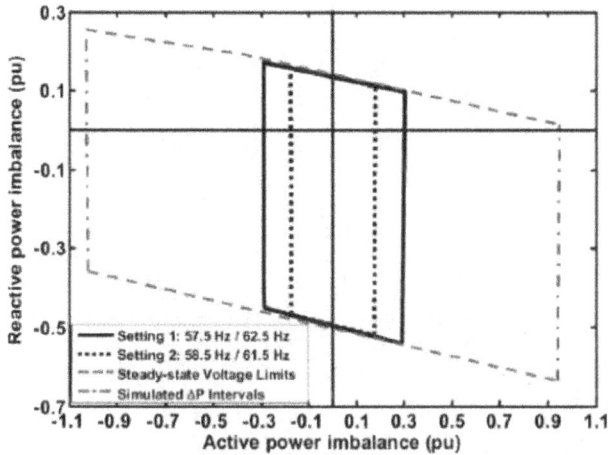

Figure 4. NDZs for different frequency relay settings [15].

6.2. Performance Curves

Performance curves represent the relationship between the islanding detection times versus the active power mismatch. This graphical tool is especially useful for synchronous DGs. Power mismatches lower than the critical power imbalance make up, a NDZ, as indicated in Figure 5 below.

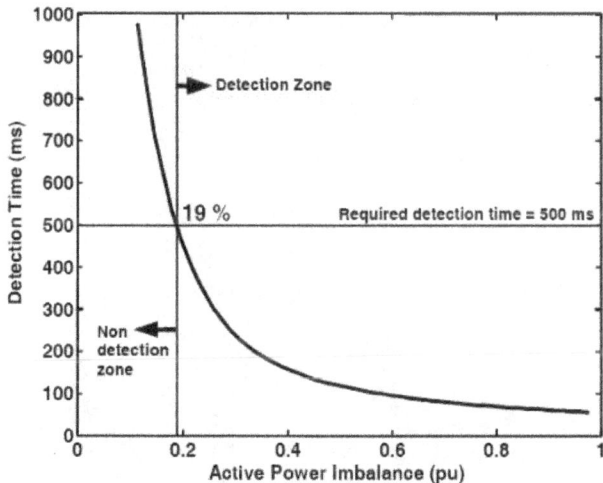

Figure 5. Performance curve of frequency-based relays [15].

In figure 5 above, at the x-axis the active power imbalance level of the islanded system of the generator is represented. Then the island detection time is represented in the y-axis. To obtain this curve the islanding occurrence simulations have to be repeated for some steps. Then, for each active power imbalance, the detection time is determined by dynamic

simulation and then the performance curve is plotted [16].

7. Methodology

First, the system that was used in this study was modelled in SIMULINK/MATLAB as shown in the figure above.

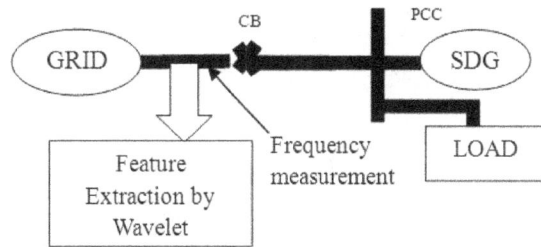

Figure 6. System used for the analysis.

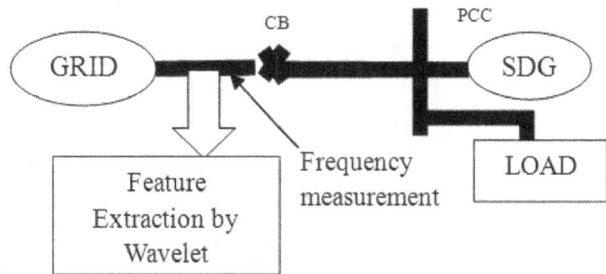

Figure 6. System used for the analysis.

Then the position of measuring the signal frequency was changed to immediately after the utility circuit breaker from the usual point of common coupling. This is meant to make frequency measurement fast and ensure the island detection time is even less than what is stipulated by IEEE 1547-2003 standard. During simulation, the grid was disconnected on the 0.2s by switching the three phase breaker off. Then the features of the measured frequency was extracted by discrete wavelet transform (DWT) found in MATLAB toolbox.

8. Results and Analysis

First, the combined three phase voltages and currents were measured and are as shown in the figure 7 below. From the figure, it clearly shows that there is remarkable changes on the side of the measured currents as compared with the measured voltages. Thus current measurement was mainly used in the rest of the analysis.

Then, the phase voltages and currents were measured separately for the sake of clarity and are as shown below in figure 8 and 9.

Then the excitation voltages and currents changes during the disturbance were monitored and are as shown in figure 10 below. It was observed that the excitation currents respond immediately as the fault occurs while the voltage changes gradually.

The current waveform of one phase was discretized during the same period and is as shown in figure 11 below.

And finally, the DWT decomposition of the same signal

was done up to level 5 and is as shown in figure 12 below.

From the figure above for the DWT, islanding of the micro grid was noted at 2.2564s.

It can be pointed out from the transform above, that the decomposition of the signal to level 2 (D2) is enough in determining the islanding condition in this case. Thus the islanding detection time is seen to be smaller when a lower decomposition level is chosen.

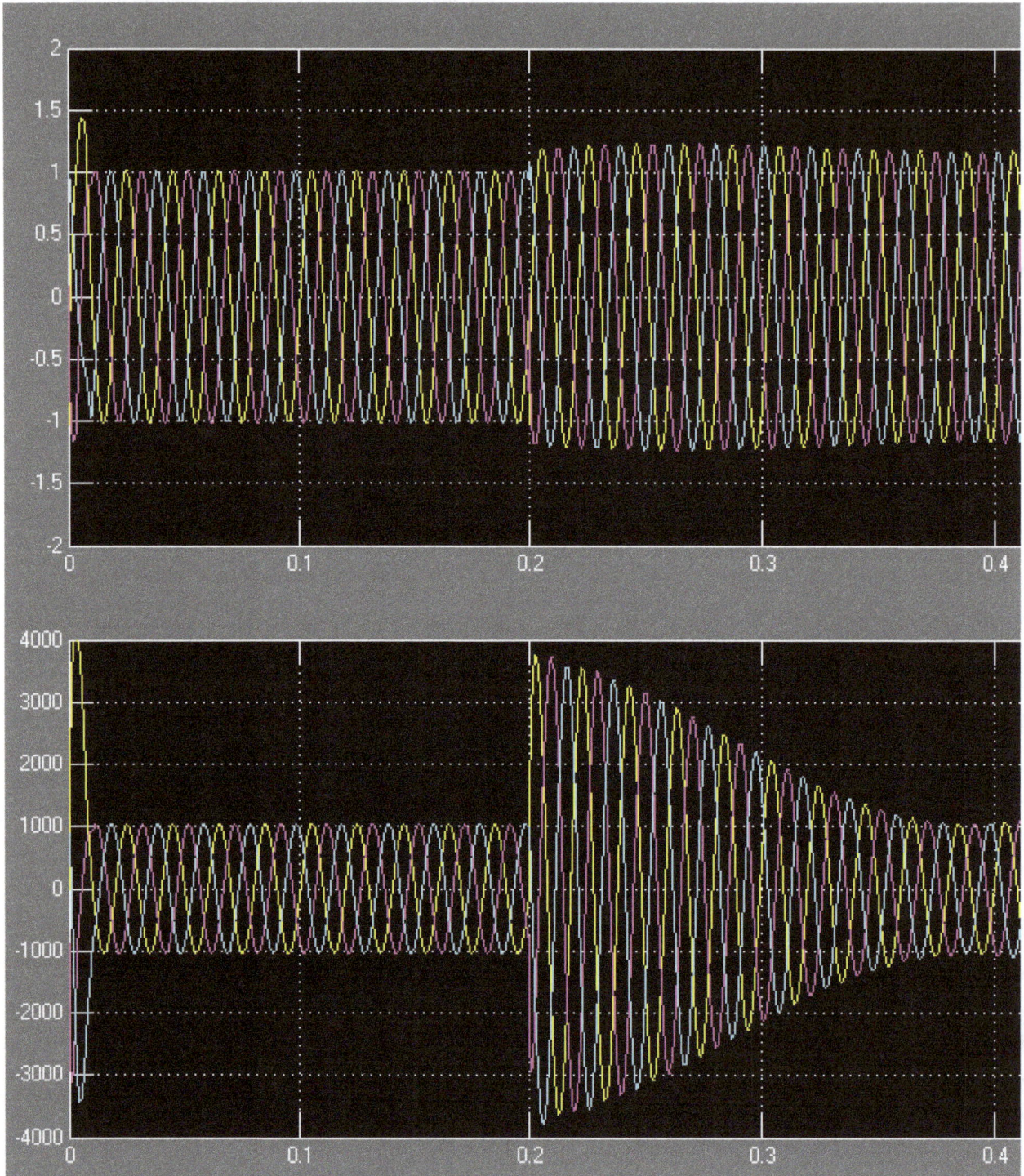

Figure 7. *Combined three phase voltage and current measurement.*

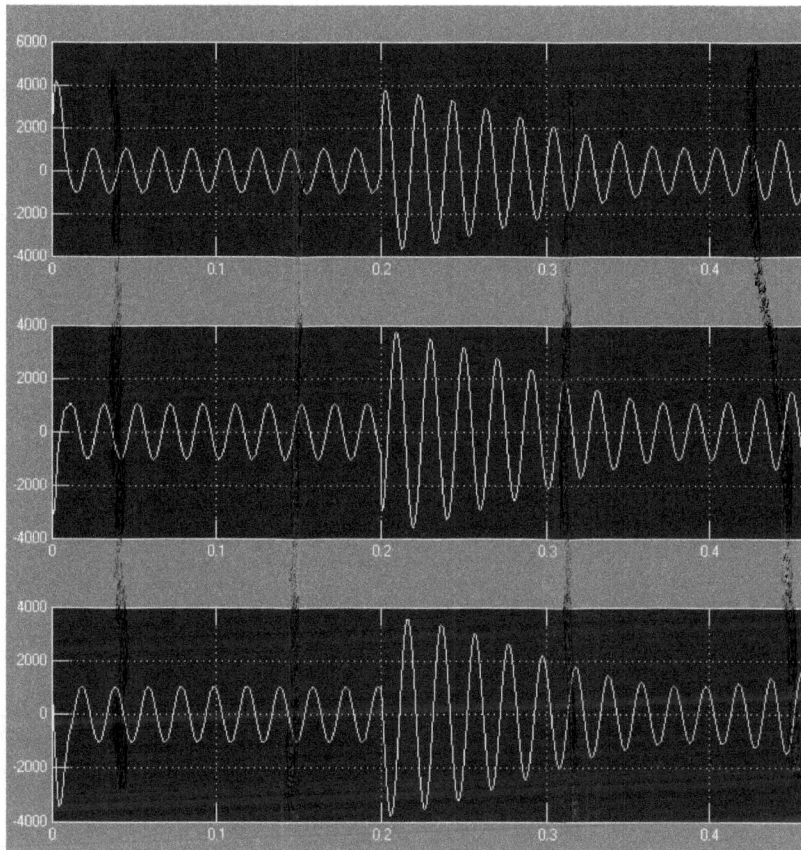

Figure 8. Three phase current measurement.

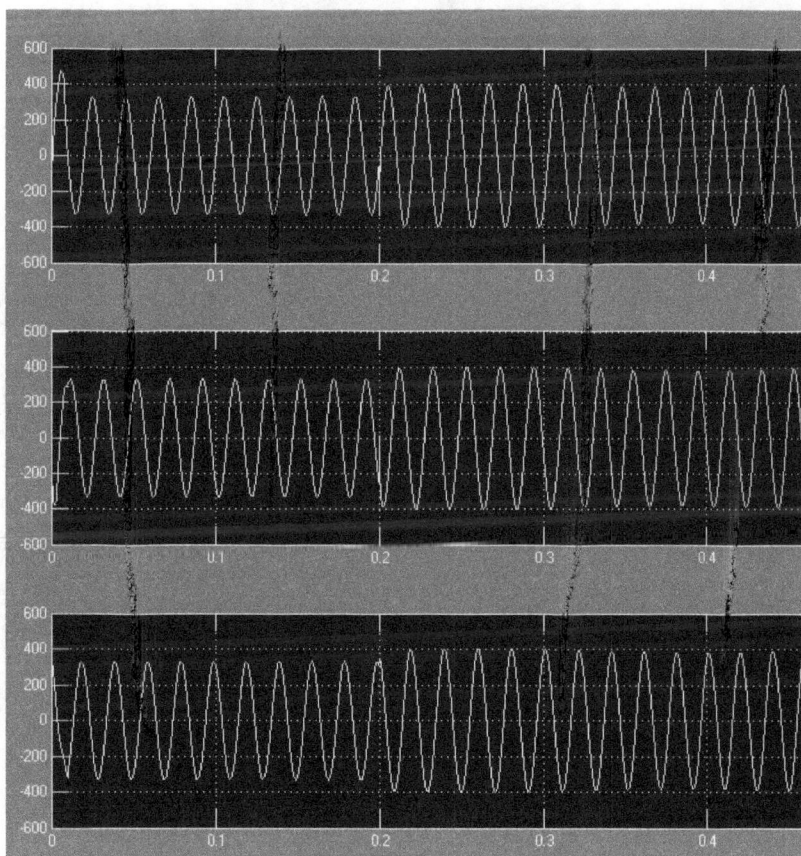

Figure 9. Voltage measurement.

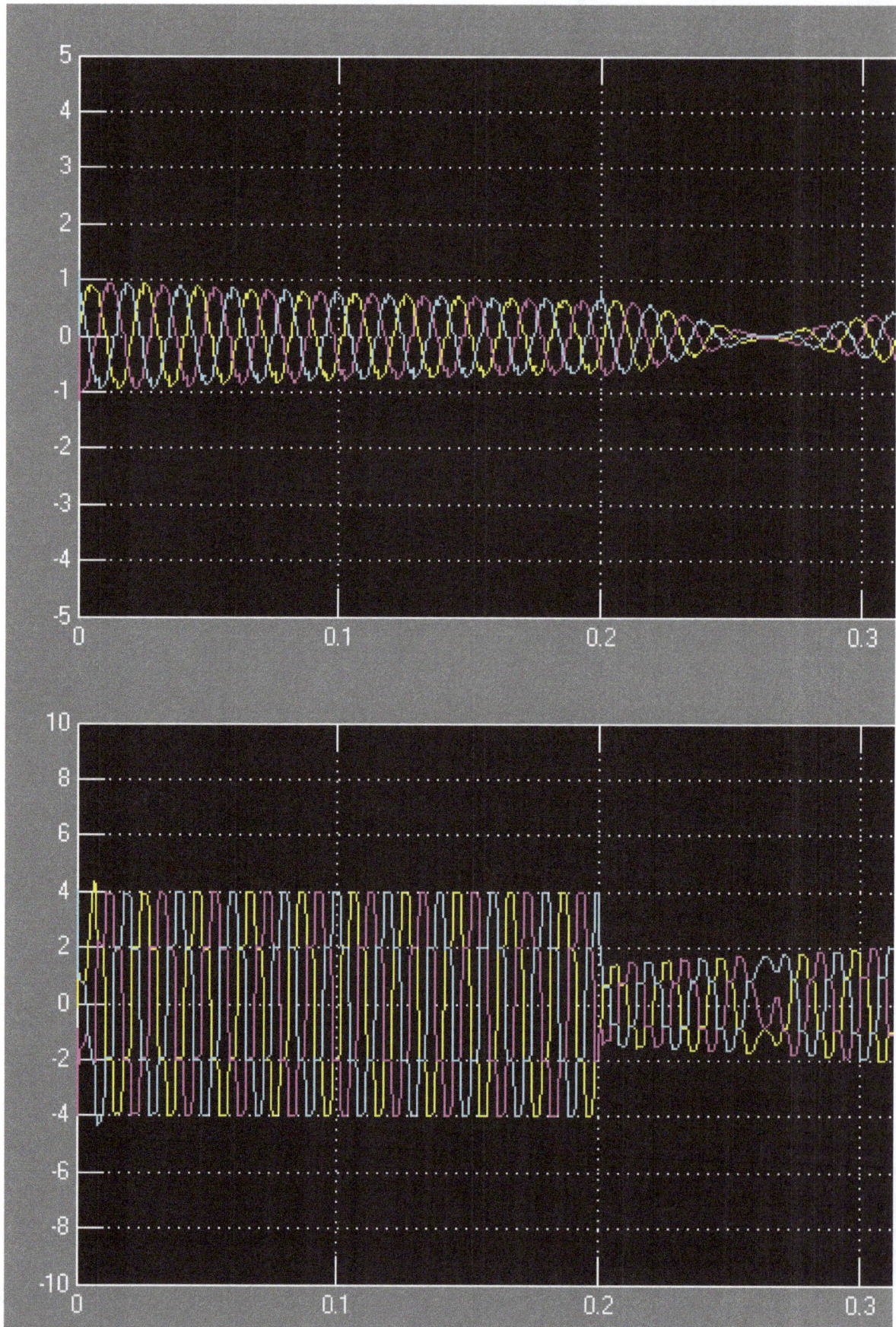

Figure 10. Combined excitation voltage and current.

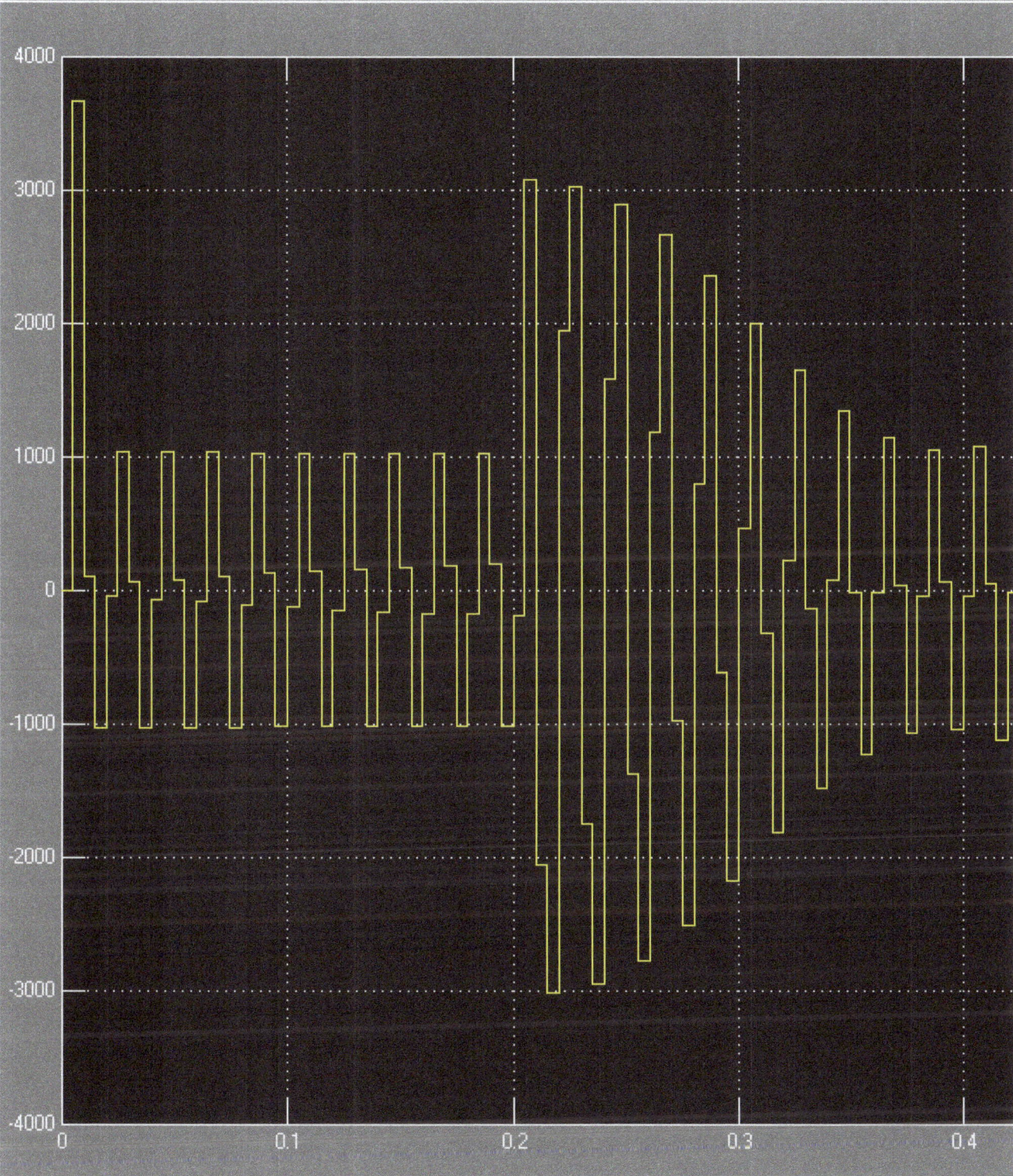

Figure 11. Discrete current signal (RED PHASE).

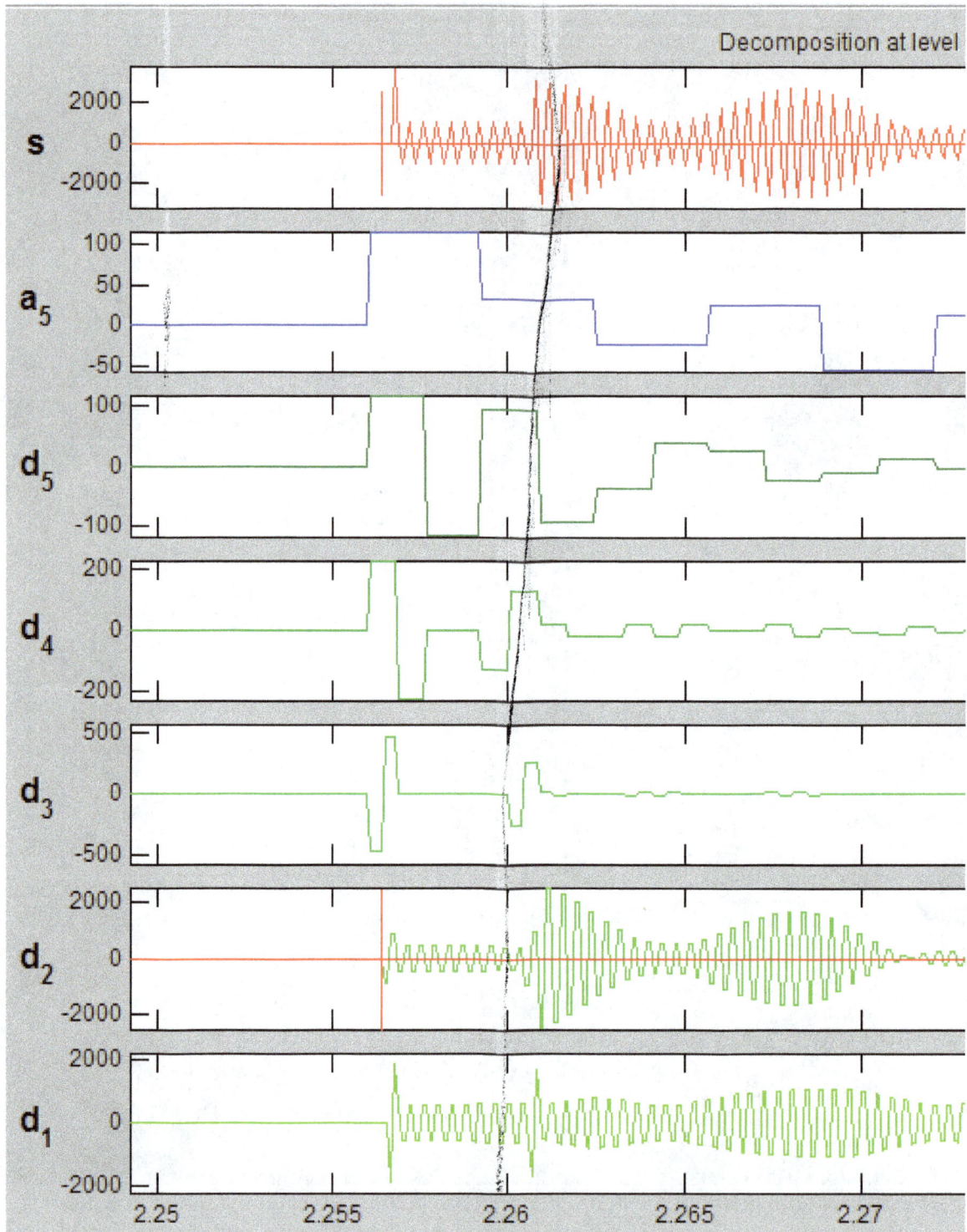

Figure 12. DWT for the analyzed signal.

9. Conclusion

Unlike inverter based distributed generators, rotating machine based DGs with fast response governors and AVRs are highly capable of sustaining an island. Therefore, islanding detection and protection for these generators is a more challenging problem in comparison with the inverter-based DG. This paper has analysed the various methods currently used in islanding detection especially for rotating machine based distributed generators especially by use of DWT. It can be seen that there is a quick response to islanding detection due to the changing of the point of measuring of the current signals from the usual PCC to just after the utility circuit breaker. Hence islanding detection time is not affected here by the circuit breaker operation time.

Though the kind of detection can be determined by human

experts from the diagram, it is desirable to process the obtained wavelet transform results further with artificial intelligence techniques Such as fuzzy logic control [17].

From this analysis, more has to be done especially in islanding detection for synchronous and induction generators so as to increase the power consumer confidence and reliability on the system.

Acknowledgement

The authors would like to express the greatest gratitude to the Technical University of Mombasa for the continued support from time to time when required.

References

[1] A. Khamis, "A review of islanding detection techniques for renewable distributed generation systems.," university kebangsaan, malaysia.

[2] J. Martin, "Distributed vs Centralised electricity generation: Are we witnessing a change of Paradigm?," may 2009.

[3] P. mahati, "Review on islanding operation of distribution system with distributed generation," pp. 1-8, 2011.

[4] T. Pujhari, "Islanding Detection In Distributed Generation," 2009.

[5] M. Ashour, "MATLAB/SIMULINK implementation and simulation of islanding detection using passive methods," in *IEEE GC conference and exhibition*, Dola, 2013.

[6] A. Khamis, "A review of islanding detection techniques for renewable distributed generation systems.," malaysia.

[7] M. Maher and G. M. A. Saad, "'Robust hybrid anti-islanding method for inverter- based distributed generation'".

[8] A. Etxegarai, I. Zamora, P. Eguia and L. Valverde, "A. Etxegarai, I. Zamora, P. Eguia, L. Valverde. Islanding detection of synchronous distributed generators.," in *International Conference on Renewable Energies and Power Quality (ICREPQ'12)*, Santiago de Compostela (Spain), 2012.

[9] Y. Sheng, Wavelet Transform."The Transforms and Applications Handbook, second edition ed., A. D., Ed., Poularikas Boca Raton: CRC Press LLC, 2000.

[10] H. MEHRDAD, S. GHODRATOLLAH and R. MORTEZA, "An intelligent-based islanding detection method using DWT and ANN.".

[11] O. Mohammed and N. Abed, "Wavelet Transform based islanding characterization method for distributed generation," in *Fourth LACCEI international Latin American and Carribean Conference for Engineering and Technology*, 2006.

[12] S. Samanta, k. El-arroudi, G. Joós and k. I., "A fuzzy rule-based approach for islanding detection in distributed generation," *IEEE Transaction on Power Delivery*, pp. 1427-33, 2010.

[13] J. Pham, N. Denboer, N. Lidula, G. Member, N. Perera and A. Rajapakse, "Hardware implementation of an islanding detection approach based on current and voltage transients.," *Electrical power and energy conference (EPEC)*, pp. 152-157, 2011.

[14] V. Mohammadreza, J. S. Mohammad and B. G. Gevork, "Islanding Detection in Multiple DG Microgird by utility side current measurement," in *Internationational Transactions on Electrical Energy Systems*, 2014.

[15] A. Etxegarai, I. Zamora, P. Eguia and L. Valverde, "Islanding detection of synchronous distributed generators," in *international Conference on Renewable Energies and Power Quality (ICREPQ'12)*, Santiago de Compostela (Spain), 2012.

[16] W. F. W. X. a. A. M. Jose C. M. Vieira, "Performance of Frequency Relays for Distributed Generation Protection," *ieee transactions on power delivery*, vol. 21, no. 3, 2006.

[17] K. Oyedoja and O. Obiyemi, "Wavelet Transform in the detection of electrical power quality disturbances," *international journal of engineering and applied sciences*, vol. 3, no. 2, April 2013.

[18] R. A. Lidula N, "A pattern recognition approach for detecting power islands using transient signals Part I: Design and implementation.," *IEEE Transactionon Power Delivery*, pp. 3070-7, 2010.

[19] S. B. Karegar H, "Wavelet transform method for islanding detection of wind turbines.," *renewable energy*, pp. 38:94-106, 2012.

[20] H. Haroonabadi, "Islanding Detection in Micro-Grids Using Sum of Voltage and Current Wavelet Coefficients Energy," *International Journal of Energy and Power Engineering*, vol. 3, no. 5, pp. 228-236, 2014.

Limitations of Inlet Air Evaporative Cooling System for Enhancing Gas Turbine Performance in Hot and Humid Climates

Majed Alhazmy[1], Badr Habeebullah[1], Rahim Jassim[2]

[1]Mechanical Engineering, King Abdulaziz University, Jeddah, Kingdom of Saudi Arabia
[2]*Saudi Electric Services Polytechnic (SESP), Baish, Jazan Province, Kingdom of Saudi Arabia*

Email address:
bhabeeb@kau.edu.sa (B. Habeebullah), mhazmy@kau.edu.sa (M. Alhazmy), rkjassim@yahoo.com (R. Jassim)

Abstract: This paper aims to investigate the evaporative cooling limitations of compressor intake-air for improving the performance of gas turbine power plants. The limitations of the evaporative cooling capability are analyzed and formulated in terms of the characteristic dimensions involving the temperature ratio, the power gain ratio (PGR), thermal efficiency change, and humidity ratio. The effects of different pressure ratios (PRs) are examined for Saudi Arabia summer weather when the turbine inlet temperature is predetermined at1373.15 K. The results of a specific example where the air evaporative cooler drops the temperature to the wet bulb temperature are presented. These indicate that the power gain ratio enhancement depends on the ambient temperature, relative humidity, evaporative cooler effectiveness, and slightly the PR. Especially for PR =10, the PGR is enhanced by 9% at 20% of relative humidity and dropped to 3.37% at 60% of relative humidity. The daily performance of the evaporative cooling method is examined for the hot humid conditions of Jeddah, Saudi Arabia. The results show that the evaporative cooler increased both the daily power output and the thermal efficiency by 2.52% and 0.112%, respectively.

Keywords: Gas Turbine, Air-Cooling, Power Enhancement, Evaporative Cooler

1. Introduction

The electric power generation sectors in many countries face two real problems which are the continuous increase in fuel prices and the incessant growth in energy demand. In order to fulfill this demand and reduce the operation cost, improving the performance of their generation units is of necessity. Nowadays, gas turbine (GT) power plants are generation units being widely used in several countries all over the world due to their low cost, quick installation, and stability of electricity grid variations. Many of these countries obviously have a wide range of climatic conditions which negatively impact the GT performance. In hot climate countries, GT power output significantly falls more in summer due to the temperature rise. At the same time, high temperature would lead to the increase in electricity demand for air conditioning. For such challenge, several attempts have been carried out to improve the GT power output as well as thermal efficiency. According to Zadpoor [1], power

augmentation can be classified into two main categories. The first category includes inlet air cooling techniques and the second involves techniques based on the injection of compressed air, steam, or water. Furthermore, a little increment of thermal efficiency could result in a significant amount of fuel being saved and a higher level of power being generated. The simplest remedy for this increment is to reduce the temperature of the inlet air. This is a reason why cooling the compressor intake air has obtained much consideration of utilities and is also a research subject in this paper.

Several methods have been using for cooling the GT intake air in which evaporation of water is one of the simplest and oldest methods. Even then the sophisticated technology available today such as mechanical compression chiller, absorption chiller, and thermal energy storage system, evaporating cooling remains the most cost-efficient method for temperature control of the gas turbine inlet air supply [2].Evaporative cooling can be achieved by many methods. In practical, two forms called direct methods involving media type evaporative cooling and spray type evaporative cooling are typically used [3].Many literature works have intensively investigated the compressor intake air cooling effect on the GT power enhancement. Johnson[4] discussed the use of evaporative cooling technique for GT installations. The calculation procedure and installation as well as operation details were also presented in his study. Ameriet al[5] applied a fog type air cooling system where fog nozzles inject water at high pressures (above 70 bars) generating micro fine droplets with sizes between 10 and 40 microns. Performance test results showed that the power output of the units has increased by 13% of the generated power, while the efficiency improvement was less than 1%. Since operation of the nozzles is critical to the fogging system, Meheret al[6] investigated the effect of nozzles type and droplets size on the performance of GT engines. Typically, the nozzles were stainless steel-316 of diameters less than 0.18 mm diameter. Because of the limited size of droplets produced by these nozzles (greater than 4 microns), humid air cannot exceed 90 to 95% of the saturation relative humidity [7-8]. Retrofitting of evaporative air coolers usually requires large ducts as the evaporation process requires low velocities. If the air velocity is high, water carry over may affect the compressor blades. For these reasons, use of evaporative cooling is limited and works better in dry air locations. Other investigations confirm the advantages of air cooling in which the most recent study is the study of Alhazmy and Najjar [5].They examined the power output and net efficiency of GT by using direct water spray process and surface cooling coils at the inlet of the compressor. For spray cooler, the drop in air temperature is 3-15°C resulting in the increase of the power output from 1 to 7%. In case of water scarcity, they suggested the use of the condensate from the waste gases to recover partially for the spray water.

Generally, the evaporative methods for cooling the compressor intake air are diverse. Each of these methods has its advantages as well as limits. In this study, an evaporative air cooler is considered and analyzed to ascertain its limitation cooling process, the capability of boosting the power output, and the thermal efficiency enhancement of GT operating for long periods in a hot and humid climate. The performance of the cooling system is then presented in a dimensionless graph where the power gain and thermal efficiency enhancement can be effortlessly evaluated for different ambient conditions and evaporative cooler effectiveness.

2. Description of Analysis

In this study, a simple open type gas turbine cycle as shown in Fig. 1 is considered. The cycle performance can be improved by cooling the compressor intake air using a direct evaporative air-cooling system. The cooling is achieved by evaporation of water spray in an evaporative cooler installed ahead of the compressor inlet manifolds. The ambient air enters the cooler at state 0and comes in the compressor at state 1.

2.1. Gas Turbine Cycle Analysis

Consider an irreversible gas turbine cycle as shown in Fig. 2; processes 1-2 and 3-4 are irreversible while processes 2-3 and 4-1 are isobaric heat addition and rejection, respectively. Processes 1-2s and 3-4s are isentropic presenting the process in an ideal cycle.

For the isentropic processes 1-2s and 3-4s, we have

$$\frac{T_{2s}}{T_1} = \frac{T_3}{T_{4s}} = \left[\frac{P_2}{P_1}\right]^{\frac{k-1}{k}} = PR^{\frac{k-1}{k}} \qquad (1)$$

where PR is the pressure ratio and k is the specific heats ratio.

2.1.1. Turbine

If a cooling system extracts its power from the turbine output as shown in Fig. 1, the thermal efficiency of the cycle is

$$\eta_{cy} = \frac{\dot{W}_{net}}{\dot{Q}_h} = \frac{\dot{W}_t - \left(\dot{W}_{comp_{air}} + \dot{W}_{el,pump}\right)}{\dot{Q}_h} \qquad (2)$$

where $\dot{W}_{el,pump}$ is the pumping power to circulate the water inside the evaporative cooler.

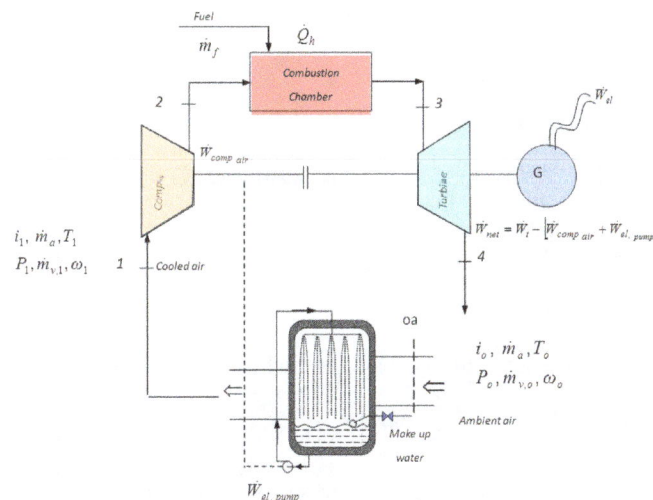

Fig. 1. *A simple open type gas turbine with a direct evaporative cooler.*

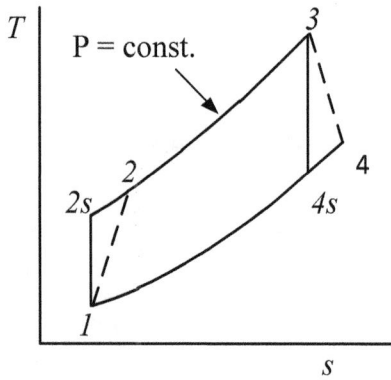

Fig. 2. *T-S diagram of an open type gas turbine cycle.*

Applying the first law of thermodynamics to the gas turbine (neglect the potential and kinetic energy terms), the power produced by the turbine is

$$\dot{W}_t = \dot{m}_t \, c_{pg} \eta_t \left(T_3 - T_{4s} \right) \tag{3}$$

where ω_1 is the humidity ratio at state 1, $f = \dot{m}_f / \dot{m}_a$ is the fuel air ratio, and $\dot{m}_t$ is the total gases mass flow rate at the turbine inlet given as

$$\dot{m}_t = \dot{m}_a + \dot{m}_v + \dot{m}_f = \dot{m}_a (1 + \omega_1 + f) \tag{4}$$

Substituting for T_{4s} and $\dot{m}_t$ from Eqs.1 and 4 into Eq. 3 yields

$$\dot{W}_t = \dot{m}_a (1 + \omega_1 + f) c_{pg} \eta_t T_3 \left(1 - \frac{1}{PR^{\frac{k-1}{k}}} \right) \tag{5}$$

The turbine isentropic efficiency can be estimated by using the practical relations recommended by Korakianitis and Wilson [9] as

$$\eta_t = 1 - \left(0.03 + \frac{PR - 1}{180} \right) \tag{6}$$

The gas turbine is almost constant volume machine at a specific rotating speed; hence, the inlet air volumetric flow rate $\dot{V}_a$ is fixed regardless of the ambient air conditions. As the air temperature rises in hot summer days, its density falls but the volumetric flow rate remains constant. Therefore, the mass flow rate reduces and consequently the power output decreases [5]. Eq.5 can be written in terms of the volumetric

flow rate at the compressor inlet state as

$$\dot{W}_t = \dot{V}_a \rho_a (1 + \omega_1 + f) c_{pg} \eta_t T_3 \left(1 - \frac{1}{PR^{\frac{k-1}{k}}} \right) \tag{7}$$

where ρ_a is the moist air density which is a function of the temperature T_1. The humidity ratio ω_1 can be calculated by using the Engineering Equation Solver (EES) software[10]. The effect of the air pressure drop across chilling coils is small and can be neglected; hence, $P_1 \cong P_0$. The air density will vary significantly with humidity ratio change from ω_0 to ω_1 and decrease in the air temperature from T_0 to T_1.

2.1.2. Air Compressor

For humid air, the compression power can be estimated from

$$\dot{W}_{comp_{air}} = \dot{m}_a \, c_{pa} \left(T_2 - T_1 \right) + \dot{m}_v \left(i_{g2} - i_{g1} \right) \tag{8}$$

where i_{g2} and i_{g1} are, respectively, the enthalpies of saturated water vapor at the compressor exit and inlet states; $\dot{m}_v = \dot{m}_a \omega_1$ is the mass of water vapor.

Relating the compressor isentropic efficiency to the changes in temperature of the dry air and assuming that the compression of water vapor behaves as an ideal gas,

$$\eta_c = \frac{T_{2s} - T_1}{T_2 - T_1} \tag{9}$$

from which T_2 is expressed in terms of T_1 and the pressure ratio PR as

$$T_2 = T_1 \left[\frac{PR^{\frac{k-1}{k}} - 1}{\eta_c} + 1 \right] \tag{10}$$

Substituting for T_2 into Eq. 8 gives the actual compressor power as

$$\dot{W}_{comp_{air}} = \dot{m}_a \left[c_{pa} \frac{T_1}{\eta_c} \left(PR^{\frac{k-1}{k}} - 1 \right) + \omega_1 \left(i_{g2} - i_{g1} \right) \right] \tag{11}$$

where η_c can be evaluated using the following empirical relation[9]:

$$\eta_c = 1 - \left(0.04 + \frac{PR - 1}{150} \right) \tag{12}$$

2.1.3. Combustion Chamber

Heat balance on the combustion chamber (as shown in Fig. 1) gives the heat rate supplied to the GT cycle as

$$\dot{Q}_h = \dot{m}_f \, NCV = \left(\dot{m}_a + \dot{m}_f \right) c_{pg} T_3 - \dot{m}_a c_{pa} T_2 + \dot{m}_v \left(i_{v3} - i_{v2} \right) \tag{13}$$

Introducing the fuel air ratio $f = \dot{m}_f / \dot{m}_a$ and substituting for T_2 in terms of T_1 from Eq. 10 give the cycle heat rate as

$$\dot{Q}_h = \dot{m}_a T_1 \left[(1+f) c_{pg} \frac{T_3}{T_1} - c_{pa} \left(\frac{PR^{\frac{k-1}{k}} - 1}{\eta_c} + 1 \right) + \frac{\omega_1}{T_1} (i_{v3} - i_{v2}) \right]$$ (14)

where f, as expressed in [5], is

$$f = \frac{c_{pg}(T_3 - 298) - c_{pa}(T_2 - 298) + \omega_1(i_{v3} - i_{v2})}{NCV - c_{pg}(T_3 - 298)}$$ (15)

i_{v2} and i_{v3} are the enthalpies of water vapor at the combustion chamber inlet and exit states, respectively. They can be calculated from [11]

$$i_{v,j} = 2501.3 + 1.8723 T_j ; \; j = 2 \, or \, 3$$ (16)

It is seen that the three terms of the gas turbine efficiency in Eq. 2 ($\dot{W}_t$, $\dot{W}_{comp,air}$, and $\dot{Q}_h$) depend on the air temperature and relative humidity at the compressor inlet whose values are affected by the type and performance of the cooling system.

2.2. Evaporative Water Spray Process

In evaporative cooling, water and intake air are brought into direct contact where the warm air stream transfers heat to sprayed water as shown in Fig. 3a. During the air-water heat

exchange process, part of the liquid water evaporates causing the temperature of the air to decrease adiabatically, presented as line *0-1* in Fig. 3b. The air humidity ratio increases from ω_0 to ω_1 approaching the saturation condition. As the air approaches the saturation limit, the evaporation process takes more time where the air cannot carry more water and further water injection is not utilized. Therefore, the evaporative cooling of air is limited by the temperature difference ($T_0 - T_1$). In practice, cooling the air to the saturation state requires water over spraying that may initiate the carryover of droplets, which causes fouling of compressor blades and/or rust of the entrance ducts. Controlling the parameters of evaporative coolers is an important key to the successful seasonal operation of coolers. The effectiveness of an evaporative air-cooler (ε_{evc}) is defined as the ratio between the actual dry bulb temperature decreases and the theoretical temperature difference if the air leaves the cooler at saturation state. Typical evaporative cooler effectiveness range is from 0.8 to 0.9 [12].

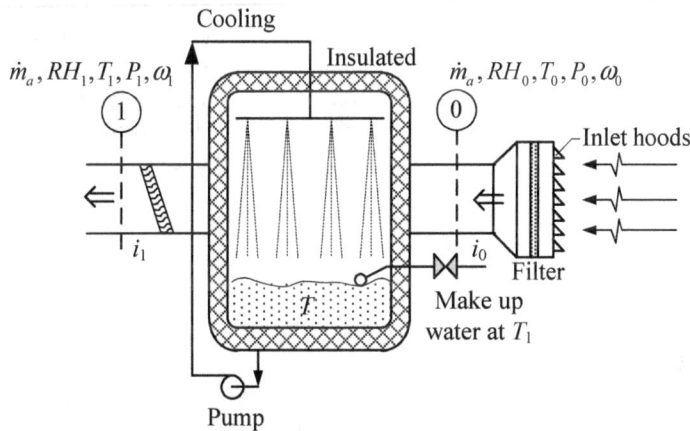

a

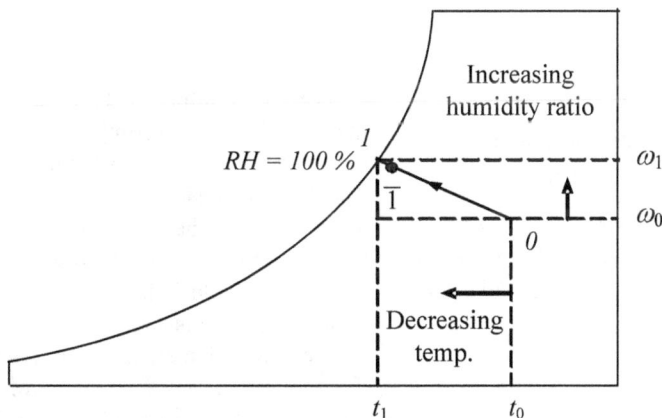

b

Fig. 3. (a Schematic of adiabatic evaporative cooler; b) adiabatic saturation process on the psychometric chart.

Fig. 3a shows a schematic of an evaporative cooler where the ambient air at T_0, ω_0, and P_0 enters the spray chamber and leaves at T_1, ω_1, and P_1. The evaporative cooler is assumed to operate in a steady adiabatic process such that the ambient moist air enters at T_0 and RH_0 and leaves at state 1. Adequate quantity of water is added to the air stream to raise its moisture content close to 100% relative humidity and decrease its temperature as shown in Fig. 3b. Applying energy balance yields [13]

$$\omega_o \left(i_{vo} - i_w \right) = c_{pa} \left(T_1 - T_o \right) + \omega_1 \left(i_{fg1} \right) \qquad (17)$$

where i_{vo} and i_w are saturated water vapor enthalpy at T_0 and saturated water liquid enthalpy at T_1, i_{fg1} is the latent heat of vaporization at state 1, and ω_0 is evaluated at the ambient conditions by using EES software.

3. Gas Turbine Coupled to an Evaporative Spray Cooler

In order to evaluate the feasibility of a cooling system coupled to a GT plant, the performance of the plant is examined with and without the cooling system. In general, the net power output of a complete system is

$$\dot{W}_{net} = \dot{W}_t - \left(\dot{W}_{comp_{air}} + \dot{W}_{el,pump} \right) \qquad (18)$$

The three terms of Eq. 18 are functions of the air properties at the compressor intake (T_1 and ω_1), which in turn depends on the performance of the cooling system. A dimensionless term that gives the advantage of using any cooling system as the power gain ratio (PGR) is defined:

$$PGR = \frac{\dot{W}_{net,with cooling} - \dot{W}_{net,without\ cooling}}{\dot{W}_{net,without\ cooling}} \times 100\% \qquad (19)$$

The PGR is a generic term that takes into account all the parameters of the GT and the associated cooling system irrespective of the cooling process. For a stand-alone GT under specific climatic conditions, PGR is equal to 0. If a cooling system is used, the PGR increases with the reduction of the intake temperature; but this increase is restricted by the physical limits of the cooling process. However, the PGR gives the percentage enhancement in power generation and the thermal efficiency of a coupled system, which is an important parameter to describe the input/output relation. Another factor that physically relates to the thermal efficiency of a stand-alone GT to that coupled cooling system as the thermal efficiency change (TEC) is given:

$$TEC = \frac{\eta_{cy,with cooling} - \eta_{cy,without\ cooling}}{\eta_{cy,without\ cooling}} \times 100\% \qquad (20)$$

However, the PGR is always positive whilst TEC can be negative, which means that the efficiency of the coupled system is less than that of a stand-alone GT even at low intake temperatures. Both PGR and TEC provide dimensionless parameters that can be easily employed and interpreted.

For evaporative cooling, the power consumed by the cooling system $\dot{W}_{el,pump}$ is the pumping power to circulate the water inside the chamber (as shown in Fig. 3a). This power is small compared to the other terms in Eq.18 and can be ignored. Therefore, the PGR for a gas turbine with evaporative cooler can be obtained by substituting Eqs.5, 11, and 18 into Eq.19. From Eqs. 2,5,11, and 18, the cycle thermal efficiency with cooling ηcy,evc in terms of the air properties at the compressor intake and the fuel air ratio can be expressed as

$$\eta_{cy,evc} = \frac{(1+\omega_1+f) c_{pg} \eta_t \dfrac{T_3}{T_1} \left(1 - \dfrac{1}{PR^{\frac{k-1}{k}}} \right) - \left[c_{pa} \dfrac{1}{\eta_c} \left(PR^{\frac{k-1}{k}} - 1 \right) + \dfrac{\omega_1}{T_1} \left(i_{g2} - i_{g1} \right) \right]}{(1+f) c_{pg} \dfrac{T_3}{T_1} - c_{pa} \left(\dfrac{PR^{\frac{k-1}{k}} - 1}{\eta_c} + 1 \right) + \dfrac{\omega_1}{T_1} \left(i_{v3} - i_{v2} \right)} \qquad (21)$$

Eq.21 for an ideal air reversible cycle when $\eta_c = \eta_t = 1$, fuel air ratio $f = 0$, $c_{pg} = c_{pa}$, and the inlet air humidity ratio $\omega_1 = 0$ gives the standard expression in which the pressure ratio is the only dependent ($\eta_{cy,rev} = 1 - 1/PR^{(k-1)/k}$). For a stand-alone (without cooling) GT the third term of the nominator vanishes and the inlet conditions T_1 and ω_1 are replaced by T_0 and ω_0.

4. Results and Discussion

In order to investigate the performance of the direct evaporative cooling method on the GT power output and thermal efficiency, a computer program has been developed to calculate the PGR and TEC for different operation conditions. The thermo physical properties were determined to obtain the accuracy from the EES software. In particular, the specific heats of air and combustion gases are both temperature dependent. Table 1 shows the range of the different operating parameters for this analysis.

The selected maximum air temperature $t_{o,max}$ which was close to 50°C is based on meteorological data recorded during the past few years in Saudi Arabia. An average design value of 44.5% for the relative humidity is selected as base data. For fixed values of T_3, T_o, and RH_o, the assumption is that the maximum cooling can be achieved if the air is cooled to the wet bulb temperature and the pumping power $\dot{W}_{el, pump}$ is neglected.

Table 1. Range of parameters.

Component	Parameter	Range
Ambient air	Max. ambient air temperature $T_{0,max}$	323.15 K
	Relative humidity RH_0	From 0% to 100%
	Volumetric air flow rate	1 m³/s
Gas turbine	Pressure ratio P_2/P_1	10
	Turbine inlet temperature T_3	323.15 K
	Turbine efficiency η_t (Eq.6)	From 0.91 to 0.93
	Air compressor efficiency η_t (Eq. 12)	From 0.88 to 0.91

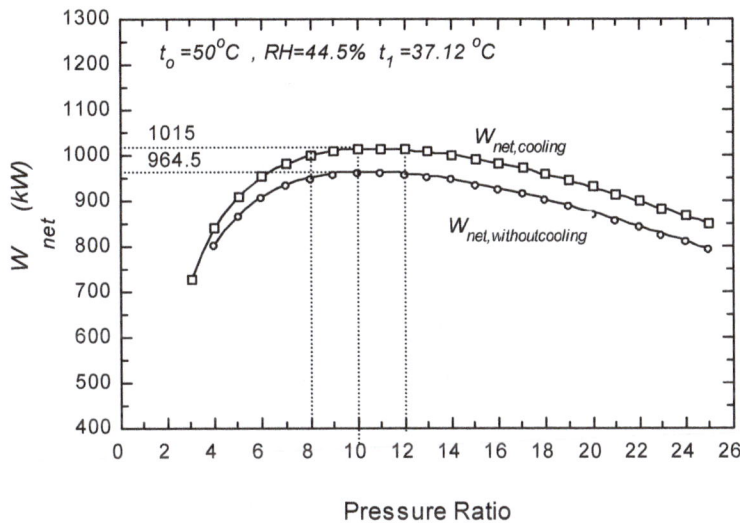

Fig. 4. Dependence of net work on the pressure ratio.

Fig. 4 shows the network is a function of PR. It can be seen that the net gas turbine power output increases together with the PR increase until it reaches a maximum at PR=10. Afterwards, the power slowly declines. The same trend is observed for the gas turbine when coupled to a cooling system. Increasing the PR causes the increase in both the turbine and compressor powers at different rates. The maximum power occurs at a PR where the rate of increase in the turbine power is greater than that required to drive the air compressor. For larger values of PR, the net work tends to decrease as a result of the percentage increase in power required to drive the air compressor as shown in Fig. 5. This figure also shows that the net output power can achieve the maximum value when the PR approximately falls into the range from 8 to 12.

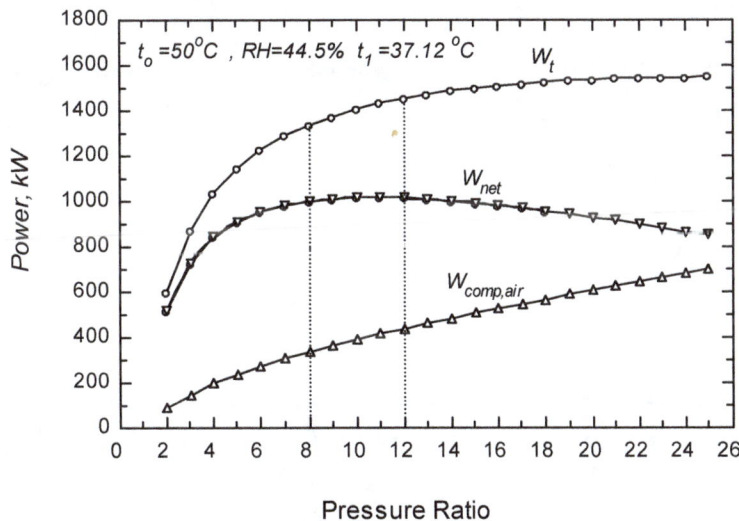

Fig. 5. The power for a gas turbine system using a direct evaporative cooler.

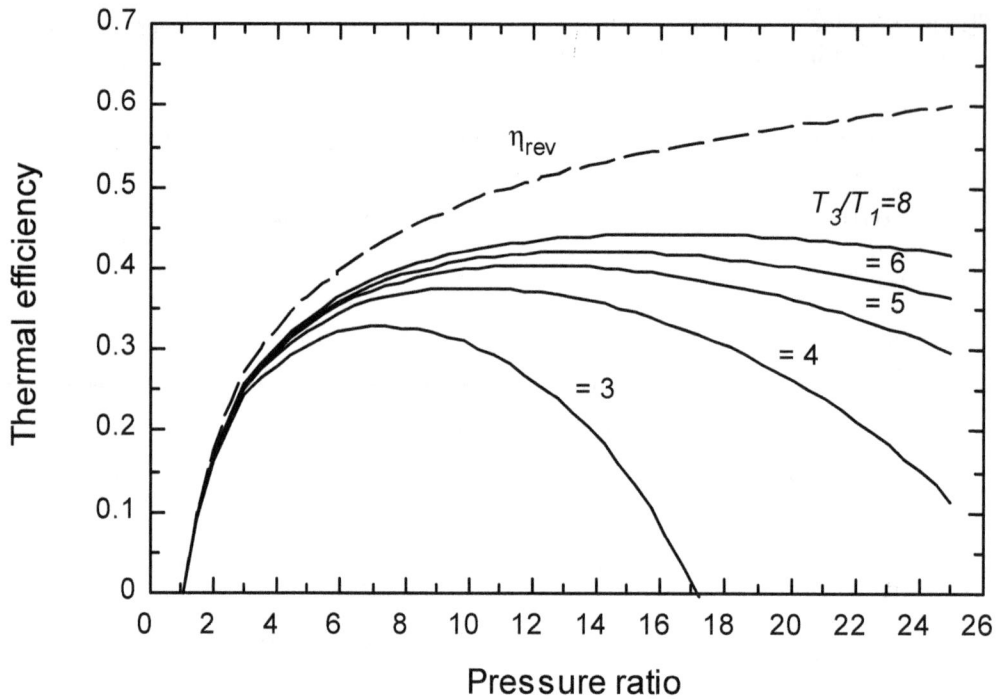

Fig. 6. *The thermal efficiency of a simple gas turbine cycle.*

Fig. 6 shows the thermal efficiency variation according to the PR for different turbine inlet absolute temperature. As the maximum temperature T_3 is kept constant (1373.15 K), an increase in the PR will increase the cycle efficiency. However, the PR is limited and will reach the maximum when the air temperature at the compressor outlet is equal to the design turbine inlet temperature. In this case, the cycle net work tends towards zero and the cycle efficiency approaches the reversible efficiency [14]. In this case study where the direct evaporative cooler is used to cool the outlet air temperature from 50°C to wet bulb temperature (37.12°C), the term T_3 / T_1 increases from 4.25 to 4.425. Consequently, the thermal efficiency increases as shown by the trends in Fig. 6. Fig. 7 shows the thermal efficiency of the gas turbine is improved by 0.24% for evaporative cooling and it is of positive magnitude.

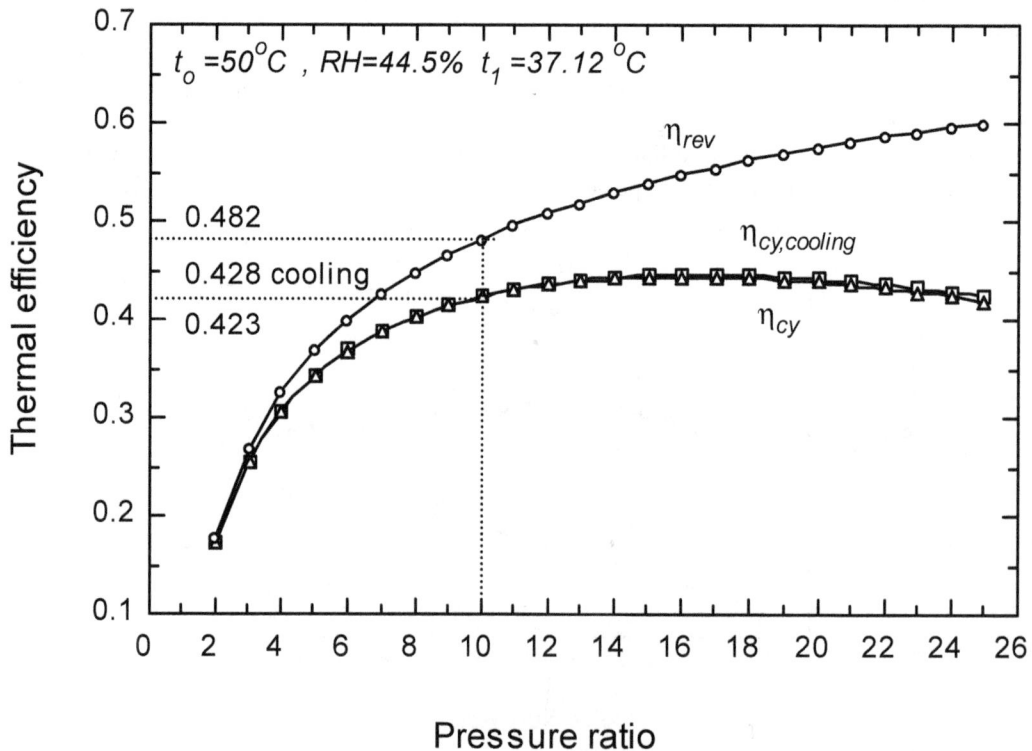

Fig. 7. *Dependence of thermal efficiency on pressure ratio of a direct evaporative cooler.*

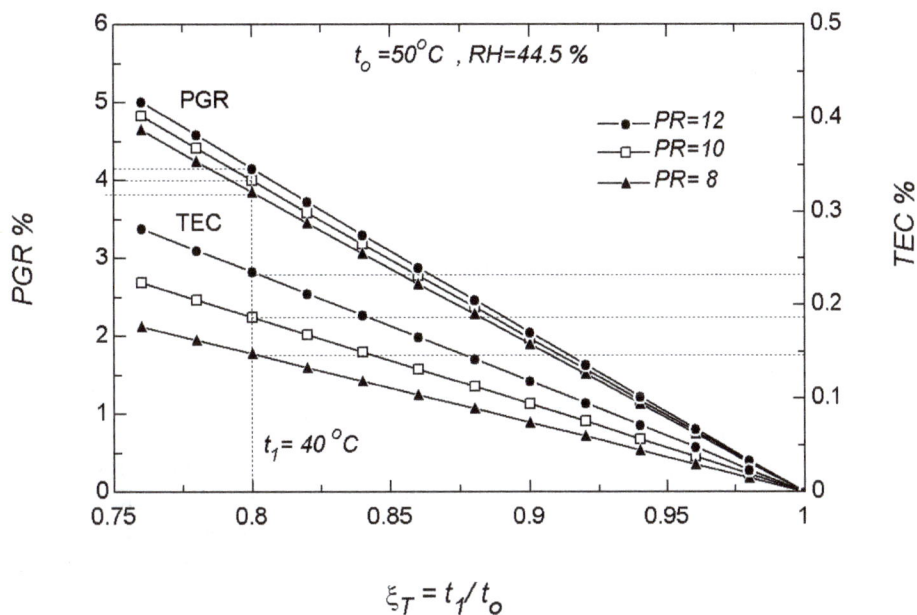

Fig. 8. *Variation of the PGR, TEC, and the temperature ratio for a gas turbine with inlet air evaporative cooling.*

By applying Eqs.19 and 20 for an evaporative air cooler system, the variation of the PGR and TEC with the temperature ratio $\xi_T = t_1/t_o$ is obtained and presented in Fig. 8. The factor ξ_T presents physically the ratio between the air temperatures at the cooler exit and the ambient temperature in Celsius (cooling process attainment). For a constant PR (PR = 10), the PGR increases with decreasing of ξ_T and a power gain factor of 4% can be obtained by cooling the air from 50°C to 40°C.

The evaporative air cooler is designed to lower the ambient air temperature to a degree close to the ambient wet bulb temperature, which limits its cooling capability in humid climate areas. By applying Eqs.19, 20, and 21 with the assumption that $\dot{W}_{el,pump}$ is zero, the PGR and TEC are computed for fixed ambient conditions as shown in Fig. 8. For temperature and relative humidity of 50°C and 44.5 %, respectively, the air temperature can be theoretically reduced in an adiabatic process to 37.12°C (wet bulb temperature) for which $\xi_T = 0.74$. As shown in Fig. 8 at $\xi_T = 0.74$ and PR of 10, the power gain reaches 4.8 % and the thermal efficiency change is 0.22%. However, the value is small but substantiates an improvement in efficiency. This result shows that there is a gain in power and improvement in thermal efficiency. This advantage is soon offset by the fact that the system cannot provide any further improvement beyond the limiting state at specified ambient conditions.

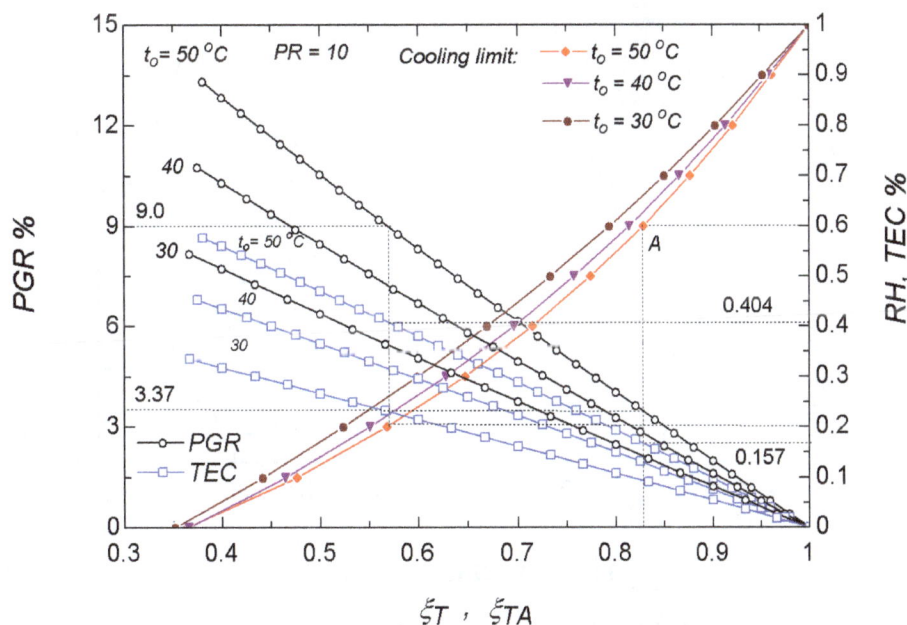

Fig. 9. *The cooling limit of air evaporative cooler.*

Fig. 9 shows the PGR and TEC for different ambient air temperatures (50°C, 40°C, and 30°C) and the whole range of RH up to 100%. The curves in the figure present the variation of the RH with ξ_T at the maximum gain, which equals t_{wb}/t_o. Therefore, these curves determine the limiting condition for evaporative cooling once the ambient temperature and relative humidity are prescribed. For example, to determine the maximum power gain ratio and thermal efficiency change for ambient conditions of 50°C and 60% RH_0, first we draw a horizontal line from 60% relative humidity until it intersects the cooling limit of $t_0 = 50^0 C$ to obtain point A on the figure. At this point, ξ_T determines the lowest air temperature at the compressor inlet; intersection with the lines of PGR and TEC (at $t_0 = 50^0 C$) gives the maximum values of the PGR and TEC as 3.37 % and 0.157 %, respectively. Further, the results in Fig. 9 are presented for an ideal evaporative cooler effectiveness $\varepsilon_{evc} = 100\%$. ε_{evc} can be computed as

$$\varepsilon_{evc} = \frac{t_o - t_{\bar{1}}}{t_o - t_1} = \frac{1 - \xi_{T,A}}{1 - \xi_T} \quad (22)$$

where $\xi_{T,A}$ is the ratio of the actual air temperature at the cooler exit to the ambient air temperature in Celsius and is equal to $t_{\bar{1}}/t_o$. From Eq.22, $\xi_{T,A}$ can be expressed as

$$\xi_{T,A} = 1 - \varepsilon_{evc}\left(1 - \xi_T\right) \quad (23)$$

To include the effectiveness ε_{evc} in the analysis, ξ_T is replaced by $\xi_{T,A}$ and the PGF and TEC can be evaluated from Fig. 9. The effect of evaporative cooler effectiveness on both the PGR and TEC is shown in Table 2 for $\varepsilon_{evc} = 0.85$.

The relative humidity of ambient air is a key parameter for the evaporative cooler cooling capability. For illustration, we consider air at constant dry bulb temperature of 40°C and 60% RH_0, for which the maximum PGR and TEC are 3.013 % and 0.137%, respectively. In the case $\xi_T = 0.814$, the air temperature can only be reduced from 40°C to 32.56°C. For an intermediate level of 40% humidity and the same dry bulb temperature, the corresponding temperature ratio ξ_T is 0.696, for which the temperature limit is 27.83°C. The corresponding power gain is 5.012%. The detailed results for the effect of the RH on the evaporative cooling capacity are summarized in Table 2, which shows a direct inverse proportionality.

Table 2. The maximum PGR and TEC with different ambient air conditions and evaporative cooler effectiveness.

Ambient air conditions		Evaporative cooler with ε_{evc} 100%				Evaporative cooler with ε_{evc} 85 %			
t_o (°C)	RH_o (%)	t_1 (°C)	ξ_T	PGR (%)	TEC (%)	$t_{\bar{1}}$ (°C)	ξ_{TA}	PGR (%)	TEC (%)
40	60	32.56	0.814	3.013	+0.137	33.68	0.842	2.55	+0.116
40	40	27.83	0.696	5.012	+0.223	29.66	0.742	4.234	+0.190
40	20	22.03	0.551	7.546	+0.328	24.73	0.618	6.359	+0.280
40	10	18.57	0.464	9.102	+0.39	21.79	0.545	7.658	+0.333

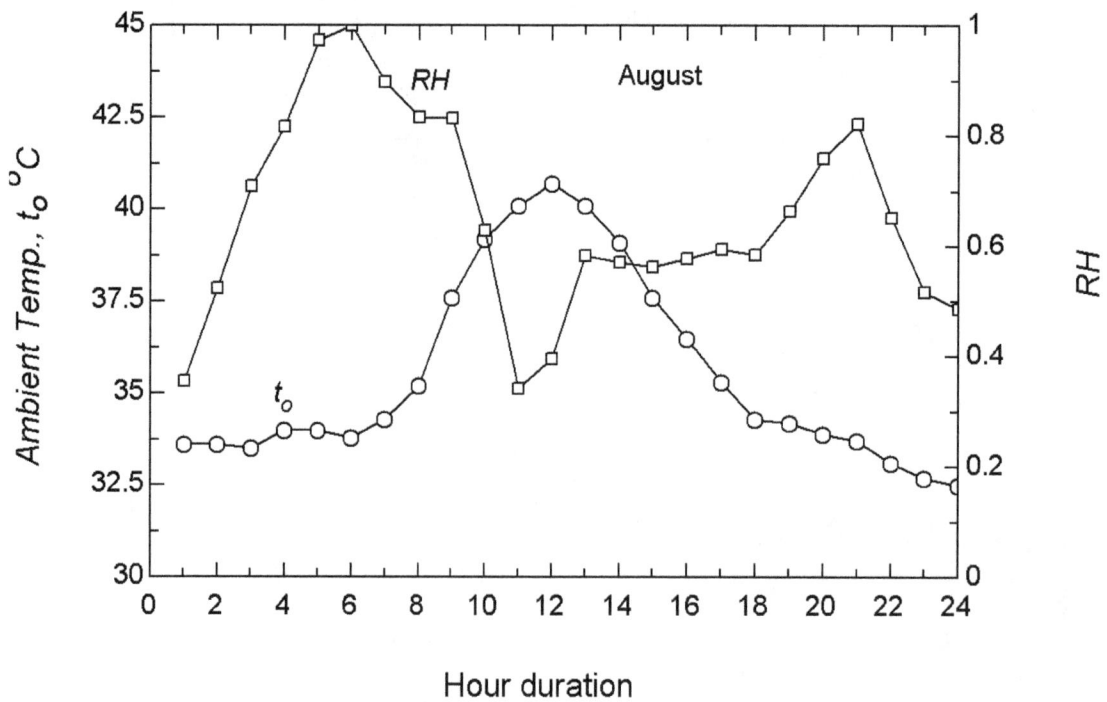

Fig. 10. August RH and ambient temperature variations.

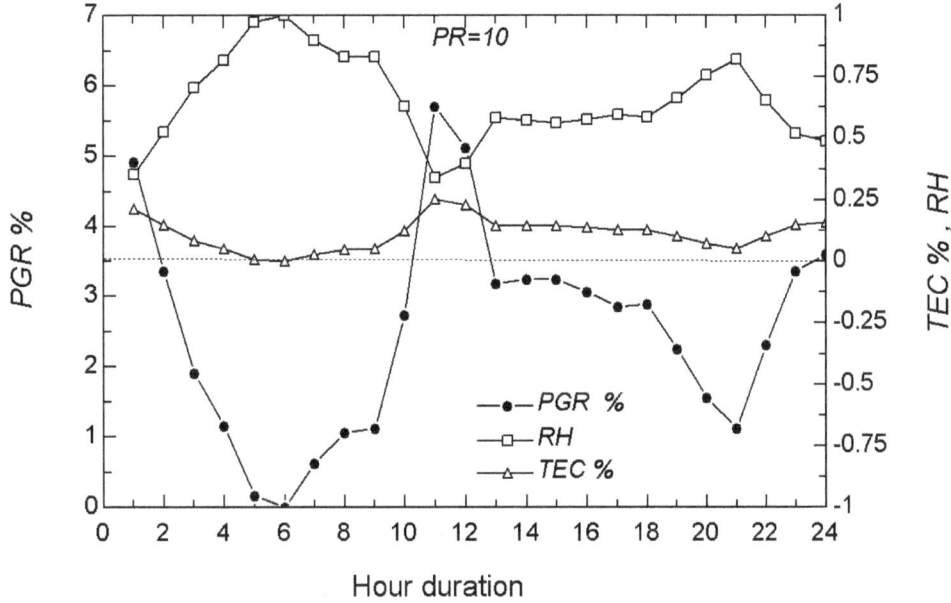

Fig. 11. *Dependence of gas turbine maximum PGRon RH of direct evaporative cooler.*

In order to examine the daily performance of the cooling system,we assume that this system reduces the intake air temperature from T0 to T1, for which the evaporative cooler effectiveness is 100% and it operates for 24 hours, applying Eqs. 19 and 20, on hourly basis, using the ambient data of Fig. 10. The variation of the PGR and TEC for the evaporative cooler is presented in Fig. 11. This figure shows that the PGR and TEC drop rapidly to reach zeros at 6AM whereξT approaches 1during early morning hours with the 100% RH. At 11AM and 40.1°C air temperature withRH of 34.1%, the gain reaches 5.706 % and the thermal efficiency improvement is 0.253%, where the humidity is the least.

The values show that the cooling system is promising when operated under favorable conditions but the overall daily performance is what counts at the end. The practical illustrative application indicates that the evaporative system provides additional energy of 2.52%.

5. Conclusions

The effect of inlet air cooling on the performance of a simple gas turbine plant is investigated. Systematic analysis of a GT cycle coupled to a cooling system is presented for an evaporative air cooling process. The performance improvement showed strong dependency on the climatic conditions and evaporative cooler effectiveness and to some degree on the gas turbine pressure ratio. The evaporative cooling is quite efficient for 40°C dry air and 10% RH ambient air. The maximum power gain and the thermal efficiency improvements are 9.102% and 0.39%, respectively. The direct evaporative cooling process is limited by the wet bulb temperature at which additional water spray would not contribute any further cooling effects. The performance of the evaporative cooling is presented in a general dimensionless working graph that directly relates the maximum power gain and thermal efficiency variation to the ambient conditions.

Finally, the performance of the air cooling system was examined for the hot and humid conditions of Jeddah, Saudi Arabia. The actual climate on August 16 is selected as base data for investigation. The daily power output and the thermal efficiency improvement were only 2.52% and 0.112%, respectively.

In addition, to improvethe thermal efficiency and power gaining, evaporative spray cooler improves the environmental impact of the GT, since increasing water vapor in the inlet air tends to lower the amount of nitrogen oxides (NOx) emissions as well as reduce the dust due to air washing.

Acknowledgments

This project was funded by the National Plan for Science, Technology and Innovation (MAARIFAH)-King Abdulaziz City for Science and Technology-the kingdom of Saudi Arabia-award number (8-ENE 288-03). The authors also, acknowledge with thanks Science and Technology Unit, King Abdulaziz University for technical support.

Nomenclature

c_p	specific heat at constant pressure (kJ kg^{-1} K^{-1})
i	specific enthalpy (kJ.kg^{-1})
k	specific heats ratio
NCV	net calorific value = 42500 (kJ kg^{-1})
P	pressure (kPa)
PR	pressure ratio = P_2/P_1
PGR	power gain ratio
$\dot{Q}_h$	heat rate (kW)
$\dot{m}$	mass flow rate (kg s^{-1})
t	temperature °C
T	absolute temperature, K
TEC	thermal efficiency change, Eq. 20

$\dot{W}$	power, kW
Greek	symbols
η	efficiency
ε_{evc}	evaporative cooler effectiveness
ξ_T	temperature ratio t_1 / t_o
ξ_T	actual temperature ratio $t_{\bar{1}} / t_o$
Subscripts	
0	ambient
a	dry air
c	compressor
cy	cycle
e	evaporator
evc	evaporative cooler
f	fuel
fg	latent heat
h	heat
rev	reversible
t	turbine
v	water vapor

References

[1] Zadpoor_AA, Golshan AH. Performance improvement of a gas turbine cycle by using adesiccant-based evaporative cooling system. Energy 2006; 31:2652-2664.

[2] Chaker M, Homji CBM, Mee III TR. Inlet fogging of gas turbine engines. Proceedings of ASME Turbo Expo; 2001 June 4-7; New Orleans, USA; 2001.

[3] AL-Hamdan OR, Saker AA.Studying the role played by evaporative cooler on the performance of GE gas turbine existed in Shuaiba North Electric Generator Power Plant. Energy Power Eng 2013; 5:391-400.

[4] Johnson RS. The theory and operation of evaporative coolers for industrial gas turbine installations. J Eng Gas Turbines Power 1989; 111:327-334.

[5] Alhazmy MM, Najjar YSH. Augmentation of gas turbine performance using air coolers. App Therm Eng 2004; 24:415-429.

[6] Meher H, Cyrus B, Mee RT, Thomas R. Inlet fogging of gas turbine engines, part B: droplet sizing analysis nozzle types, measurement and testing. Proc ASME Turbo Exo 2002; 2002 June 3-6; Amsterdam, The Netherlands.

[7] Bettocchi R, Spina PR, Moberti F. Gas turbine inlet air cooling using non-adiabatic saturation process. ASME Cogen-Turbo Power Conf; 1995 August 23-25; Vienna, Austria .ASME; 1995.p.1-10.

[8] Cyrus B, Mee RT. Gas turbine power augmentation by fogging of inlet air. Proc 28th Turbomach Symp. 1999.p.95-114.

[9] Korakianities T, Wilson DG. Models for predicting the performance of Brayton-cycle engines.Eng Gas Turbine Power 1994; 116:381-388.

[10] Klein KA, Alvarado FL. EES-Engineering Equation Solver, F-Chart Software.Middleton, WI.

[11] Dossat RJ. Principles of refrigeration, New York: John Wiley and Sons; 1997.

[12] Cortes CPE, Willems D. Gas turbine inlet cooling techniques: an overview of current technology. Proc Power GEN 2003; 2003 Dec. 9-11;Nevada, USA.

[13] McQuiston FC, Parker JD, SpilterJD. Heating, ventilating and air conditioning: design and analysis, 5th ed. New York: Willey; 2000.

[14] Li KW, Priddy A P. Power plant system design. New York: John Wiley & Sons; 1985.

3

A Co-Integration Analysis Between Electricity Consumption and Economic Development in Hebei Province

Huiru Zhao[1], Yaowen Fan[1], Nana Li[1], Fuqiang Li[2], Yuou Hu[2]

[1]School of Economics and Management, North China Electric Power University, Beijing, China
[2]North China Grid Company Limited, Beijing, China

Email address:
nancyli1007@163.com (Nana Li)

Abstract: Electricity is a convenient and clean energy which can provide strong support for the development of all walks of life. The power development should maintain coordination with the regional economic development, in which power development may advance sometimes. Hebei province is a typical resource-based area in China, and its power consumption is closely related to economic development, which makes it important to study the relationship between electricity and economic development in Hebei. In order to select the economic factors affecting electricity demand, the gray correlation analysis is used to analyse the correlation among different factors. And then, a long-term equilibrium model between electricity consumption and economic factors is proposed through co-integration analysis. The analysis result showed that the electricity consumption, GDP, the level of residential consumption, efficiency levels and economic structures have a long-run equilibrium relationship in Hebei Province, in which the economic structure has the strongest impact on the electricity consumption, followed by GDP, energy consumption intensity and the residential consumption level. Currently, the impact of the economic restructuring on electricity demand in Hebei cannot be overlooked. Furthermore, this model can be used to give a reference to Hebei Electric Power Planning.

Keywords: Electricity Consumption, Economic Development, Co-Integration Theory

1. Introduction

In recent years, Hebei has made considerable progress in economy through taking advantage of the unique natural resources, geographical location and other aspects. The GDP of Hebei has reached 2.83 trillion yuan in 2013 and the average annual growth rate from 1978 to 2013 is 10.78%.Meanwhile, Hebei's economic development is inseparable from the rapid growing electricity consumption. In 2012, Hebei's total electricity consumption has reached 307.773 billion kWh, accounting for 6.21 percent of China's power consumption and its per capita electricity consumption is 4223.01 kwh.

In recent years, as the non-stationary of the time series of electricity and economy, the co-integration theory related with non-stationary time series has become the main method for the relationship analysis between power and economy. Engel-Granger method is implied to estimate the relationship between the US energy and economy in 1974-1990[1]. This method has also been applied to the Group of Seven and 16 newly industrialized countries except China[2]. Analysis of the relationship between the Greek GDP and energy consumption in 1960-1996 has been made using vector error correction model[3]. In the related researches for China, Johansen Method has been used for Taiwan in 1955-1993 and 1954-1997 [4-5], this method has also been used to analyse the relationship between electricity consumption and economic development between 1952-2001 and 1970-2001in China[6-7]. Hebei is located in north central China, and it is a typical resource-based province whose electricity consumption condition is quite different with whole China obviously. Thus this paper uses gray relational analysis model to select economic factors which affect the electricity demand most in Hebei. Then co-integration analysis model has been proposed to illustrate the relationship between electric demand and economic growth which could be helpful for future electricity planning.

2. Theory and Methodology

2.1. Grey Correlation Analysis

The basic idea of gray correlation analysis is to determine the degree of association between factors based on the similarity between the curve. Grey correlation analysis usually compares the data series which reflects changes in each factor. The correlation level between factors is judged by the degree of association which is obtained by comparing the association between factor curves. There are six steps in the gray correlation analysis:

(1)Determine the reference sequence and comparative sequence $X_0 = [X_0(1), X_0(2), \cdots X_0(n)]$ is the reference sequence and $X_i = [X_i(1), X_i(2), \cdots X_i(n)](i = 1, 2, \cdots, l)$ is the comparative sequence.

(2)Calculate sequence X_0 and gray absolute correlation ϕ_{0i} of sequence X_i.

Gray absolute correlation is only associated with geometric shape of sequence X_0 and X_i, which has nothing to do with their relative location. The specific formula is shown as follows:

$$\phi_{0i} = \frac{1 + |S_0| + |S_i|}{1 + |S_0| + |S_i| + |S_i - S_0|} \tag{1}$$

Where, $k = 1, 2, \cdots n-1$.

$$|S_0| = \left| \sum_{k=2}^{n-1} X_0^0(k) + \frac{1}{2} X_0^0(n) \right| \tag{2}$$

$$|S_i| = \left| \sum_{k=2}^{n-1} X_i^0(k) + \frac{1}{2} X_i^0(n) \right| \tag{3}$$

$$\begin{cases} X_0^0(k) = X_0(k) - X_0(1) \\ X_i^0(k) = X_i(k) - X_i(1) \end{cases} \tag{4}$$

(3) Calculate the gray relative correlation π_{0i} between X_0 and X_i.

Gray relative correlation denotes the relationship between the changing rate of sequence with respect to the starting point. Namely, the closer the changing rate of X_0 and X_i, the greater the gray relative correlation is.

$$\pi_{0i} = \frac{1 + |S_0'| + |S_i'|}{1 + |S_0'| + |S_i'| + |S_i' - S_0'|} \tag{5}$$

Where, $k = 1, 2, \cdots n-1$.

$$|S_0'| = \left| \sum_{k=2}^{n-1} X_0^{0'}(k) + \frac{1}{2} X_0^{0'}(n) \right| \tag{6}$$

$$|S_i'| = \left| \sum_{k=2}^{n-1} X_i^{0'}(k) + \frac{1}{2} X_i^{0'}(n) \right| \tag{7}$$

$$\begin{cases} X_0^{0'}(k) = \dfrac{X_0(k)}{X_0(1)} \\ X_i^{0'}(k) = \dfrac{X_i(k)}{X_i(1)} \end{cases} \tag{8}$$

(4) Calculate the grey synthetic correlation degree ρ_{0i} between X_0 and X_i.

$$\rho_{0i} = \theta \phi_{0i} + (1-\theta)\pi_{0i} \tag{9}$$

Where, $\theta \in [0,1]$, in general, θ can be 0.5.

(5) Select the key factors according to the calculation results.

Determine the threshold value for ρ and select the key factors. If synthetic correlation degree ρ_{0i} is bigger than ρ, X_i should be the key indicator.

2.2. Co-Integration Analysis

Co-integration was firstly put forward by Granger in 1981and Engle and Granger proposed the theorem and its concrete operational framework [8-9].In general, yt~I(d) shows that yt becomes stationary after d times differencing. If variables are all I(1), their linear combination o usually satisfies I(1). Further, stationary linear combination series can be called co-integrated which implies the existence of a long-term equilibrium relationship between the variables.

When examine the stationary of each history series, Augmented Dickey and Fuller(ADF) approach is usually used[10]. ADF test is an improved form of DF (Dickey and Fuller) unit root test in which the statistical test function is presented as:

$$\Delta X_t = \alpha + \beta t + \delta X_{t-1} + \sum_{i=1}^{p} \theta_i \Delta X_{t-1} + \varepsilon_t \tag{10}$$

Where ε_t is the random error. It is an independent identically distributed (i.i.d.) white noise process. For a given level of significance, the time series is stable if the ADF test value is less than the critical value.

The next step after the unit root test is JJ (Johansen and Juselius) co-integration test. If two or more variables are I(d), their linear combination is co-integration, which can be defined as said long-run equilibrium relationship[11]. There are two ways to test the co-integration: TRACE and MAX are commonly two output results in this test.

3. Quantitative Analysis of the Relationship Between Power and Economy in Hebei

Studies have shown that there is a certain relationship between the electricity demand and economy development, which has been demonstrated by scholars [6-7]. But the economic factors impacting the electricity demand have not gotten consistent conclusions. Economic growth, industrial structure, population, urbanization rate, export and technological advance may affect the electricity demand. Therefore, the gray correlation model has been used to select the economic factors which affect the electricity demand most. The variables are picked and defined as follows.

Electricity Consumption (Q).The total electricity consumption of Hebei Province in 2012 has reached 307.773 billion kWh, accounting for 6.21 percent of China's power consumption, and accounting for 0.02% of world electricity consumption. Meanwhile, per capita electricity consumption of Hebei has reached 4223.01 kwh. Hebei Electric Power Consumption data is shown as Figure 1.

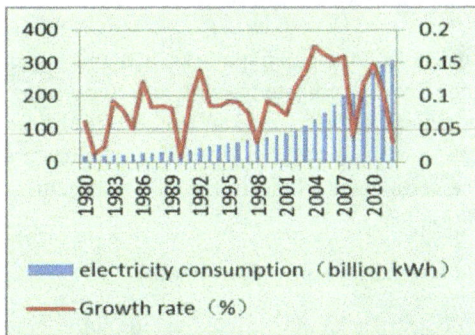

Figure 1. The electricity consumption in Hebei.

Gross domestic product (GDP).GDP refers to the gross domestic product in Hebei, which reflects the level of economic development. The GDP of Hebei is 2.657501 trillion yuan in 2012, accounting for 5.12 percent of China's GDP. The data of GDP is shown as Figure2.

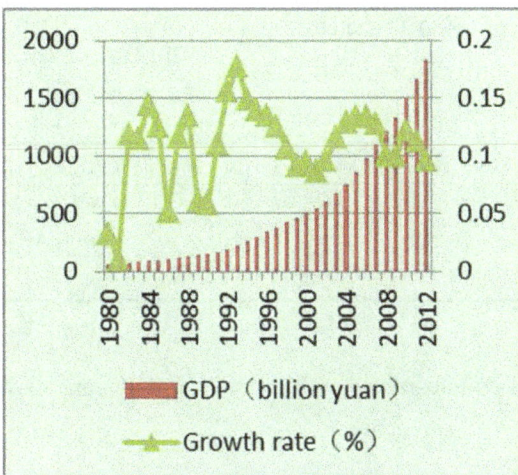

Figure 2. The GDP in Hebei.

Residential consumption level (RCL). RCL reflects the living standards of local residents. When in economic expansion, RCL is considerably high. From 1978 to 2012, the level of consumption in Hebei province increase with the growth of GDP, and reached 10,749 yuan (nominal value) in 2012. The data is shown as Figure 3.

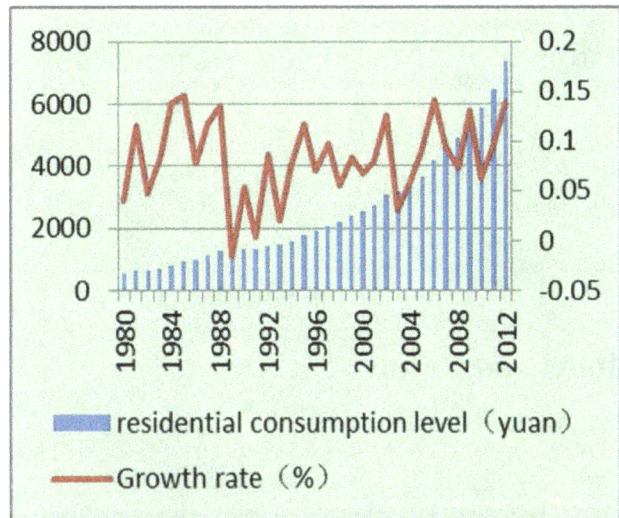

Figure 3. The residential consumption level in Hebei.

Efficiency level (EF). EF refers to energy consumption intensity. In 1980-2011, the energy consumption intensity of Hebei is reduced from 5.00 tons of standard coal to 1.66 tons of standard coal. The data is shown as Figure 4.

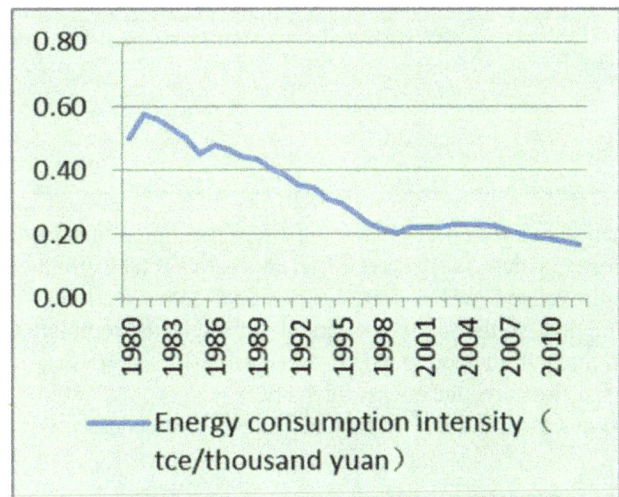

Figure 4. The energy consumption intensity in Hebei.

Economic structure (STRU).STRU refers to the tertiary industry proportion of GDP. The proportion of the first industry is relatively small while the proportion of the secondary industry is relatively stable with the highest proportion. Compared with 1978, the proportion of primary industry dropped from 28.52% to 12.37% in 2012, the proportion of secondary industry has remained at about 50% level, the proportion of tertiary industry increased to 21.02% from 35.47% in 2012. The data is shown as Figure 5.

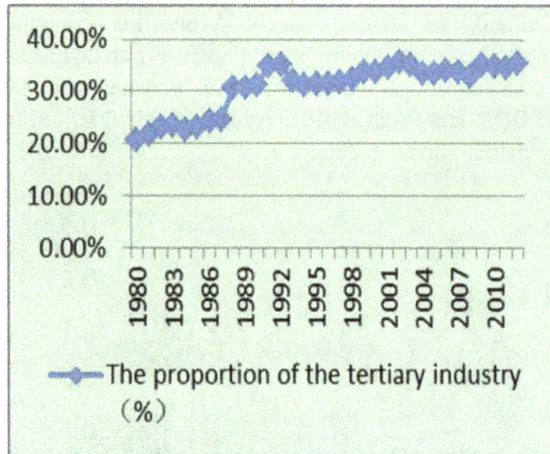

Figure 5. *The proportion of the tertiary industry.*

4. Model and Results

4.1. The Gray Correlation Analysis

To fully characterize the level of economic development , the GDP, residential consumption level, exports level, industry structure (tertiary proportion) and technical progress (energy intensity) are chosen to select the key factors based on the gray correlation analysis. Data interval is 1980-2012 and the results are shown in Table 1.

Table 1. *Gray correlation analysis result*

Indicators	Correlation	Rank
GDP	0.727617259	2
RCL	0.813460967	1
EXP	0.700820384	5
STRU	0.758933167	2
EF	0.721373249	4

When the indicators increase, the co-integration model requires more data. Also, spreadsheet shows that total exports in multivariate model has little effect on electricity consumption, so the following variables are selected to build the co-integration model: GDP, the level of consumption, industrial structure, and energy intensity.

In order to eliminate fluctuations and heteroscedasticity,

part of the data are processed logarithmic.

4.2. Results of Unit-Root Tests

To check the stationary of variables used in the model, ADF test is adopted in this paper, results are shown in Table 2.

Table 2. *Unit root test of variables*

Series	Trends and intercept	ADF value	P-Value
lnGDP	(C,C,3)	-2.492644	0.3290
$\Delta \ln GDP$	(C,C,5)	-4.994957	0.0024***
lnQ	(C,C,0)	-1.913891	0.6240
$\Delta \ln Q$	(C,C,0)	-3.828068	0.0283**
ln RCL	(C,C,0)	-1.913891	0.6240
$\Delta \ln RCL$	(C,C,0)	-3.828068	0.0283**
ln STRU	(C,C,0)	-1.868447	0.6472
$\Delta \ln STRU$	(C,C,5)	-3.744667	0.0369**
ln EF	(C,C,2)	-2.107893	0.5208
$\Delta \ln EF$	(C,C,0)	-5.266468	0.0009***

Δ indicates the first difference of a time series.***,** and *indicate results are statistically significant at 1%,5% and 10% levels, respectively.

As shown in Table 2, the null hypothesis of a unit root could not be rejected. Results indicate that the null hypothesis of a unit root was significantly rejected in the second difference. Therefore we conclude that all five variables are at first-difference stationary, which fulfills the requirements of the co-integration test.

4.3. Select the Lag Intervals for VAR model

We need to construct an unrestricted vector autoregressive (VAR) that is composed of lnQ, lnGDP, STRU, lnRCL, and EF to analyze the co-integration relationship among the variables. Therefore, we need to select the optimal lag order. In this paper, we choose a lag of 2 by using the Schwarz information criterion (SC), sequential modified LR test (LR), Final prediction error (FPE), Akaike information criterion (AIC) and Hannan-Quinn information criterion (HQ).As shown in table 3, lag interval of 2 is conformably chosen by the criteria of LR, AIC and HQ.

Table 3. *The lag intervals*

Lag	LogL	LR	FPE	AIC	SC	HQ
0	157.37	NA	3.70e-11	-9.83	-9.60	-9.76
1	371.98	346.14	1.85e-16	-22.06	-20.68*	-21.61
2	407.19	45.44*	1.10e-16*	-22.72*	-20.18	-21.9*

*indicates lag order selected by the criterion.

4.4. Johansen Co-Integration Test

We use trace statistic and the maximum eigenvalue statistic to estimate the number of co-integration relationships and the

normalized co-integrating coefficient. The results are shown as table 4.

Table 4. *Co-integration test results*

Hypothesized No. of CE(s)	Eigenvalue	Trace Statistic	0.05Critical Value	Prob.**
None	0.859236	104.7236	69.81889	0.0000
At most 1	0.486274	43.94271	47.85613	0.1112
At most 2	0.421657	23.29472	29.79707	0.2319
At most 3	0.163145	6.319511	15.49471	0.6578
At most 4	0.025422	0.798281	3.841466	0.3716
Hypothesized No. of CE(s)	Hypothesized No. of CE(s)	Eigenvalue	Trace Statistic	0.05Critical Value
None	0.859236	60.78087	33.87687	0.0000
At most 1	0.486274	20.64799	27.58434	0.2981
At most 2	0.421657	16.97521	21.13162	0.1732
At most 3	0.163145	5.521231	14.26460	0.6751
At most 4	0.025422	0.798281	3.841466	0.3716

Trace method and the max eigenvalue method show the presence of a co-integration relationship between the variables, indicating that GDP, consumer level (RCL), industry structure (STRU) ,energy intensity(EF)hold a long-term stable relationship with electricity consumption.

4.5. The Long-Term Equilibrium Model of Economy and Electricity Demand

The Co-integration coefficients in time interval 1980-2012 is shown as table 5.

Table 5. *Co-integration coefficients*

Co-integrating Equation(s):	Log likelihood		464.3303	
Normalized co-integrating coefficients (standard error in parentheses)				
$\ln Q$	$\ln GDP$	$\ln STRU$	$\ln RCL$	$\ln EF$
1.000000	-0.836429	0.6096	-0.5197	-0.6689
	(0.09965)	(0.29864)	(0.11907)	(0.08141)

$$\ln Q = 0.84\ln GDP + 0.51\ln RCL - 0.61\ln STRU + 0.67\ln EF \quad (11)$$

The results show that 1% of GDP growth can cause 0.84% growth in electricity demand, 1 point change in tertiary industry proportion accounted for 0.61% reduction in power consumption, energy consumption intensity increases 1 percent, electricity consumption increase 0.67%. The impact of changes in the GDP is maximum among all key factors affecting the demand for electricity.

5. Conclusion and Policy Implications

Through qualitative analysis and gray correlation analysis, GDP, the residential consumption level, industrial structure, technological advancement have an impact on electricity demand in Hebei Province. The co-integration analysis shows that long term equilibrium relationship among GDP, the residential consumption level, industrial structure, energy intensity (EF) is exist. 1% of GDP growth can cause 0.84% growth in electricity demand, 1 point change in tertiary industry proportion accounted for 0.61% reduction in power consumption, energy consumption intensity increases 1 percent, electricity consumption increase 0.67%. The impact of changes in GDP is maximum in all the factors affecting the demand for electricity. The model also shows when future electricity demand forecasting is being made, economic growth, economic structure, living standards and technical progress factors should all be take into consideration comprehensively and the impact of structural changes needs extra attention.

Acknowledgments

This study is supported by the Humanity and Social Science project of the Ministry of Education of China (Project number: 11YJA790217), the National Natural Science Foundation of China (Project number: 71373076), and Science and Technology Project of State Grid Corporation of China (Contract number: SGHB0000DKJS1400116).

References

[1] Stern D I. Energy and Growth in the USA: multivariate approach[J]. Energy Economics, 15(2), pp.137-150,1993.

[2] Soytas U, Sari R. Energy consumption and GDP: causality relationship in G7 countries and emerging markets[J]. Energy Economics, 5(1), pp.33-37, 2003.

[3] Hondroyiannis G, Lolos S, Papapetrou E. Energy consumption and economic growth: assessing the evidence from Greece[J]. Energy Economics, 24(4), pp.319-336, 2002.

[4] Cheng B L, Lai T W. An investigation of co-integration and causality between energy consumption and economic activity in Taiwan [J]. Energy Economics, 19(4), pp.435-444, 1997.

[5] Yang H Y. A Note on the causal relationship between energy consumption and GDP in Taiwan [J]. Energy Economics.22(4), pp.309-317, 2000.

[6] Lin Boqiang. Structural change, efficiency improvement and electricity demand forecasting[J]. Economic Research, 38(5), pp. 57-65, 2003.

[7]   Shiu A, Lam P L. Electricity consumption and economy growth in China [J]. Energy Policy, 32(1), pp. 47-54, 2004.

[8]   Granger C.W.J. Some properties of time series data and their use in econometric model specification [J]. Journal of econometrics, 16(1), pp. 121-130, 1981.

[9]   Engle R F, Granger C W J. Co-integration and error correction: representation, estimation, and testing [J]. Econometrica: journal of the Econometric Society, pp. 251-276, 1987.

[10]   Dickey D A, Fuller W A. Likelihood ratio statistics for autoregressive time series with a unit root [J]. Econometrica: Journal of the Econometric Society, pp.1057-1072, 1981.

[11]   Johansen S, Juselius K. Maximum likelihood estimation and inference on co-integration—with applications to the demand for money [J]. Oxford Bulletin of Economics and statistics, 52(2), pp. 169-210, 1990.

4

Monthly Stream Flow Predition in Pungwe River for Small Hydropower Plant Using Wavelet Method

Miguel Meque Uamusse[1, 3, *], **Petro Ndalila**[2], **Alberto JúlioTsamba**[3], **Frede de Oliveira Carvalho**[4], **Kenneth Person**[1]

[1]Department of Water Resource Lund University, Lund, Sweden
[2]Department of Mechanical Engineering, Mbeya University of Science and Technology, Mbea, Tanzania
[3]Faculdade de Engenharia, Universidade Eduardo Mondlane, Maputo, Mozambique
[4]Departamento de Engenharia Química, Universidade Federal de Alagoas, Brazil

Email address:
miguelmeque@gmail.com (M. M. Uamusse)

Abstract: This study investigates the effect of discrete wavelet transform data pre-processing method on neural network-based on monthly streamflow prediction models to produce energy from small Hydro power plant in along of Pungwe river basin in Mozambique. The study used data from Vanduzi gauging a station which is along of Pungwe river basin. Eight different single-step-ahead neural monthly stream flow prediction models were developed. Coupled simulation between MATLAB and wavelet neural network was used to solve the problem. Different models were teste with the same sample and found that, Artificial Neural Network had the best performance. The main objective of this research project is to analyze the monthly stream flow prediction in the Pungwe river, so as to make suitable decisions in dry or wet spells also to resolve probable conflicts about water recourses.

Keywords: Renewable Energy, Hydropower, Wavelet Artificial Neural Network, Monthly Flow Prediction

1. Introduction

According currently energy crisis, energy consumptions is become a big problem in Mozambique especially the big and concentrated cities such Maputo, Beira and Nampula. For the reason is to bring renewable energy hydropower technology to replace current energy mix in Mozambique. One of the best energy alternative in Mozambique is the conversion of Hydropower energy because the source is abundant, is clean energy, is renewable, and it can creat more job opportunity (EDM, 2015; IEA, 2009; Cuamba et al.,2010; Uamusse et al.,2014 and FUNAE).

To study about level of river flow and its fluctuations time at different times of the year is one of the significant factor to achieve sustainable development for water resource issues and energy planning Hydropower energy is one of promising clean energy technologies, however this technology has some challenges compare with biomass, solar, and wind energy. The most challenging issue here is high capital investment cost.

Steam flow perdition is essential in all activities involving the operation and optimization of water resources management and planning issues, than for this reason, the development of Mathematical model to be able provide more reliable long-term forecasting has attracted the attention of hydrologist throuth time (Santos and Silva,2013; Seo et al,. 2014).

In this article, two-hybrid models are studied, Wavelet Neural Fuzzy and Wavelet Artificial Neural Network (WANN) for flow prediction of the Pungwe River. Water flow prediction can play an important role in water resources management over river basin, in this work different methods have been use to predict Monthly flow. The main objective of this research project is to analyze the monthly stream flow prediction in the Pungwe river, so as to make suitable decisions in dry or wet spells also to resolve probable conflicts about water recourses.

Nowadays many studies of rivers flows and aplications of artificial networks to varios aspects of hydrological modelling has ben constituted interesting to some scientists(Nourani et al., 2013; Santos and Silva 2013;Nayak *et al.*,2013, Muhanmadi *et al.*, 2008; Sreekath et al 2009; Partal and Kisi, 2007 and Seo *et al.*, 2014). The interest of this Hydrological modeling is to improve performance of artificial neural networks (ANN) to predict seasonal time series. Several structures of ANN presented to predict seasonal time series.

Several scientific paper of Hydrology about streamflow prediction methods, time series analysis have been demonstrated the good performance of WANN, ARIMA and WANFIS models and inclusion richer information and good systematic way to modeling flow(Santos and Silva 2014; Rezaeianzadeh et all.,2014; Nejad and Nourani 2012; Kisi, 2007 and Kisi, 2008).

Nonetheless to use wavelet transforms for time series treatment and ANN although flexibility and applicabiliy are still present difficulties (Neurani *et all 2013*, Santos and Silva, 2013). The application of some methods to pre-process imput data has been highlighted as an efficient alternative to improve the performance of ANN models like wavelet which has recently received attentions because provide useful decomposition os original time series into high to low frequence components.

The knowledge of stream flow system along the Pungwe River will help to plan how many Hydro power plants can be built along the river as well as other consumption such as; domestic, animal, and irrigation, so as avoid possible conflicts of sharing water along the river basin. Monthly modeling and prediction of Pungwe river flow was used by wavelet neural network (WNN) method in combination with neural networks (NN) and Neural Fuzzy Model using monthly flow data from one gauging stations in Pungwe river basin in Mozambique. Other Motivation of this research investigation is Renewable energy has been considered as one of the key elements to address climate change issues due to global warming and because the fossil fuels contribute to increase greenhouse gases (Kisi,2006, Kisi 2008, Nourani *et al,*. 2014; Naghizadeh *et al,*. 2012).

2. Description of the Study Area

The Pungwe river basin is located between latitudes 18S and 20S and the longitudes 33E and 35E and covers an total catchment area of some 31150 km^2 according to the new estimate, with a perimeter of 318.08 km. The Pungwe river is born in eastern highlands in Zimbabwe and flows eastwards through the Mozambican provinces of Manica and Sofala on its way to the Indian Ocean at Beira. The Zimbabwean part of the basin is 1 460.7 km^2 and the Mozambican part is 29 689.8 km^2.

The climate of Pungwe River basin stretches over two climate types, the is tropical of 6 months dry season and the remaining 6 months is rainy season annually. The average annual precipitation is 2000 mm the months of December to February with an average monthly rainfall of 300 mm in the month of February. In the mountains as shown in Figure 1 there is normally rain every month of the year, with a concentration in November-April, while the area east of the mountains has a pronounced concentration of the rain for the warm season November-April, and mainly no precipitation at all from May to October. In the eastern region, near Beira, the climate is classified as tropical humid, with a temperature variation from 22° in July to 29°C in January. The mean rainfall varies from 300 mm in January to 20 mm in July.

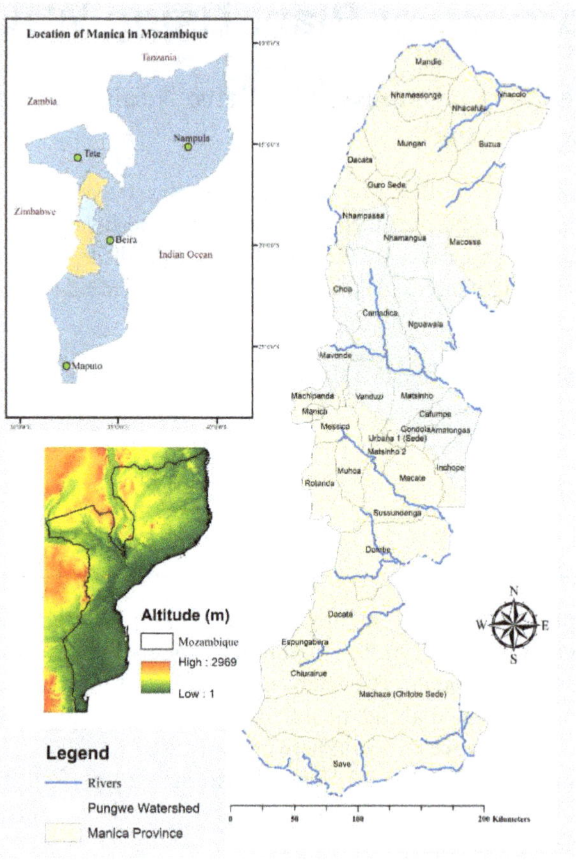

Figure 1. *Location of Pungwe river basin in Manica, Mozambique.*

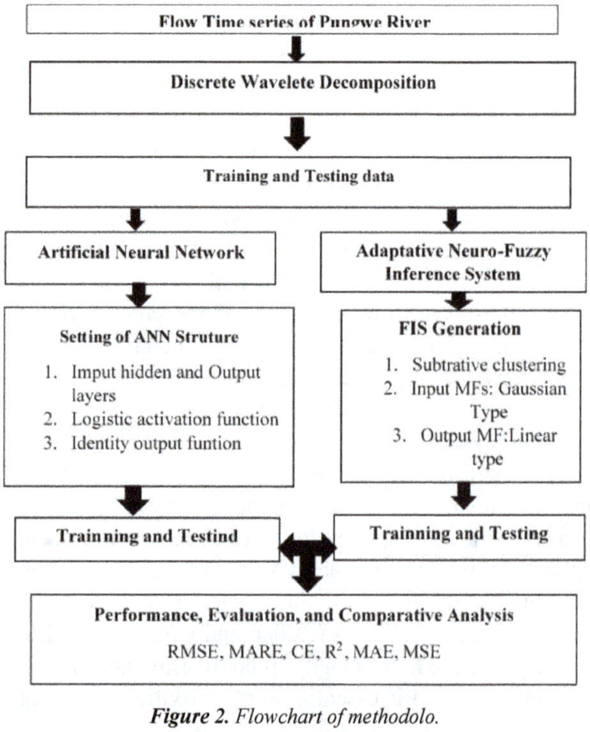

Figure 2. *Flowchart of methodolo.*

3. Methodology

For the purpose of this research and to achieve the objectives the following activities we carried out Literature :(Santos and Silva, 2014; Khan, 2012; Nourani *et al.*, 2013; Santos and Silva 2013; Nayak et al.,2014; Muhanmadi et al.,2010; Sreekath *et al,.* 2009; Partal and Kisi 2007, Seo *et al.*, 2014), materials, Matlab technology and data collection for 588 month and 49 years.

The Figure 2 its shows a flowchart for flow predition using artificial neural networw and adaptive neuro-fuzzy interference system.

3.1. Wavelet Analysis and Neural Network Model

The Wavelet transformation is a mathematical technique that transforms a signal in the time domain to frequency domain and by integral calculation, this transformation is described by equation 1. Wavelet neural network, a time series decomposes into the higher and low frequency components, multiple levels of details, sub-time series, which provide an interpretation of the original time series structure and history in both the time and frequency domains using a few coefficients (Badrzadeh, 2013; Alizdeh et al,. 2015, Nayaka *et al,.* 2013 and Solgi *et al,.* 2014, Chiu,1994).

$$W(a,b) = \int_{-\infty}^{+\infty} x(t)\phi_{a,b}(t)\,dt \qquad (1)$$

$$\phi_{a,b}(t) = \frac{1}{\sqrt{|a|}}\left(\frac{t-b}{a}\right) \qquad (2)$$

Where *x (t)* represents the temporal signal, $\varphi(t)$ is a mother wavelet function, in time and frequency domain and *a* is scale parameter; *b* is a position parameter and W (a, b) is the wavelet coefficients.

$$\int_{-\infty}^{\infty} \phi(t)\,dt = 0 \qquad (3)$$

$$0 < \int_{-\infty}^{\infty} |\phi(t)|^2\,dt < \infty$$

Wavelet transformation is more effective instrumental than the Fourier transform during the studying non-stationary time series. The main advantage of wavelet transformation is the ability to simultaneously obtain information on the location and frequency of a signal, while Fourier Transformations its separates a time series in to sine waves of various frequencies. There are two types of wavelet transform where are Continuous wavelet Transform (CWT) and discrete wavelet transform (DWT).

The CWT calculations requires a significant amount of computation period and resoiurce. Considering the discrete nature of observed data of flow in the time series. DWT the original signal time series, passes through two complementary filter and emerges as one approximation to detail components, The DWT is most preferred in hydrological prediction(Kim and valdes,2003, Neurani et al.,

2011; Honey et all, 2013; Dibike and solomatie,2001; kisi,2006)

In practical applications, hydrology researchers have access to a discrete time signal rather than continuous time signal (Santos and Silva 2014, Khan 2012, Nourani et al., 2013).

The same authors advise for the purpose of prognosis hybrid model ANN and ANFIS following the scheme represented in the figure below. The Figure 3 show the schematic diagram of hybrid wavelet neural network model with one input and the time series signal was decomposed by wavelet transform and this sub signal as input into neural network.

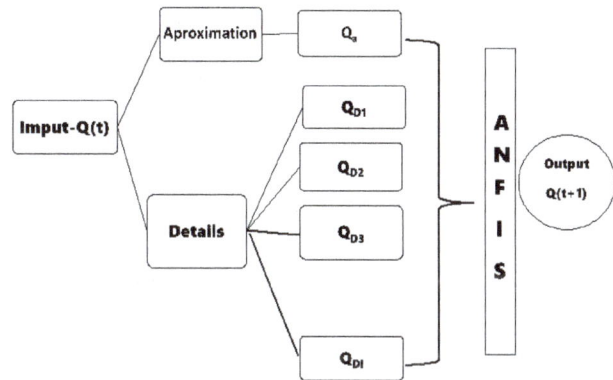

Figure 3. *Schematic diagram of hybrid WANN model.*

In terms of development strategies of objective of this work, one question options presented is about kind of wavelet transform used (CWT,DWT) and combination with the ANNN and ANFIS. The table 1. summarizes the possibilities of the building models.

Table 1. *Wavelet transform strategy used.*

Model Struture	Wavelet transform	Model ANFIS
Model 1- NN	Without Pré-Treatment	Neural Network
Model 2-ANFIS	Without Pré-Treatment	Neural-Fuzzy
Model 3-CWTNN	CWT	Neurais Network
Model 4-DWTNN	DWT	Neural Network
Model 5-CWTANFIS	CWT	Neural-Fuzzy
Model 6-DWTANFIS	DWT	Neural-Fuzzy

3.2. Artificial Neural Network (ANN)

Artificial neural network are computational and mathematical model with a wide ranges of applications have great ability in forecasting modeling for nonlinear hydrological time series (flow, and precipitation).

The ANN Procedure through the combination of neural networks to prediction components of frequency up to 5 layers, subsequently combining the simulated network values to reconstruction of the original signal by wavelet technique reconstruction, This Model is basically works with inverse decomposition process as shown in figure 5. For Artificial neural networks, the backpropagation algorithm automatically acquires the knowledge, but the learning process is relatively slow and analysis of the trained network is difficult (Jain *et. al.*, 1999; Badrzadeh et al., 2013).

Input Leyer   hidden leyer   output leyer

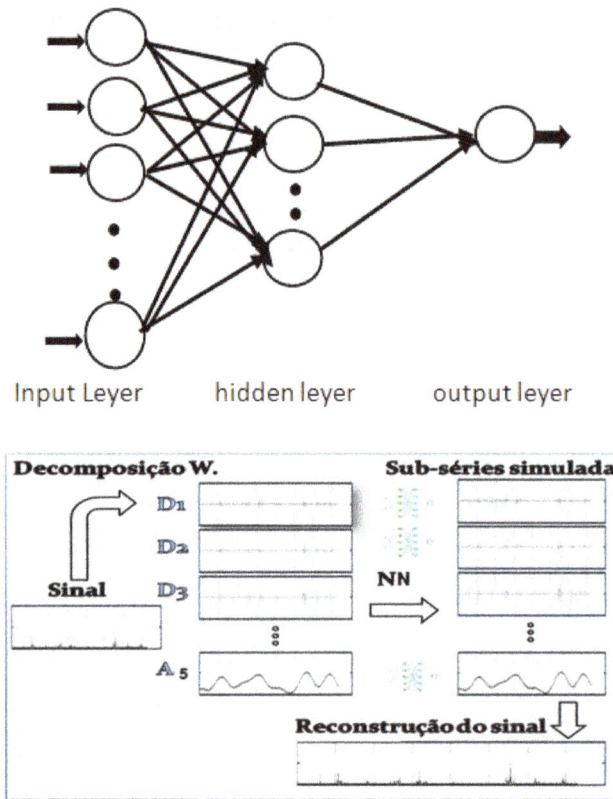

Figure 5. Artificial Neural network Model Arquiteture.

3.3. Wavelet Artificial Neuro Fuzzy Inference System(WANFIS)

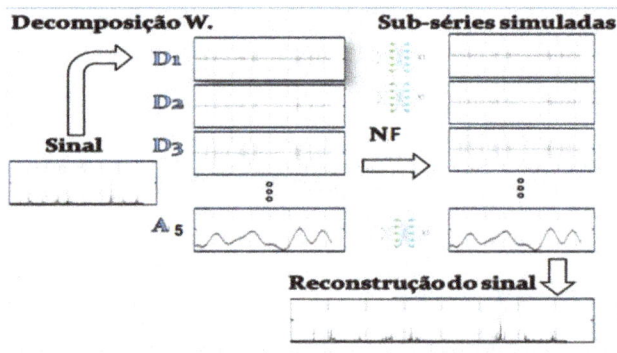

Figure 4. Neuro fuzzy Model Arquiteture.

The WANFIS model was develop by using the wavelet sub series as ANFIS input, information of input and output data is convert into linguistically interpretable. In This Model, the Procedure is the same way like Artificial Neuro Network prediction components up to five layers of frequency, than combining the simulated network values for the reconstruction of the original signal using wavelet technical as shown in figure 4. Fuzzy logic systems, which can reason with imprecise information, are good at explaining their decisions but they cannot automatically acquire the rules they use to make those decisions For example, while neural networks are good at recognizing patterns, they are not good at explaining how they reach their decisions (Jang, 1993; Chiu, 1994; Neurani et al., 2011).

3.4. The Model Performance

The performance of different forecasting models was accessed in terms of goodness to fit once each of the model structures is calibrated using the training, validation data set and testing data. Degree of correlation was measure (R^2) in equation 6. The equation 4 is coefficient of correlation(CE) was used to compare the goodness to fit between the measured flow and the simulated flow, mean-squared error (MSE) in equation 5 were used to evaluates the variance of error (Nourani *et al.*, 2013; Santos and Silva 2013; Nayak *et al.*, 2014, Muhanmadi *et al.*, 2010;Sreekath et al 2009, Partal and Kisi 2007). To calculate the root-mean-square error (RMSE), mean absolute error (MAE) and mean absolute relative error (MARE) will use remaining equations in the same literature.

$$CE = \frac{\sum_{l}^{n}\left(Q_{obs} - \overline{Q_{obs}}\right)\left(Q_{pre} - \overline{Q_{pred}}\right)}{\sqrt{\sum_{l}^{n}\left(Q_{obs} - \overline{Q_{obs}}\right)^2}\sqrt{\sum_{l}^{n}\left(Q_{pred} - \overline{Q_{pred}}\right)^2}} \quad (4)$$

$$MSE = \sqrt{\frac{\Sigma\left(Q_{obs} - Q_{pre}\right)^2}{n}} \quad (5)$$

$$EFF \text{ or } R^2 = 1 - \frac{\sum_{l=1}^{N}\left(Q_{obs} - Q_{Pre}\right)^2}{\sum_{l=1}^{N}\left(Q_{obs} - \overline{Q_{obs}}\right)^2} \quad (6)$$

$$MAE = \frac{\sum_{l}^{n}\left|\left(Q_{obs} - Q_{pred}\right)\right|}{n} \quad (7)$$

$$MARE = \frac{1}{n}\sum_{l}^{n}\frac{\left|Q_{obs} - \overline{Q_{obs}}\right|}{Q_{obs}} \quad (8)$$

$$RMSE = \sqrt{\frac{\sum_{l}^{n}\left(Q_{obs} - QPred\right)^2}{n}} \quad (10)$$

Where Q_{obs} is corresponds the steam flow observed value, Q_{pre} is corresponds the predicted flow rate value, and $\overline{Q}$ is the average value of flow rate. MSE is mean square error and CE is coefficient of efficiency and n is number of data point used and n is number of observed data or sample of size.

The parameters used to evaluate and the validation model and the best conjunction wavelet transform with neural networks or neuro fuzzy predict are: The mean square error (MSE), defined by (Equation 5) and linear regression (equation 6) between actual the real data and simulated by the network.

4. Results and Discussions

The realiability of experimental setup is established by

comparing the performance of ANN and wavelet hybrid models of 588 month and 49 years. Due to the temporal serie of hydrological process its recomended to use 45 yers for first part for training equivalente of 80 percent than, the rest 4 yers for verificatior or test equivalente 20 percent of time series by figure 6.

Table 2 and figure 8 below shows training and test results of forecasting of flow in pungwe river basin bybusing WANFIS end WANN model with corresponding the Root-Mean-Squere (RMSE), Degree of correlation (R^2) and Nash-Sutcliffe of efficiency (NSE) and coeficient of correlation (R).

Table 2. Performance results of the WANN and WANFIS model with diferent situation.

	Decomposition		Training		Testing	
	WANN	WANFIS	WANN	WANFIS	WANN	WANFIS
R^2	Sym3,3	Db3,4	0.8282	0.9061	0.6894	0.8706
RMSE	Sym3,2	Db3,4	190.1096	139.57	179,60	123.68
NSE	Sym3,2	Db3,4	0,843	0.861	0.4682	0.5261
R^2	Coifl,3	Db5,3	0.6150	0.729	0.5382	0.5454
RMSE	Coifl,3	Db5,3	199.42	178.83	172.57	164.212
NSE	Coifl,3	Db5,3	0.8047	0.79	0.7813	-0.421

The value of degree correlation(R^2) in WANFIS in Db3,4 is higher than WANN where in general the model shoul be greater than 0.5 and is considered as an acetable macth to the real system. Other situation using Nash-Sutclifficient of coeffient of efficiency I can see is varies between 0.46 to 0.843 which it mean the model efficient corresponds a perfect prediction.

Table 3 shows training and test results of forecasting of flow in pungwe river basin bybusing ANFIS end ANN model with corresponding the Root-Mean-Squere(RMSE), Degree of correlation(R^2), Nash-Sutcliffe of efficiency(NSE) and coeficient of correlation(R), Coefficient of Correlation(CE) and Mean Absolute Error(MAE).

Table 3. Performance results of the ANN and ANFIS model with diferent situation.

	Time		Training		Testing	
	ANN	ANFIS	ANN	ANFIS	ANN	ANFIS
R^2	2,5	2,5	0.8710	0.978	0.7621	0.982
RMSE	2,5	2,5	128.61	185.43	105.06	146.06
NSE	2,5	2,5	0.9293	0.7813	0.8667	0.5831
MAE	2,5	2,5	134.3	107.07	122.05	97.45
CE	2,5	2,5	0.6770	0.872	0.567	0.856
R^2	2,9	2,9	0.9326	0.971	0.9254	0.976
RMSE	2,9	2,9	115.64	132.76	85.4510	115.67
NSE	2,9	2,9	0.9109	0.8021	0.6420	0.7081
MAE	2,9	2,9	118.85	197.14	97.020	154.3
CE	2,9	2,9	0.9016	0.8902	0.8632	0.8871

Table 3 shows the performance and summarize the values of ANFIS and ANN models for difernt layers, and we verify that the result are litle similar. The Root-Mean-Squere(RMSE) for both model was good betwen the 85.45 to 128.61 m³/s. Based on the results of table 2 and table 3, it was noticed that

the number of decomposition levels had considerable impact on the results. Since the random parts of original time series were mainly in the first resolution level, obviously the prediction errors were also mainly in the first resolution level.

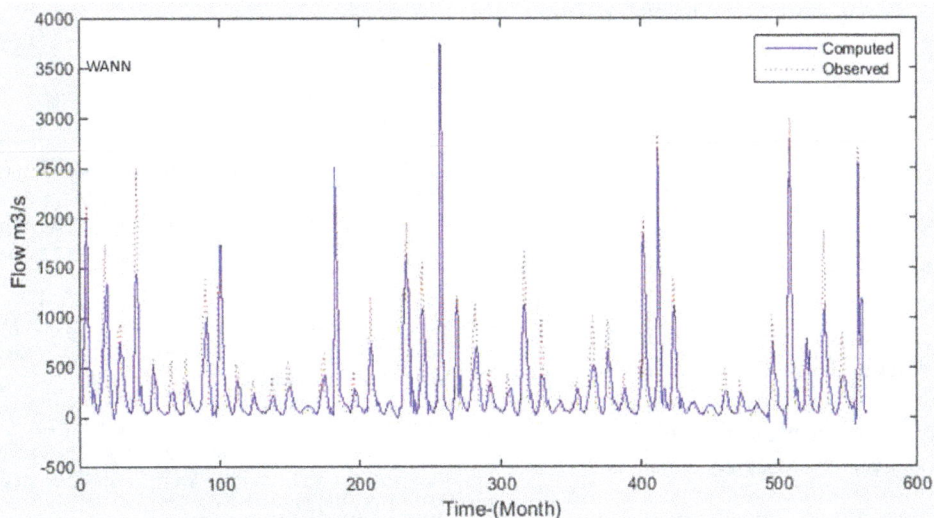

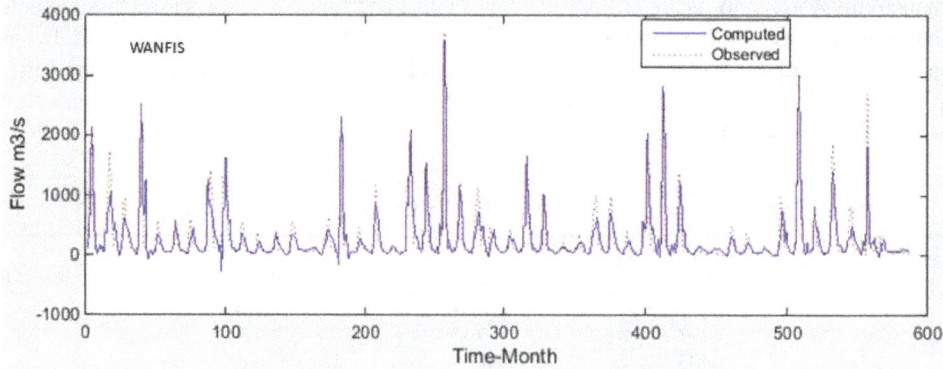

Figure 6. Time series between obserced and computed WANN and WANFIS.

The figure 8 represent the scattter plots of observed flow data and and simulation during the validation time. The different between the regration line and 45° are close or de

regretion value for all situation is aproximatley 1 and give clear that this data are corelected and it can be useful to built hydropower plant.

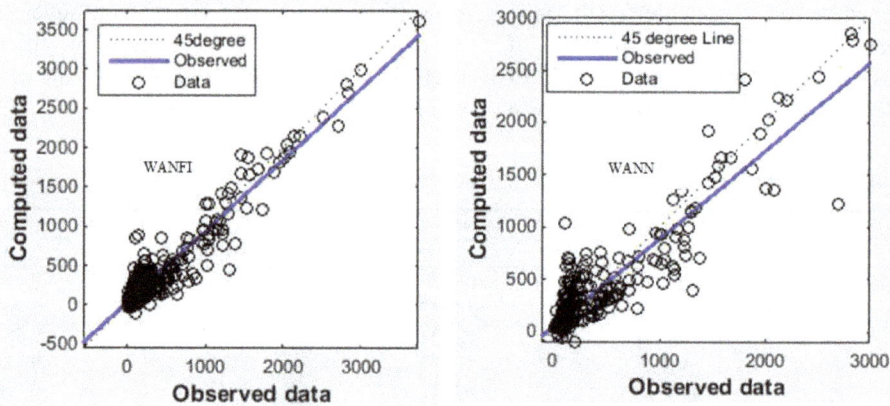

Figure 8. Regression of WAFIS and WANN model.

5. Conclusions

The dataset used in this modelling experiment was a time series of average monthly river discharge during the period 1960–1994, which was collect from the Ara centro gauging hydrological station in the pungwe river basin Manica, and this study applied time series models for different matlab program like ANN,WANN, ANFIS, WANFIS at Pungwe basin station in Manica, Mozambique including the statistics tools ,RMSE, MAE, R^2, were adopted to evaluate models performance. It was found that for almost all lead times WANN and WAFIS model has given better and consistent results compared to conventional ANN and ANFIS model. Also, the effect of decomposition level on WANN models efficiency was studied.

It was observed through this paper that the hybrid model is really have high efficiency because it possessed a high value linear regression 0.9548 combined with a good values of Nash-Sutcliffe of efficiency(NSE) between 0.46 to 0.86.

Regards the flow prediction I can conclude that the flow presented in this river has a strong correlation and is good site to build a Small hydro power plant in Manica, than the climatic changes is not creat any problem about water to generate power.

Acknowledgement

The authors would like to thank the Department of Water Resources at Lund University and the Chemical department in Eduardo Mondlane University for help to carry out this beautifull Project.

References

[1]   EDM (Electricidad de Mocambique), 2015. Stastical Summary. www.edm.co.moz.

[2]   IEA, 2009. IEA Energy Statistics - Energy Balances for Mozambique. Available at: http://www.iea.org/stats/balancetable.asp Accessed May, 2015].

[3]   Cuamba B.C., Uthui R. Chenene M.L. et al. (unpubl.) Identification of areas with likely good wind regimes for energy applications in Mozambique. Eduardo Mondlane University, Maputo

[4]   Santos, C. A. G., Silva, G. B. L., (2014) Daily streamflow forecasting using a wavelet transform and artificial neural network hybrid models. Journal des Sciences Hydrologiques, 59 (2) 312–324.

[5] Solgi, A., Radmanesh, F., Zarei, H., Nourani, V., (2014), Hybrid Models Performance Assessment to Predict Flow of Gamasyab River International journal of Advanced Biological and Biomedical Research Volume 2, Issue 5, 2014: 1837-1846.

[6] O. Kisi.,(2008). "Stream flow forecasting using neuro-wavelet technique," *Hydrological Processes*, vol. 22, no. 20, 4142–4152.

[7] FUNAE (Fundo de Energia), 2015. Annual Report. www.funae.co.moz.

[8] V. Nourani, M. T. Alami, and M. H. Aminfar, (2009), "A combined neural-wavelet model for prediction of Ligvanchai watershed precipitation," *Engineering Applications of Artificial Intelligence*, Vol. 22, no. 3, 466–472.

[9] Nourani V, Hosseini Baghanam A, Adamowski J, Gebremichael M, (2013), Using self-organizingmaps and wavelet transforms for space–time preprocessing of satellite precipitation and runoff data in neural network based rainfall–runoff modeling. J Hydrol 476:228–243.

[10] Sreekanth, P., Geethanjali, D.N., Sreedevi, P.D., Ahmed, S., Kumar, N.R., Jayanthi, P.D.K., (2009), Forecasting groundwater level using artificial neural networks. Current Science 96 (7), 933–939.

[11] Mohammadi, K., (2008), Groundwater table estimation using MODFLOW and artificial neural networks. Water Science and Technology Library 68 (2), 127– 138.

[12] Nourani, V., Hosseini, A., Adamowski, J., Kisi,O, (2014), Applications of hybrid wavelet–Artificial Intelligence models in hydrology: A review, Journal of Hydrology 514 , 358–377.

[13] Partal, T., Kisi, Ö. (2007), Wavelet and neuro-fuzzy conjunction model for precipitation forecasting. Jornal of Hydrology, 342,199-212.

[14] Krishna, B.; Satyaji Rao, Y. R., Naya, P.C. (2011) Times Series Modeling of River Flow Using Wavelet Neural Networks, Journal of water resource and protection, 3, 50-59.

[15] Nayak P.C., Venkatesh B., Krishna, B., Sharad, K J., (2013) Rainfall-runoff modeling using conceptual, data driven, and wavelet based computing approach. Journal of Hydrology 493 57–67.

[16] Rezaeianzadeh, M., Tabari, H, Yazdi, A. A., Isik, S., and Kalin, L. (2014). "Flood flow forecasting using ANN, ANFIS and regression models." Neural Computing and Applications, Vol. 25, Issue 1, pp. 25-37, DOI: 10.1007/s00521-013-1443-6.

[17] Nejad, F. H., and Nourani, V. (2012). "Elevation of wavelet denoising performance via an ANN-based streamflow forecasting model." International Journal of Computer Science and Management Research, Vol. 1, Issue 4, pp. 764-770.

[18] Kisi, O. (2006). "Streamflow forecasting using different artificial neural network algorithms." Journal of Hydrologic Engineering, Vol. 12, Issue 5, pp. 532-539, DOI: 10.1061/(ASCE)1084-0699(2007)12: 5(532).

[19] Kim, T.W., Valdes, J.B., 2003. Nonlinear model for drought forecasting based on a conjunction of wavelet transforms and neural networks. Journal of Hydrologic Engineering 6, 319–328.

[20] Nourani, V., Kisi, Ö., Komasi, M., 2011. Two hybrid artificial intelligence approaches for modeling rainfall-runoff process. Journal of Hydrology 402, 41–59.

[21] Nourani, V., Baghanam, A.H., Adamowski, J., Gebremichael, M., 2013. Using selforganizing maps and wavelet transforms for space–time pre-processing of satellite precipitation and runoff data in neural network based rainfall-runoff modeling. Journal of Hydrology 476, 228–243.

[22] Dibike, Y.B., Solomatine, D.P., 2001. River flow forecasting using artificial neural networks. Physics and Chemistry of the Earth, Part B: Hydrology, Oceans and Atmosphere 26, 1–7.

[23] Badrzadeh, H., Sarukkalige R., Jayawardena,A.W., 2013. Impact of multi-resolution analysis of artificial intelligence models inputs on multi-step ahead river flow forecasting. Journal of Hydrology 507,75–85.

[24] Jang, J.S., 1993. ANFIS: adaptive-network-based fuzzy inference system. IEEE Transactions on Systems, Man and Cybernetics 23, 665.

[25] Chiu, S., 1994. Fuzzy model identification based on cluster estimation. Journal of Intelligent & Fuzzy Systems 2.

[26] Uamusse, M., Persson, K. and Tsamba, A. (2014) Gasification of Cashew Nut Shell Using Gasifier Stovein Mozambique. Journal of Power and Energy Engineering, 2, 11-18. doi: 10.4236/jpee.2014.27002.

5

Finding Better Solutions to Reduce Computational Effort of Large-Scale Engineering Eddy Current Fields

Dexin Xie[1], Zhanxin Zhu[2], Dongyang Wu[1], Jian Wang[2]

[1]School of Electrical Engineering, Shenyang University of Technology, Shenyang, China
[2]TBEA Shenyang Transformer Co., Ltd., Shenyang, China

Email address:

xiedx2010@163.com (Dexin Xie), zzxin111@163.com (Zhanxin Zhu), shineast_521@163.com (Dongyang Wu)

Abstract: In the finite element analysis of the engineering eddy current fields in electrical machines and transformers there are the problems such as the huge scale of computation, too long computing time and poor precision which could not meet the demand of engineering accuracy. The current research situation and difficulties of these problems are analyzed in this paper mainly from the aspect of computation methodology. The methods to deal with these problems, e.g., homogenization models of the laminated iron core, the sub-domain perturbation finite element method, domain decomposition method, and EBE (Element by Element) parallel finite element method are described. Their advantages and limitations are discussed, and the authors' suggestions for the further research strategies are also included.

Keywords: Engineering Eddy Current Fields, Huge Scale of Computation, Homogenization of Laminated Iron Core, Finite Element Method, Sub-Problem Perturbation Finite Element Method, EBE Parallel Finite Element Computation

1. Introduction

The electromagnetic fields in most of the electrical devices, such as electrical machines and transformers, are classified as quasi-static field. The quasi-static electromagnetic field in which conductive materials are included is called eddy current field too. It is of great significance to calculate the distribution of the eddy current field with its losses induced in the conductive materials accurately for optimal design and safe operation of the electrical devices. As an example, there are many metal structural components in power transformer. The eddy current losses in the components induced by the variation of leakage magnetic field are one of the heating sources for temperature rise. Although decreasing the whole losses in the structural components is important, the local loss concentration due to non-uniform distribution of the losses is the direct reason of local over-heating and operating faults, which deserve to pay close attention. Even if the computational technology has got rapid development in nowadays, to improve the computation precision is still a difficult task for numerical analysis of eddy current fields in super-huge type of power transformers. The features of this kind of computation include very small skin depth of ferromagnetic material, nonlinear and anisotropic electromagnetic characteristics of the structure components, structural discontinuity of the laminations, three dimensional feature of the spacial configuration, non-sinusoidal variation of the field physical quantities, and so on. Therefore, although there have been some commercial softwares of electromagnetic field analysis used in manufacture enterprises of power transformers extensively for performance verification and aided design, the calculated results are not agree well with the experimental ones. The crux of the problem is in the aspects of the conflict between the huge computational scale and high precision required, the difficulty of magnetic characteristic modeling for laminated iron core, the higher harmonics of exciting electric current and magnetic field being not easy to included, and lack of the data which can describe the electromagnetic characteristics of materials of the structural components accurately and completely. The aim of this paper is to analyze the current study situation of these problems, discuss the research strategies, and propose the further research directions from the authors' point of view.

2. Simulation of the Material Characteristics of Laminated Iron Core

2.1. Difficulty of the Magnetic Characteristic Modeling of Laminated Iron Core

Proper modeling of material characteristics is the basis to improve the accuracy of electromagnetic field computation. However, it is not easy to simulate the electromagnetic characteristics of laminated iron core. The iron core and magnetic shield of power transformer consist of grain-oriented silicon steel sheets laminated, and the thickness d of each sheet is around 0.3 mm or even less, see Fig. 1(a). In the plane of the sheet and the angles with the rolling direction ranging from 0^0 to 90^0, the magnetic characteristics including permeability and loss per kg are angle-dependent anisotropic, which need to be depicted by the so-called two dimensional magnetic characteristic model. M. Enokizono and N. Soda proposed the well-known E&S model [1] in 2000 and kept improving it. With this model not only the loss in silicon steel sheets resulted from alternating magnetic field can be calculated, but the losses due to local rotating magnetic field can also be computed. In order to incorporate the E&S model into finite element (FE) analysis the 2D magnetic characteristic test device [2] has to be used to obtain a great deal of data for different angles in the sheets, different exciting levels and moments of a sinusoidal period, as the computation basis. However, this model has been confined to 2D computation [3, 4], not being extended to 3D yet. The reason is that at first the specialized 2D test device has not been used commonly, and more importantly this model involves time-discretization and iterations, which will cause the calculational scale over-increasing thus lead to unworkable computation effort if the model is used in 3D analysis directly. Furthermore, for 3D analysis more factors have to be considered, e. g., the material nonuniformity of the laminated iron core, the increasing of the unknowns, the ill-conditioned degree of the coefficient matrix, and so on.

It is known that for the magnetic field in iron core under a sinusoidal excitation when the direction of the magnetic field is parallel to the surface of the sheet the distribution of its magnitude and phase are nonuniform along the thickness of the sheet. At this condition the 1D classical analytic solutions of flux density $\dot{B}_z(x)$ and eddy current density $\dot{J}_y(x)$ are given as [5, 6]

$$\dot{B}_z(x) = \dot{B}_0 \frac{\text{ch}\sqrt{j\varpi\mu\sigma}x}{\text{ch}\sqrt{j\varpi\mu\sigma}\dfrac{d}{2}} \qquad (1)$$

$$\dot{J}_y(x) = \dot{J}_0 \frac{\text{sh}\sqrt{j\varpi\mu\sigma}x}{\text{sh}\sqrt{j\varpi\mu\sigma}\dfrac{d}{2}} \qquad (2)$$

where x stands for the coordinate along the direction

perpendicular to the sheet, $\dot{B}_0$ and $\dot{J}_0$ are the amplitude of $\dot{B}_z(x)$ and $\dot{J}_y(x)$ at $x = \pm d/2$ respectively. ϖ, μ, σ are angle frequency, permeability and conductivity respectively. Fig. 1(b) shows the magnitude of flux density and eddy current density versus x respectively, in which the component of magnetic field perpendicular the sheet is neglected, although it exists in practical operating condition.

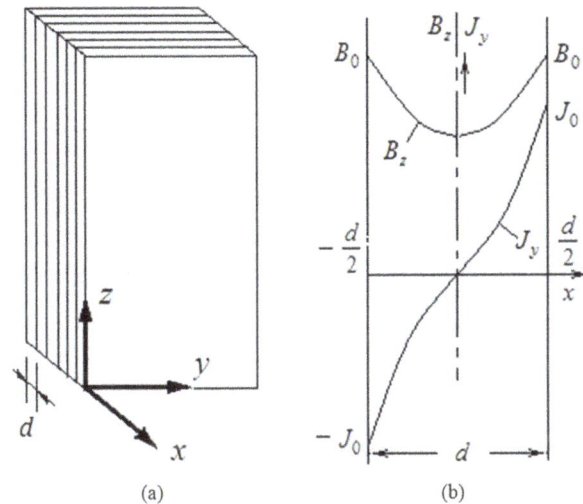

Figure 1. (a) Sketch map of laminated iron core (b) B_z and J_y versus x.

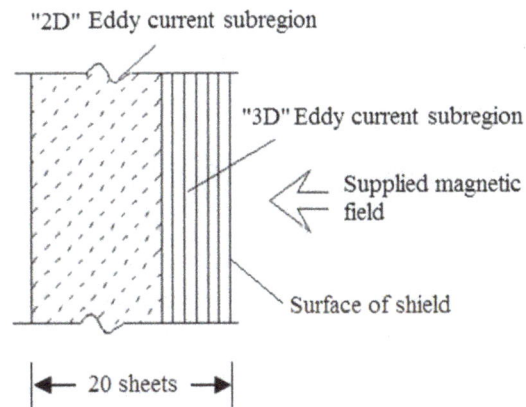

Figure 2. FEM model of the eddy current in laminated sheets of magnetic shield [7].

Considering the nonuniformity of the magnetic field in the thickness of the sheet, the sheet itself should be meshed in 3D FE analysis. It has been done so as in [7], in which the TEAM Problem P21-M1/M2 is chosen as a numerical model and the single sheet of magnetic shield is meshed. The magnetic shield is made of 20 sheets with a thickness of 0.3mm for each in the model, and the supplied magnetic field is perpendicular to the surface of the sheets, as shown in Fig. 2. According to the results of experiment, the distribution of magnetic field in the 6 sheets which are near the surface of the shield is three dimensional, while in the other sheets further from the surface the magnetic field direction is parallel to the sheet surface basically. Therefore, the laminated region is divided into two sub-regions of "2D" and "3D" in 3D FE analysis, see Fig. 2. In the "3D" sub-region each sheet is meshed into 3 layers, and the insulation layer between two sheets is also meshed with one layer, while for the "2D" sub-region the electromagnetic characteristics are given using more rough meshes with the conventional homogenization of iron core material, which has been used extensively and will be illustrated in Section 2.2. The calculated results of this model satisfied the demand of engineering precision basically. However, the model is too small compare with the practical huge transformer products. For the latter, the so refined meshes lead to over huge computational scale so that the computation task cannot be fulfilled.

2.2. Conventional Rough Homogenization of Laminated Iron Core

To avoid too large scale of computation in FE analysis the simplified modeling method for the electromagnetic characteristics of the laminated iron core has been used commonly, that is, taking the laminated stack as a continuum and its electromagnetic characteristics are given uniformly as follows, i. e. [8]

$$\mu = \begin{bmatrix} \mu_x & 0 & 0 \\ 0 & \mu_y & 0 \\ 0 & 0 & \mu_z \end{bmatrix} \quad (3)$$

$$\sigma = \begin{bmatrix} \sigma_x & 0 & 0 \\ 0 & \sigma_y & 0 \\ 0 & 0 & \sigma_z \end{bmatrix} \quad (4)$$

where μ and σ are permeability and conductivity respectively, and their components can be determined, depending on the connection in series or in parallel of the silicon steel sheets and insulation layers according to the magnetic and electric circuit theory. After proper simplifying the values of the components can be given as $\sigma_x = 0$, $\sigma_y = \sigma_z = \sigma_s$, $\mu_x = \mu_0/(1-k)$, $\mu_y = k\mu_{sy}$, $\mu_z = k\mu_{sz}$ for the laminated direction shown in Fig.1(a), where k is the lamination coefficient, μ_0 is the permeability of vacuum, and the subscript s stands for silicon steel sheet. This method has been used extensively, and some commercial softwares of electromagnetic field analysis also use the method in a similar way. However, for the computation of eddy current field and its losses due to leakage magnetic field in practical transformer products this modeling of the homogenization is too rough to obtain satisfied results.

2.3. New Development of Magnetic Characteristic Modeling of Laminated Iron Core

Reference [9] proposes a more refined homogenized method for laminated iron core, with which the eddy current density in silicon steel sheet is divided into two components, that is, the component 1 induced by the alternating flux density parallel to the sheet and the component 2 due to the

flux density perpendicular to the sheet. To incorporate the method into 3D FE analysis of magnetic vector potential and electric scalar potential scheme, for the component 2, the electromagnetic characteristic modeling is the same as that described in Section 2.2, the conventional homogenization, while for the component 1, an addition item is added into the Galerkin weak formulation.

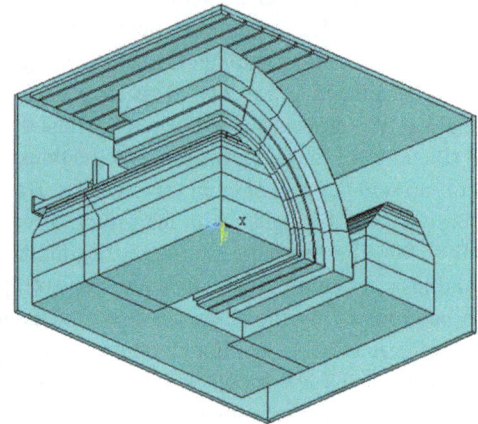

(a)

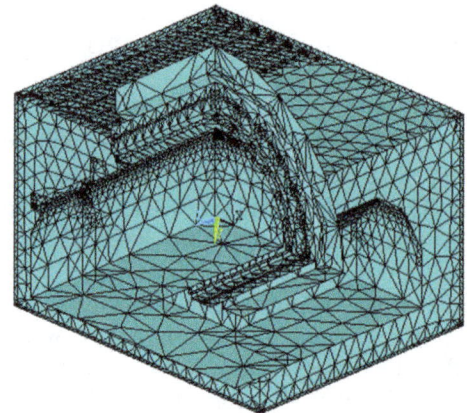

(b)

Figure 3. Single phase transformer model of 380MVA/500 KV (a) Structural diagram (b) Meshes of the transformer model.

For the low frequency case, the addition item is given as

$$\left(\frac{\sigma d^2}{12} \partial_t \mathrm{curl}A, \mathrm{curl}N \right)_{\overline{V}} = -\left(J_\beta, N \right)_{\overline{V}} \quad (5)$$

where N is the basis function, A is the magnetic vector potential, J_β corresponds to the eddy current density generated by the flux density parallel to the sheet, which is regarded as a kind of source electric current density, and $\overline{V}$ is the laminated region. The deducing of (5) is on the condition that the skin effect in a sheet is neglected, i. e. the distribution of flux density along the thickness of a sheet is uniform which equals to the average flux density, and the distribution of J_β is linear. That is, contrasting with J_y in Fig. 1(b), J_β is a straight line instead of a curve in this case. The method is used in a numerical model of laminated stack with FE computation in frequency domain and a satisfied result is obtained [9]. For more complicated calculated model, a practical product model of a 380MVA/500 KV single phase power transformer with the engineering frequency of 50 Hz is used by the authors, and its structure and FE meshes are shown in Fig. 3. The method proposed in [9] is extended to time domain and nonlinear permeability is considered [10, 11], the calculated total losses due to leakage magnetic field are agree with the experimental values and the error is around 6%. Figure 4 shows parts of the numerical results, i. e. the flux density distribution on the symmetry plane and the eddy current density distribution on the oil tank inner surface of the transformer model. It is obvious that with this method the good computed results are obtained with relative less and acceptable computational effort.

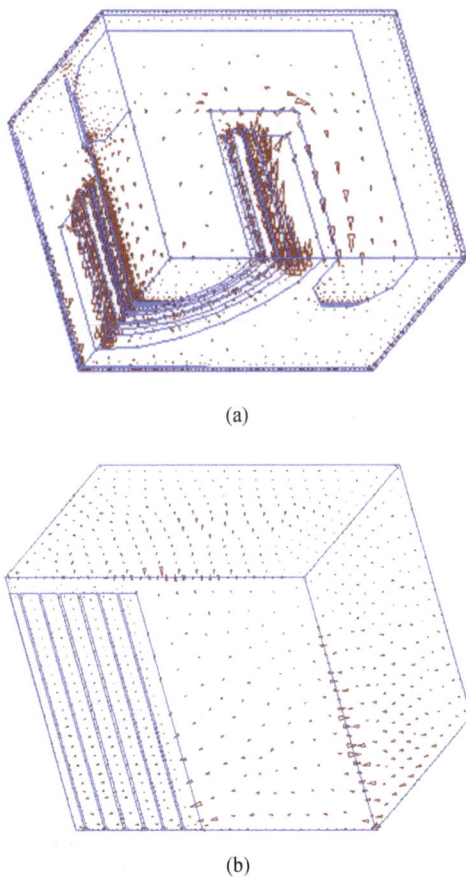

(a)

(b)

Figure 4. (a) Flux density distribution of the symmetry plane of the transformer, (b) Eddy current density distribution of the oil tank inner surface of the transformer.

For higher frequency, the skin effect cannot be neglected. A more accurate homogenization method is presented by [9]. Based on the classical analytic solutions of magnetic flux density and eddy current density, i.e. (1) and (2), a more complicated addition item is put into the weak formulation to simulate the eddy current generated by the flux density parallel to the sheet. Furthermore, the method has been extended from frequency domain to time domain, and the

insulating layers of finite width between the laminated sheets are considered too [12]. The test numerical model is a linear 3D axisymmetrical one and a good result is achieved. A quantitative conclusion is drawn by [12] that the simplified model for the low frequency case is valid up to a frequency for which the skin depth is equal to the half-thickness of the laminations, while the accurate model is valid for any frequency. However, till now the method hasn't been used in practical engineering model yet.

3. Transformation of Global Solution to Combination of Partial Solutions

Since the computational scale is very large for a global solution, trying to substitute combination of a certain partial solutions with less computational effort for global solution may be a novel strategy, among which the sub-problem perturbation FE method and domain decomposition method are both typical examples.

3.1. Sub-Problem Perturbation FE Method

The so-called perturbation FE method is proposed initially in [13]. The method is applied to eddy current nondestructive evaluation problems. To probe the flaw of metallic tube using external magnetic field created by exciting coil a 3D eddy current field calculation has to be carried out. However, compared with the dimension of the coil and tube, the size of the flow is too small, which results in the conflict of computation scale and accuracy. In fact, the computation of the tube eddy current field is much easier for the case without the flaw, e. g., through analytical solution, or adopting FE calculation of axisymmetric filed based on the symmetry of the tube, but when computing the eddy current field with the flaw, the known solution without the flaw cannot be used. To solve the problem an ingenious method is presented by [13]. The main idea is that the practical eddy current field with the flaw involves three fields, denoted here as field 0 for the tube with the flaw, field 1 for that without the flaw, and field 2 the perturbed field created by the flaw. The governing equations of field 2 can be obtained by subtracting the governing equations of field 1 from that of field 0. The 3 governing equation systems are all Maxwell equations, however, there is an additional item in the right side of the equations of field 2. Taking the curl equation of magnetic field as an example, that is

$$\nabla \times \boldsymbol{H}_2 = \sigma \boldsymbol{E}_2 + \boldsymbol{J}^i \qquad (6)$$

Equation (6) is the perturbation equation of magnetic field, where $\boldsymbol{J}^i = (\sigma_2 - \sigma_1)\boldsymbol{E}_1$, called as incident electric current density. It is $\boldsymbol{J}^i$ which causes the perturbation to field 1, and the perturbation results from the conductivity change of the flow region. $\boldsymbol{H}$ and $\boldsymbol{E}$ are magnetic field intensity and electric field intensity respectively, subscript 1 and 2 correspond to field 1 and 2 respectively, σ_2 is the conductivity of the flaw region, and σ_1 is that of the region outside of the flaw.

Obviously the field of the tube with flaw, the field 0, is equal to the sum of the field 1 and field 2 for the case of linear electromagnetic characteristics. The advantage of this perturbation FE method is that the solution of field 1 can be obtained easily; the computation of field 2 can be carried out using conventional FE method and is independent of field 1. Furthermore, the computational domain and mesh size of the field 2 are much less than that of field 1, because the affected area of the flaw is not very large, in general beyond 5~6 skin depth the flaw field may considered zero.

Getting the hint of perturbation FE method in nondestructive testing, reference [14] extends the method to the computation of the skin and proximity effects in conductors of any properties and shapes, in both frequency and time domains. Because two sub-problems have to be calculated, the developed method could be called as sub-problem perturbation FE method. The main points are as follows. The practical eddy current field (field 0) is regarded as the superposition of reference field (field 1) and perturbation field (field 2). The field 1 and field 2 are calculated respectively, and their computational effort is much less than that of computing the field 0 directly. The calculation methods in two cases in which the conductive or magnetic material is included in the solved domain, are proposed in [14], which will be described in Section 3.1.1 and 3.1.2 respectively and briefly.

3.1.1. Case of Electric-Conductive (Non-Magnetic) Material Included

Reference field (field 1): Set the solid conductive material as ideal conductor, i.e., the conductivity of which equals to infinite, so that the conductor can be excluded from the solved region. Provide that the normal component of magnetic flux density is equal to zero at the boundary of the region.

Perturbation field (field 2): Include the conductive domain in the solved region, in which the practical conductivity is assigned. Consider the difference of the conductivity from that of ideal conductor as the perturbation to field 1. At the interface of conductive and non-conductive domains set the tangential component of magnetic intensity as that of the calculated results of field 1, which is taken as the source of the perturbation field.

3.1.2. Case of Magnetic (Non-Electric-Conductive) Material Included

Reference field (field 1): Set the solid magnetic material as ideal magnetic conductor, i.e., the permeability of which equals to infinite, so that which is removed from the solved region. At the boundary adjacent to the ideal magnetic conductor the tangential component of magnetic intensity is set as zero.

Perturbation field (field 2): Include the magnetic material in the solved region, in which the practical permeability is assigned. Assign the normal component of flux density as that of the calculated results of field 1 at the interface of the magnetic and non-magnetic domains, which is taken as the source of the perturbation field.

3.1.3. Discussion

In the two cases described in Section 3.1.1 and Section 3.1.2 the solutions of the practical problem are all equal to the superposition of the solutions field 1 and field 2. For the calculation of field 1 the computational scale is reduced because that the conductive or magnetic region is removed so that the pressure of mesh generation is decreased. For the calculation of field 2 the solved region can be reduced as explained in Section 3.1. Fig. 5 and Fig. 6 are two numerical examples given by [14] for the two cases respectively.

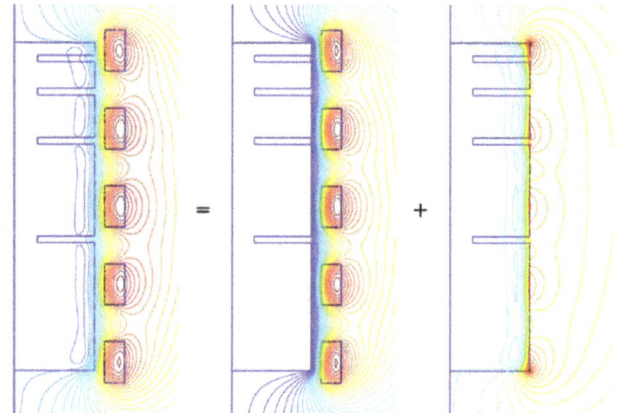

Figure. 5. *Magnetic flux lines of the system with a conductive non-magnetic core [14]. Left: for the conventional FE solution, Middle: the reference solution, Right: the perturbation solution.*

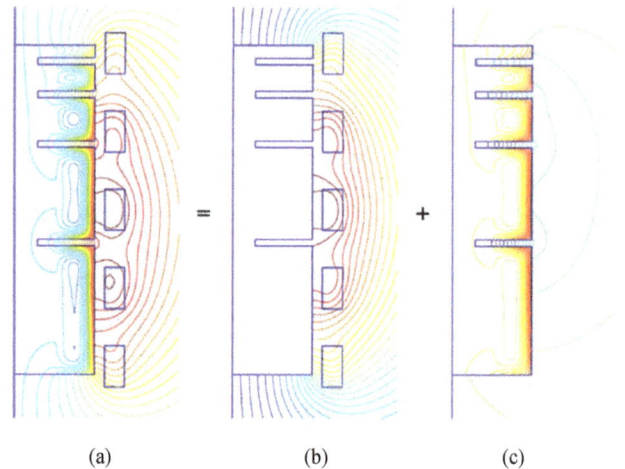

(a)   (b)   (c)

Figure 6. *Magnetic flux lines of the system with a magnetic conductive core (non-electric conductive) [12]. (a) for the conventional FE solution, (b) the reference solution, (c) the perturbation solution*

Compared with the well-known Surface Impedance Method [15-17], Sub-problem Perturbation FE Method can describe the field inside the conductors in more detail. However, it worth noting that there is still room for studying the method further, e. g., in fact the silicon steel is not only magnetic but also electric conductive material, therefore when considering both the magnetic and electric parameters at the same time, how to deal with the problem? Besides, the principle of superposition is valid only in the linear cases, then how to incorporate the nonlinearity of the material parameters in the calculation? Even though there are the questions to be answer, the method offers a very good idea to solve complicated problems.

3.2. Domain Decomposition Method

Domain decomposition method [18] is the one with which the solved domain of a definite-solution problem of partial differential equation is divided into two or more than two subdomains. The method itself can be classified overlapping domain decomposition method and non- overlapping domain decomposition method according to that the subdomains are overlapping or not. The solution of each subdomain is conducted independently, and the discretized meshes of subdomains are independent of each other. The interaction of the subdomains is dealt with by means of the iterations of certain interface conditions. In this way, the solution of the original problem for global domain is transformed into the solutions of subdomains, so that the computational scale is reduced greatly.

3.2.1. Overlapping Domain Decomposition Method

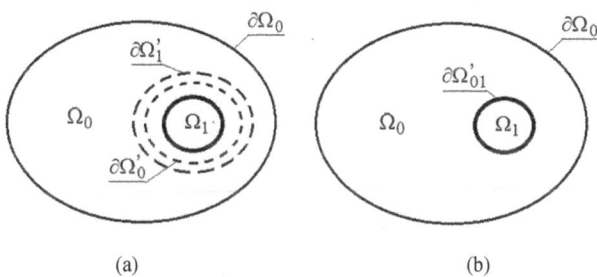

(a)   (b)

Figure 7. Sketch map of domain decomposition (a) Overlapping domain decomposition method; (b) Non-overlapping domain decomposition method.

Overlapping domain decomposition method (ODDM) is based on the so-called Schwarz alternating method. There are various choices for the domain division strategy due to the difference of the concrete problems. One of the division modes is shown in Fig. 7(a), where Ω_0 is the global domain of the original problem, and $\Omega_1 \in \Omega_0$ is the interested domain. In the calculating process of the domain decomposition, denote the subdomain between boundary $\partial\Omega_0$ and $\partial\Omega'_0$ as subdomain 1 while the subdomain surrounded by boundary $\partial\Omega'_1$ as subdomain 2, and the calculation of the two subdomains are conducted separately. The numerical boundary conditions are given alternately at the boundary $\partial\Omega'_0$ and $\partial\Omega'_1$ till the iterative solutions at these boundaries satisfy the convergent condition. A typical calculating procedure is given as follows.

1. Carry out the FE calculation in global domain Ω_0 with coarse meshes then retrieve the discrete solution at $\partial\Omega'_1$.

2. Calculate the FE solution in subdomain 2 with refined meshes. The Dirichlet condition at the outside boundary $\partial\Omega'_1$ of the subdomain 2 is provided by step 1 at the first iteration and by step 3 at and after the second iteration.

3. Calculate the FE solution in subdomain 1 with coarse meshes. The boundary condition at the inside boundary $\partial\Omega'_0$ is provided by step 2.

4. Calculate the iteration error of the solution at $\partial\Omega'_0$ and $\partial\Omega'_1$ respectively. If the error criterion is met then stop, or go to step 2.

It should be pointed that the setting of step 1 is to accelerate the convergence, which is not a necessary step. If the whole computation begins with the step 2, the boundary condition at $\partial\Omega'_1$ could be set arbitrarily.

3.2.2. Non-Overlapping Domain Decomposition Method

The sketch map of non-overlapping domain decomposition method (NODDM) is shown in Fig. 7(b). It is obvious from the figure that there is no overlap between the two subdomains, Ω_0 and Ω_1. Their interface, $\partial\Omega'_{01} = \partial\Omega'_0 \cup \partial\Omega'_1$, can be regarded as the limit case of the overlapping area reduced in Fig. 7(a). The calculation steps of the NODDM are similar as that of the ODDM. However, in the calculation of the former the Dirichlet and Neumann boundary conditions are alternately adopted in general. For a same problem, the convergence speed of the ODDM is faster than that of the NODDM. Furthermore, the lager the overlapping area, the faster the convergence, but the greater the calculation scale of each subdomain. Therefore, it is necessary to balance the convergence speed and the calculation scale of the subdomains properly and choose the division of overlapping or non-overlapping subdomains reasonably.

3.2.3. An Example to Compare the Computation Time

To compare the computation time of the domain decomposition method with conventional FE method, a small model, the P21[a]-0 of TEAM-based Benchmark Family [7] is calculated using the ODDM [11]. Table 1 lists the discretization data of the model for two subdomains, and Table 2 shows the comparison of calculated results of the model with the ODDM and conventional FE Method. It can be seen from Table 2 that at similar accuracy level the computation time with ODDM reduced to a quarter of that with the conventional method.

3.2.4. Discussion

The domain decomposition method has been applied in fluid dynamics problems, wave and Laplace equation problems more commonly [19-21], but not been used to eddy current field computation very often. For the FE analysis of nonlinear and anisotropic eddy current field the use of the method is restricted to the model with relatively simple and regular structure [11]. It should be noted that when using the method to analyze 3D eddy current field of large scale transformer products the boundary positions of subdomains have to be chosen carefully to avoid the abrupt change of electric and magnetic parameters in the boundary that could result in poor convergence performance. Therefore, the division of subdomains for practical products with complicated structure becomes troublesome, which is an obstacle to the application of the domain decomposition method in this field.

Table 1. Discretization Data of P21ᵃ-0 Model for Two Subdomains.

Name of Subdomain	Subdomain 2 (Interested)	Subdomain 1
Type of element	Hexahedron of 8 nodes	Hexahedron of 8 nodes
Number of elements	18816	20844
Number of nodes	21315	18161
Number of unknowns	71820	54204
Non-Zero Elements of Matrix	3066446	2019909

Table 2. Comparison of Calculated Results of P21ᵃ-0 Model with the ODDM and Conventional FE Method.

	Measured Losses (W)	Calculated Losses (W)	Error (%)	Computation Time (h)
Conventional	9.17	9.233	0.687	2.0
ODDM	9.17	9.245	0.818	0.5

4. New Development of Parallel Algorithm-EBE Method

Parallel algorithm is a powerful tool to deal with the large scale computation in engineering domain. The parallel methods for FE analysis include mainly the parallel solution of FE equations, the parallel of domain decomposition, and the parallel based on element level, i.e., Element by Element (EBE) parallel FE method.

The parallel solution of FE algebraic equations has been adopted in different areas, but its parallel efficiency is not very high. Especially for the large scale computation of an engineering problem the requirement of memory rises steeply, then when conducting the conventional CPU-based calculation the date interchange have to be performed, which makes the computation time for that increased greatly, so that the calculating speed declines [22-23]. The limitations of the domain decomposition method have been stated in Section 2.2.

The EBE FE method based on the element-level is an effective method to improve the degree of parallelism at the algorithm-level. With this method the main calculation can be performed independently and parallelly for each element, and only in limited stages the correlation and transmission of the element data should be carried out. Therefore, it doesn't need to create and store the global coefficient matrix, thus the requirement of memory is reduced greatly. The more great the computational scale, the more obvious the effect by using the method, so that this method is especially suitable for the numerical computation of engineering electromagnetic field problems with complicate structure and huge computational scale. The method is proposed initially by [24] in1983, and is applied to the domains of heat conduction, solid mechanics and structural mechanics gradually [25-26].

At present the research of EBE method has come to a stage of combining algorithm, software and hardware. In recent years, based on the Graphic Processing Units (GPU), the General Purpose Computation on GPUs (GPGPU) has developed rapidly. Furthermore, the arising of the Compute Unified Device Architecture (CUDA) [27] supplies a reliable programming environment for the realization of the GPGPU, which provides a certain condition for the parallel computation of large-scale engineering problems. Currently, the GPU has been applied to electrical system, graphic processing, mechanics [28-32], etc..

For the application of the EBE algorithm to electromagnetic field it is still confined to the problems of static field [33] now. Because of the specificity of eddy current problems, the EBE parallel FE Method with the GPU computing platform is not applied in the solution of the eddy current fields yet so far. The key point is that the implementation of EBE method is combined with Conjugate Gradient (CG) method, but the convergence performance is poor when using the CG method to solve the FE equations of eddy current field. The main calculation of the CG method involves the product of matrix and vector, which is particularly suitable for parallel computation, but the use of the method should be based on the condition that the coefficient matrix of the solved equations is positive definite and symmetric. For the boundary value problem of Laplacian equation, the EBE method can be used successfully because that the condition is satisfied. As an example, it is shown that for a 2D coaxial-cable model of static electric field using the EBE parallel calculation with the multi-core GPU of the first and third generations the computational speed is accelerated 14 and 111 times respectively than that of the serial calculation with CPU. However, for the eddy current problem, although the coefficient matrix of FE discretized equations is symmetric, which is not positive definite, so that the precondition of applying CG method is dissatisfied. Furthermore, when solving 3D eddy current field by using $A, \phi - A$ or $T, \psi - \psi$ scheme which is adopted commonly at present, the coefficient matrix of the resultant FE equations is ill-conditioned. In this case if the CG method is still used to solve the equations, the convergence performance will get worse greatly, or even a stable solution cannot be obtained. This is the major obstacle to use EBE method for solution of eddy current field. To overcome the obstacle it is necessary to start with mathematical model, preprocessing of parallel CG method, etc. The attempt of this aspect is now in progress [34].

5 Conclusions

This paper starts from the current problems in large scale eddy current field FE computation of electrical machines and transformers, then analyzes the present research situation in this domain, expounds the main difficulties to solve the problems, introduces several methods for reducing computational effort and improving calculating accuracy, e.g., the material performance homogenization model of laminated iron core, sub-problem perturbation FE method, domain decomposition method, and EBE parallel FE method, at the same time indicates the advantages and limitations of the different methods. In the methods above mentioned the EBE parallel FE method is a powerful tool to implement the accurate computation of 3D eddy current field by the authors' opinion, but in order to realize the computation it is still necessary to solve a set of problems. Therefore, more efforts still need to be put into exploring further the ways to fulfill the computation task rapidly and effectively by drawing lessens from and synthesizing the research results with the above methods.

References

[1] N. Soda, and M. Enokizono. "Improvement of T-Joint part constructions in three-phase transformer cores by using direct loss analysis with E&S model". IEEE Trans. Magn., Vol. 36, No. 4, pp. 1285-1288, 2000.

[2] E. Enokizono, T. Suzuki, J. Sievert, and J. Xu. "Rotating power loss of silicon steel sheet". IEEE Trans. Magn., Vol. 26, No. 5, pp. 2562-2564, 1990.

[3] Yanli Zhang, Houjian He, Dexin Xie, and Chang-seop Koh. "Study on vector magnetic hysteresis model of electrical steel sheets based on two-dimensional magnetic property measurement," Proceedings of the CSEE, Vol. 30, No. 3, pp. 130-135, 2010. (in Chinese)

[4] Yanli Zhang, Houjian He, Dexin Xie, and Chang-seop Koh. "Finite element analysis of magnetic field in transformer core coupled with improved vector hysteresis model and its experimental verification," Proceedings of the CSEE, Vo. 30, No.21, pp. 109-113, 2010. (in Chinese)

[5] K. Simonyi. Theoretische Electrotechnik, VEB Deutscher, Verlag, 1956.

[6] Yunqiu Tang. Electromagnetic Field in Electrical Machines. (Second edition), Beijing: Science Press, 1998. (in Chinese)

[7] Zhiguang Cheng, Norio Takahashi, Behzad Forghani et al. Electromagnetic and Thermal Field Modeling and Application in Electrical Engineering. Beijing, Science Press, 2009. (in Chinese)

[8] Dexin Xie et al. Finite Element Analysis of Three Dimensional Eddy Current Field. Second edition. Beijing: China Machine Press, 2007. (in Chinese)

[9] Patrick D,and Johan G. "A 3-d magnetic vector potential formulation taking eddy currents in lamination stacks into account," IEEE Transactions Magn.,Vol. 39, No. 3, pp. 1424-1427, 2003.

[10] Zhanxin Zhu, Dexin Xie, and Yanli Zhang. "Time domain analysis of 3D leakage magnetic field and structural parts loss of large power transformer," Proceedings of the CSEE, Vol. 32, No. 9, pp. 156-160, 2012. (in Chinese)

[11] Zhanxin Zhu. Research on Calculation Method of 3D Eddy Current Field Structural Part Losses in Large Power Transformer, Doctoral Dissertation, Shenyang: Shenyang University of Technology, 2012. (In Chinese)

[12] J. Gyselinck and P. Dular. "A Time-domain homogenization technique for laminated iron cores in 3-D finite-element models," IEEE Trans. Magn., Vol. 40, No. 2, pp. 856-859, 2004.

[13] Zsolt Badics, Yoshihiro Matsumoto, Kazuhiko Aoki, Fumio Nakayasu, Mitsuru Uesaka, and Kenzo Miya. "An affective 3-D finite element scheme for computing electromagnetic field distortions due to defects in eddy-current nondestructive evaluation," IEEE Trans. Magn., Vol. 33, No. 2, pp. 1012-1020, 1997.

[14] Patrick Dular, Ruth V. Sabariego, and Laurent Krähenbühl. "Subdomain perturbation finite-element method for skin and proximity effects," IEEE Trans. Magn., Vol. 44, No. 6, pp.738-741, 2008.

[15] S. A. Schelkunoff, "The impedance concept and its application to problems of reflection, shielding and power absorption," Bell System Technical Journal, pp. 17-48, 1938.

[16] E. M. Deeley, Serface "Impedance near edges and corners in three-dimensional media," IEEE Trans. Magn.,Vol. 26, No. 2, pp. 712-714, 1990.

[17] Yong-Gyu Park et al. "Three dimensional eddy current computation using the surface impedance method considering geometric singularity," IEEE Trans. Magn.,Vol. 31, No. 3, pp. 1400-1403, 1995.

[18] Tao Lv, Jimin Shi, and Zhenbao Lin. Domain Decomposition Algorithms—New Technology of Numerical Solution of Partial Differential Equation. Beijing: Science Press, 1997. (in Chinese)

[19] R. Glowinski, Q. V. Dinh, J. Periaux. "Domain decomposition methods for nonlinear problems in fluid dynamics," Computer Methods in Applied Mechanics and Engineering, Vol. 40, No. 1, pp. 27-109, 1983.

[20] Zhu Z. H., Ji H., and Hong W. "An efficient algorithm for the parameter extraction of 3D interconnect structures in the VLSI circuits domain decomposition method," IEEE Transactions on Microwave Theory and Techniques, Vol. 45, No. 9, pp. 1179-1184, 1997.

[21] Hanqing Zhu. Study on the Applications of Domain Decomposition Method in Electromagnetic Problems, Doctoral Dissertation, Chengdu: University of Electronic Science and Technology of China, 2002. (in Chinese)

[22] Mifune T, Iwashita T, and Shimasaki M. "A fast solver for FEM analysis using the parallelized algebraic multi-grid method," IEEE Trans. Magn., Vol. 38, No. 2, pp. 369-372, 2002.

[23] Steve McFee, Qingying Wu, Mark Dorica, et al. "Parallel and distributed processing for h-p adaptive finite-element analysis: a comparison of simulated and empirical studies," IEEE Trans. Magn., Vol. 20, No. 2, pp. 928-933, 2004.

[24] Hughus T J R, Levit I, and Winget J. "An element-by-element solution algorithm for problems of structural and solid mechanics," Computer Methods in Applied Mechanics and Engineering, Vol. 36, pp.241-254, 1983.

[25] Shunxu Wang, Boguo Sun, and Shuquan Zhou. "A mixed EBE parallel algorithm for transient heat conduction problems," Journal of Huaihai Institute of Technology, Vol. 8, No. 3, pp. 7-9, 1999.

[26] Yaoru Liu, Weiyuan Zhou, and Qiang Yang. "A distributed memory parallel element by element scheme based on Jacobi-conditioned conjugate gradient for 3D finite element analysis," Finite Elements in Analysis And Design, Vol. 43, pp. 494-503, 2007.

[27] Shu Zhang, Yanli Chu. CUDA of GPU High Performance Computation, Beijing: China Water Power press, 2009. (in Chinese)

[28] Han Jiang and Quanyuan Jiang. "A two-level parallel transient stability algorithm for AC/DC power system based on GPU platform," Power System Protection and Control, Vol. 40, No.21, pp. 102-108, 2012. (in Chinese)

[29] Youquan Liu, Kangxue Yin, and Enhua Yin. "Fast GMRES-GPU solver for large scale sparse linear systems. Journal of Computer-Aided Design & Computer Graphics," Vol. 23, No. 4, pp. 553-560, 2011. (in Chinese)

[30] Xiaohu Liu, Yaoguo Hu, and Wei Fu. "Solving large finite element system by GPU computation," Chinese Journal of Computational Mechanics, Vol. 29, No. 1, pp.146-152, 2012. (in Chinese)

[31] Barnat, J., and Bauch, P. "Employing multiple CUDA devices to accelerate LTL model checking," IEEE 16th International Conference on Parallel and Distributed System, pp. 259-266, 2010.

[32] Thurley, M. J. and Danell, V. "Fast Morphological Image Processing Open-Source Extensions for GPU Processing with CUDA," IEEE Journal of Selected Topics in Signal Processing, Vol.6, No. 7, pp. 849-855, 2012.

[33] David M. Fernández, Maryam Mehri Dehnavi, and Warren J. Gross. "Alternate Parallel Processing Approach for FEM," IEEE Trans. Magn., Vol.48, NO.2, pp. 299-402, 2012.

[34] Renyuan Tang, Dongyang Wu, and Dexin Xie. "Research on the key problem of element by element parallel FEM applied to engineering eddy current analysis," Transaction of China Electrotechnical Society, Vol.29, No.5, pp. 1-9, 2014.

6

Effect of Leading Edge Radius and Blending Distance from Leading Edge on the Aerodynamic Performance of Small Wind Turbine Blade Airfoils

Mahasidha R. Birajdar, Sandip A. Kale

Mechanical Engineering Department, Trinity College of Engineering and Research, Pune, India

Email address:

mrbirajdar88@gmail.com (M. R. Birajdar), sakale2050@gmail.com (S. A. Kale)

Abstract: The aerodynamic performance of a wind turbine depends upon shape of blade profile blade airfoils. Today, small wind turbine industries are extensively focusing on blade performance, reliability, materials and cost. The wind turbine blade designers are required to give emphasis on accurate analysis of flows around the blade and loads on wind turbine blades. Low Reynolds number airfoils suited for small wind turbine applications must be designed to have a high degree of tolerance in avoiding high leading suction peaks and high adverse pressure gradients that lead to flow separation. This paper presents a study to investigate the effect of leading edge radius and leading edge blending on the aerodynamic performance of wind turbine airfoils. In the present work NACA 4412 airfoil is considered as base airfoils. In this work six modified airfoils having different new to the old ratio of leading edge radii are considered for performance analysis. The performance of these six profiles is compared with basic airfoil performance. In this paper, the effect of blending distance from leading edge of airfoil on aerodynamic performance is also determined. Different five blending distances from leading edge are analyzed and compared with basic profile. The performance analysis of airfoils is carried out using Blade Element Momentum. In the present analysis, chord length of airfoils and Reynolds number are kept constant.

Keywords: Airfoil, Aerodynamic Performance, Leading Edge Radius, Leading Edge Blending

1. Introduction

A condition for an efficient conversion of the wind energy into mechanical energy with wind turbines is the optimal design of the rotor blades. Quick, reliable and simple methods for predictions of the aerodynamic characteristics and simulation of the flow conditions around a rotor blade are essential for this design work [1-3]. Wind flow around blade is complicated in nature. In addition to these, complicated wind turbine flows, wide range of operating conditions, rotating lifting surfaces, transitional blade flows and low Mach numbers are the major challenges in performance analysis. The wind turbine blade designers are required to force accurate analysis of flows around the blade and loads on wind turbine blades. The wind tunnel testing of the wind turbine blade airfoils with or without flow control and load control is time consuming and expensive. The two dimensional computational tools are based on viscous flow theory having, steady flow, smooth surface and limited flow

separation condition. The three dimensional computational tools are majorly based on Blade Element Momentum (BEM) theory. Q-blade is a one of the examples of these 3D tools [4-6]. The passive flow control and load control methods are leads to improve the performance of the turbine, to mitigate the loads on the structure and reduce the stress levels in the structure. Passive control techniques includes the laminar flow control, Passive porosity, Riblets, Vortex generators, Stall strips, Gurney flaps, Serrated trailing edges, Aero elastic tailoring, Special purpose airfoils such as restrained maximum lift, high lift, blunt trailing edge, modified leading edge etc. Passive load control is extensively used in wind turbine design, for the most part focused on power production [7-8].

Numerical Analysis of new airfoils for small wind turbine blade is carried out successfully by Birajdar et. al [1]. Two new blade airfoils are designed for small wind turbine and comparison of new airfoils and blade performance using different techniques is carried out. It is remarked that Q-blade is a reliable tool for analysis of wind turbine airfoils and blade [2]. In the wind turbine airfoils the parabolic leading edge affects the performance of airfoils. The blunt leading edge portion is fair into a pressure surface characterized by leading edge radius i.e. leading convex portion [3].

The thin airfoils are chosen for low Reynolds number application to decrease the suction peak near the leading edge of the airfoil to decrease the adverse pressure gradients on the upper surface of the airfoils. The low Reynolds number airfoils operate below a Reynolds number of 500,000, where the flow across the upper surface of the airfoil is predominantly laminar. Airfoils within this Reynolds number range suffer from laminar separation bubble and are susceptible to laminar flow separation that occurs when the laminar separated flow does not reattach to the surface, resulting in a loss in aerodynamic performance. Low Reynolds number airfoils suited for small wind turbine applications must be designed to have a high degree of tolerance in avoiding high leading suction peaks and high adverse pressure gradients (APG) that lead to flow separation. A small degree of roughness needs to be associated with airfoils operating at low Reynolds number conditions as explained by Lissaman, where the introduction of turbulators or trip wire devices, promote early transition from laminar to turbulent flow to eliminate laminar separation bubbles and delay the possible chances of separation from the upper surfaces at higher angles of attack [3].

The use of specifically sized trip wires has been employed near the leading edges of low Reynolds number airfoils to show this effect as studied by Giguere and Selig, where the devices 'trip' laminar flow into the high energy turbulent flow able to negotiate the adverse pressure gradients (APG). Roughness can easily be introduced to airfoils at low Reynolds number as it does not appear significant in relation to boundary layer thickness whereas the opposite happens at high Reynolds number. Since the boundary layer thickness is inversely proportional to Reynolds number, a small amount of roughness would appear noticeable with decreasing boundary layer thickness as Reynolds number is increased since the physical size of the introduced roughness stays the same [8-23].

Wind turbine blades are exposed to precipitation that occurs in a variety of forms and myriad abrasive airborne particles that can, over time, erode their surfaces, particularly at the leading edge. These airborne particles can cause significant blade erosion damage that reduces aerodynamic performance and hence, energy capture. Moreover in some environments, insect debris and other airborne particles can accrete on the leading edges of wind turbine blades. Leading edge blade erosion and debris accretion and contamination can dramatically reduce blade performance particularly in the high-speed rotor tip region that is crucial to optimum blade performance and energy capture. The erosion process on wind turbine blades typically starts with the formation of small pits near the leading edge, which increase in density with time and combine to form gouges. If left to the forces of nature, the gouges, then grow in size and density, and combine to cause delamination near the leading edge [4].

This paper presents the study to investigate the effect of leading edge radius and leading edge blending on the aerodynamic performance of wind turbine airfoils. The objective of this study was to test a wind turbine airfoil with shape modifications to simulate the leading edge radius going through the evolutionary stages of development. The goal was to develop a baseline understanding of the aerodynamic effects of various types and magnitudes of leading edge radius and to quantify their relative impact on airfoil performance. The ultimate aim of conducting the study was to examine the potential detrimental effects of leading edge radius and leading edge blending on the wind turbine airfoils.

2. Wind Turbine Blade Airfoils

2.1. Airfoil Nomenclature

The basic nomenclatures of airfoils are shown in Figure.1 and basic explanation is given as follows

Leading edge: The front edge of the airfoil is called leading edge. The leading edge is the part of the airfoil that first contacts the incoming air and the principal edge of an airfoil section.

Mean Camber line: The locus of the points that lie half way between the upper and lower surfaces is called the mean camber line.

Chamber: The maximum distance between the chord line and the mean camber line is called the camber. Camber is generally designed into an airfoil to increase the maximum lift coefficient.

Trailing edge: The back of the airfoil is called trailing edge.

Chord line: The straight line drawn from leading edge to trailing edge [1].

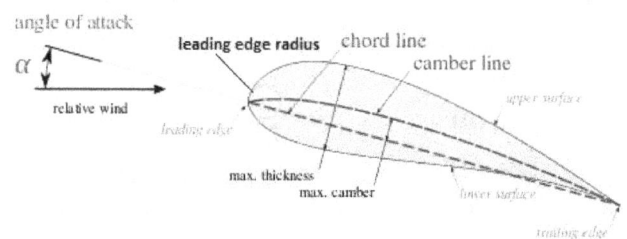

Figure 1. Airfoil nomenclature.

Airflow over any surface creates two types of aerodynamic forces drag forces, in the direction of the airflow, and lift forces, perpendicular to the airflow. Either or both of these can be used to generate the forces needed to rotate the blades of a wind turbine. The airfoil shape is designed to create a differential pressure between the upper and lower surfaces, leading to a net force in the direction perpendicular to the wind direction [2].

2.2. Airfoil Selection

In the present work NACA 4412 airfoil is considered as

base airfoils as shown in Figure. 2. The NACA 4412 is a four digit airfoil. The first digit expresses the camber in percent chord, the second digit gives the location of the maximum camber point in tenths of chord, and the last two digits gives the thickness in percent chord. Thus 4412 has a maximum camber of 4% of chord located at 40% chord back from the leading edge and is 12% thick [24].

Figure 2. *NACA 4412 airfoil.*

2.3. Aerodynamic Analysis

Efficient wind turbines operate with the higher lift force and low drag force. If the angle of attack is less, then the lift force will be high and the drag force will be lower. But, if the angle of attack increases beyond a certain value, the lift force decreases and the drag forces increases. Hence, angle of attack plays a vital role in designing a blade. In this work, angle of attack is given as input in the Q-blade tool and the values of C_L and C_D were obtained. The analysis of airfoil is carried out by using Q-blade. Q-blade is a three dimensional computational tool is majorly based on Blade Element Momentum (BEM) theory [1]. The airfoil design has also taken place using Q-blade tool. Therefore an airfoil design requirements include information regarding C_{Lmax} as well as the operating range over which low drag is achieved.

These requirements can be translated into specific characteristics to be embodied in the pressure distribution. The low drag points require extended runs of laminar flow on the lower and upper surfaces, respectively, while the high lift requirement is achieved by limiting the leading edge suction peak behavior, each of which must be achieved at the corresponding design lift coefficient [6].

3. Effect of Leading Edge Radius

The front edge of the airfoil is called leading edge. The leading edge is the point at the front of the airfoil that has a maximum curvature means minimum radius. This minimum radius of the leading edge is called the leading edge radius. In this paper NACA 4412 airfoil is considered as base airfoils. In this work seven modified airfoils having different new to the old ratio of leading edge radii are considered for performance analysis.

Table 1. *Modified airfoil for different radii ratio.*

Airfoil	New to the old ratio of leading edge radii
AFR1	0.6
AFR2	0.8
NACA4412	1.0
AFR3	1.2
AFR4	1.4
AFR5	1.6
AFR6	1.8
AFR7	2.0

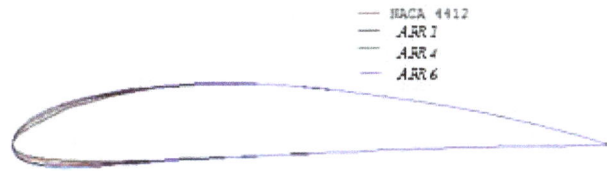

Figure 3. *NACA 4412 airfoil with varying leading edge radii.*

During this analysis the blending distance from leading edge is kept constant as 10 % of chord. The profiles with change in leading radius are named as AFR1 to AFR7. Here, AFR stands for an airfoil. Table 1.shows different radii and profile names analyzed. Figure 3 shows NACA 4412 and airfoils with few new to the old ratio of leading edge radii. The base airfoil NACA 4412 has radii ratio of 1. The effect of these variations is investigated and plotted in fig. 4.

The performance analysis of airfoils is carried out at constant Reynolds number of 250000 and constant chord length. Figure 4 describes the performance variations for different new to the old ratio of leading edge radius. At minimum new to the old ratio of leading edge radius i.e. less than one, the starting lift to drag ratio is higher compared to maximum new to the old ratio of leading edge radius i.e. greater than one. But the performance curve of these airfoils falls down at lower angle of attack compared to other. This shows that range of maximum performance is greater for higher new to the old ratio of leading edge radius. The wider performance curve provides lesser fluctuations in power output. The maximum lift coefficient to drag coefficient ratio is obtained for AFR1 airfoils, whereas the minimum lift to drag ratio is obtained for AFR7airfoils.

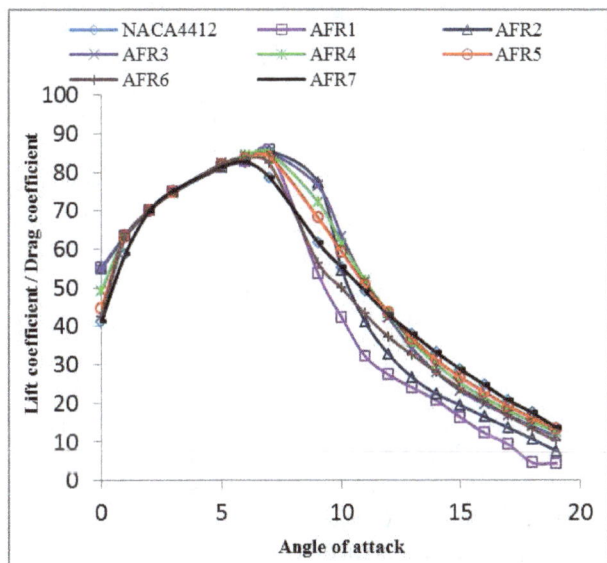

Figure 4. *Effect of leading edge radius on NACA 4412 airfoil.*

4. Effect of Blending Distance from Leading Edge

In the present work, the focus has been on designing airfoils that can be used along the entire blade span of small horizontal axis wind turbine. The NACA 4412 airfoil is considered as base airfoils. The changes in airfoils are carried out by changing the blending distance from the leading edge. For NACA 4412 airfoils the blending distance from leading is 10 percent of chord. Similarly AFB1, AFB2, AFB3, AFB4 and AFB5having the blending distance from leading is 06%, 08%, 20%, 30%, 40% of chord respectively. The nomenclature AFB stands for Airfoil blending. Figure 5 shows the cross section of NACA 4412 airfoils with varying blending distance from the leading edge.

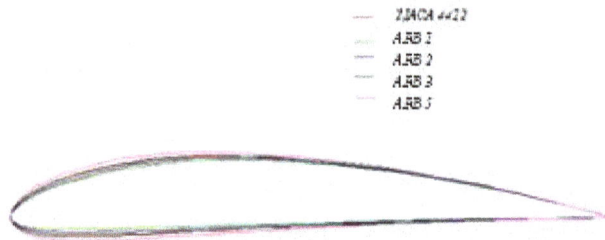

Figure 5. NACA 4412 airfoils with varying blending distance from leading edge.

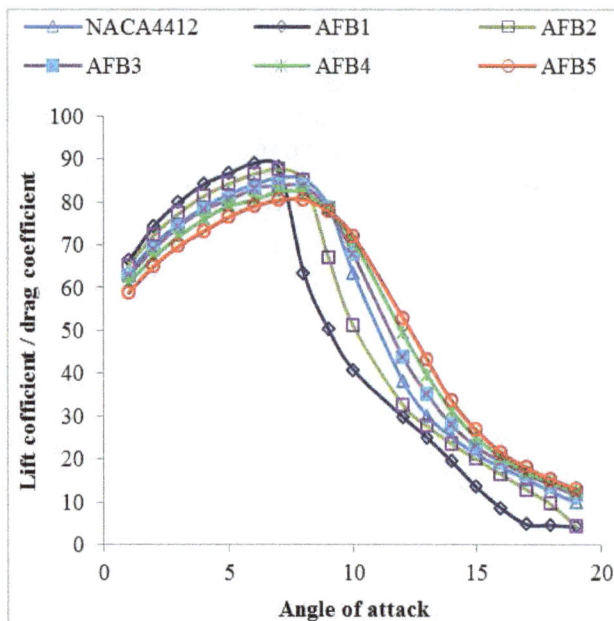

Figure 6. Effect of blending distance from leading edge on NACA 4412 airfoils.

The designed airfoils should provide optimum performance over a broad range of operating conditions. The analysis of airfoils is carried out at constant Reynolds number of 250000 and constant chord length. Figure 6 describes the performance variations of NACA 4412 airfoils with varying blending distance from the leading edge. At the lesser blending distance from leading edge higher lift coefficient/ drag coefficient ratio is obtained. With the increase in blending distance from leading edge the maximum lift to drag ratio is decreased. For the AFB1 airfoil the maximum lift to drag ratio 90 is higher than other airfoils, but the range of performance curve is very narrow. Whereas, for AFB5 airfoil the maximum lift to drag ratio is 79 is lesser

than other airfoils, but the range of performance curve is too broad. The optimum lift to drag ratio is obtained for NACA 4412 airfoil with a wide range of performance curve and at the 10% blending distance from the leading edge.

5. Conclusion

For different seven leading edge radii and five blending distance from the leading edge on an aerodynamic performance of small wind turbine blade airfoils is determined through Q-blade simulation. From results obtained, some concluding remarks are as follows:

- With the increase in leading edge radius, the performance of the airfoil is decreases, but range of performance becomes broader.
- With the decrease in leading edge radius, the performance of the airfoil is increases, but range of performance becomes narrow.
- The ratio of lift to drag coefficient increases with reduction in blending distance from leading edge and decreases with increase in blending distance from leading edge for up to angle of attack value 7. After that the ratio of lift to drag coefficient increases with increase in blending distance from leading edge and increases with increase in blending distance from leading edge.

References

[1] Mahasidha R. Birajdar, Sandip A. Kale, S. N. Sapali, "Numerical Analysis of New Airfoils for Small Wind Turbine Blade", Journal of Alternate Energy Sources and Technologies, ISSN: 2230-7982, Volume 6, Issue 1, 2015.

[2] Mahasidha R. Birajdar, Sandip A. Kale, "Comparison of new designed small wind turbine airfoils and blade performance using different techniques", International Journal of Applied engineering research, ISSN: 0973-4562, Volume 10, No.71, 2015.

[3] Ronit K. Singh, M. Ra.uddinAhmeda, Mohammad AsidZullah, Young-Ho Lee, "Design of a low Reynolds number airfoil for small horizontal axis wind turbines", Renewable Energy 42 (2012) 66-76.

[4] AgrimSareen, Chinmay A. Sapre and Michael S. Selig, "Effects of leading edge erosion on wind turbine blade performance", Wind Energ.2014; 17:1531–1542.

[5] C.P. (Case) van Dam, "Blade Aerodynamics - Passive and Active Load Control for Wind Turbine Blades", University of California, Davis.

[6] Giguere P, Selig MS. "New airfoils for small horizontal axis wind turbines", Wind Engineering 1998; 120:111.

[7] Jasinski WJ, Noe SC, Selig MS, Bragg MB.Wind turbine performance under icing conditions. ASME Journal of Solar Energy Engineering 1998; 120: 60–65.

[8] Giguère P, Selig MS. Aerodynamic effects of leading-edge tape on airfoils at low Reynolds numbers. Wind Energy 1999; 2: 125–136.

[9] Van Rooij RPJOM, Timmer WA. Roughness sensitivity considerations for thick rotor blade airfoils. AIAA Paper 2003–352, Reno, NV, August 2003.

[10] Fuglsang P, Bak C. Development of the Risk wind turbine airfoils. Wind Energy 2004; 7(2): 145–162.

[11] Giguere P, Selig MS. Low Reynolds number airfoils for small horizontal axis wind turbines. Wind Engineering 1997; 21:367-80.

[12] Miley SJ. A catalog of low Reynolds number airfoil data for wind turbine applications. College Station, Texas: Department of Aerospace Engineering, Texas A&M University; 1982.

[13] Elizondo J, Martínez J, Probst O. Experimental study of a small wind turbine for low- and medium-wind regimes. International Journal of Energy Research 2009; 33:309-26.

[14] Lissaman PBS. Low-Reynolds-number airfoils. Annual Reviews of Fluid Mechanics 1983; 15:223-39.

[15] L.J.Vermeer et.al, in the paper "Wind turbine wake aerodynamics" Progress in Aerospace Sciences 39 (2003) 467–510.

[16] Lucas I. Lago a, Fernando L. Ponta, Alejandro D. Otero Analysis of alternative adaptive geometrical configurations for the NREL-5 MW wind turbine blade, Renewable Energy 59 (2013) 13-22

[17] Clausen PD, Wood DH. Research and development issues for small wind turbines. Renewable Energy 1999; 16: 922-7.

[18] Peacock AD, Jenkins D, Ahadzi M, Berry, Turan S. Micro wind turbines in the UK domestic sector, energy and buildings.

[19] Wright AK, Wood DH. The starting and low wind speed behavior of a small horizontal axis wind turbine. Journal of Wind Engineering and Industrial Aerodynamics 2004; 92: 1265-79.

[20] Tangler J. L. et al., "Wind Turbine Post-Stall Airfoil Performance Characteristics Guidelines for Blade Element Momentum Methods," National Renewable Energy Laboratory, Technical Report 2004 NREL/ CP-500-36900.

[21] Habali SM, Saleh IA. Local design, testing and manufacturing of small mixed airfoil wind turbine blades of glass .ber reinforced plastics. Part I: design of the blade and root. Energy Conversion & Management 2000; 41: 249-80.

[22] Peter J. Schubel, Richard J. Crossley in the paper "Wind Turbine Blade Design" Energies, 2012, 5, 3425-3449.

[23] Ahmed MR, Narayan S, Zullah MA, Lee YH. Experimental and numerical studies on a low Reynolds number airfoil for wind turbine blades. Journal of Fluid Science and Technology 2011; 6: 357-71.

[24] S A Kale, R N Varma, "Aerodynamic Design of a Horizontal Axis Micro Wind Turbine Blade Using NACA 4412 Profile", International Journal of Renewable Energy Research, Vol. 4, Issue 1, 69-72 p.

7

Highly Porous Polymer Electrolytes Based on PVdF-HFP / PEMA with Propylene Carbonate/Diethyl Carbonate for Lithium Battery Applications

P. Sivakumar[*], M. Gunasekaran

Department of Physics, Periyar E.V.R. College, Tiruchirappalli, India

Email address:

ssilabp6pevrca@gmail.com (P. Sivakumar), 8012072465p6@gmail.com (M. Gunasekaran)

Abstract: The development of new materials is a vital to meet the challenges faced by battery technologies. Ionic conducting solid polymer electrolytes could reduce the risk of explosion with non-flammability and high thermal stability. The use of solid polymer electrolyte is the additional strength of the electrodes performances to increase the number of cycle for the rechargeable batteries. In the present study, preparation of PVdF-HFP/PEMA blend based solid polymer electrolytes enclosure of two different plasticizers such as propylene carbonate (PC) and diethyl carbonate (DEC) at different concentrations and the accumulation of lithium perchlorate as salt. To confirm the structural changes and complex formations, the prepared electrolytes were subjected into XRD and FTIR analyses, and the porous nature of the electrolytes was identified using scanning electron microscopy. AC impedance studies were performed at various temperatures from 303 K to 363K for the prepared samples. The results suggest that the PC/DEC (1:1) based electrolyte exhibited the higher ionic conductivity is 0.00477 S/cm at room temperature and 0.00843 S/cm at 363K. The temperature dependence of ionic conductivity also complies with the VTF relation.

Keywords: Solid Polymer Electrolyte, FTIR, XRD, SEM, Conductivity

1. Introduction

One of the most feasible challenges as a sustainable energy conversion and storage systems is the Rechargeable Lithium Ion Battery (LIB). The LIB is based on a cathode and an anode, which has the property of reversible insertion and extraction of lithium ions. Transfer of lithium ions is enabled by the addition of an organic liquid electrolyte and a mechanical separator between the anode (negative electrode) and the cathode (positive electrode). When the lithium ion is inserted and extracted in the cathode and the anode, electrical energy is generated by electrochemical oxidation and reduction process. The electrolyte between the anode and cathode has to be an ionic conductor, electronic insulator and is responsible for the transport of lithium ions. The optimal electrolyte should combine the conduction properties of liquid and the mechanical stability of solid with high chemical stability. Even though liquid electrolytes are commonly used, due to high ionic conductivity, application of polymer and ionic liquid electrolytes also attract interest as they might improve the safety of lithium batteries [1]. A membrane (separator) is an important component of a battery, as it prevents short circuit by separating the anode from the cathode. In the LIB, the membrane is required to be capable of battery shutdown at a temperature below that at which thermal runaway occurs, and the shutdown should not result in loss of mechanical integrity. Otherwise, the electrodes could come into direct contact and the resulting chemical reactions cause thermal runaway. Shutdown is an important trait of a good membrane for the safety of lithium batteries. The promising membranes are those with high electrolyte permeability and mechanical strength, as well as good thermal, chemical, and electrochemical stability. In order to concentrate the above parameters, polymer electrolytes have been improved for better performance in the electrochemical characteristics devices with stretchy natures and concurrence with safety concern. Many kinds of P. Sivakumar et al: Highly Porous Polymer Electrolytes based on PVdF-HFP/PEMA with Propylene Carbonate/Diethyl Carbonate for Lithium Battery

Applications. polymers have been preferred as the matrix of poly (ethylene oxide) (PEO), poly (ethylene glycol) (PEG), poly (vinyl chloride) (PVC), poly (vinylidene fluoride) (PVdF), poly (acrylonitrile) (PAN), poly (vinyl acetate) (PVAc), poly (vinyl pyrrolidone) (PVP), poly (methyl methacrylate) (PMMA), etc. [2-9]. Among them, PVdF-HFP has good electrochemical stability, affinity to electrolyte solution, high dielectric constant (ε = 8.4) and also it is chemical copolymer, which contains both amorphous (HFP) and crystalline (PVdF) phase, which provides plasticity and mechanical strength. PVdF-HFP based membranes tend to be opaque [10], while PEMA based membranes are transparent. Blend host matrices also help to increase the ionic conductivity [11]. In order to further increase the ionic conductivity, an attempt has been taken to incorporate the $LiClO_4$ as salt, PC and DEC as plasticizers with PVdF-HFP/PEMA based blend polymer. The prepared electrolytes are subjected into various studies, such as a.c. Impedance measurement, FTIR, XRD, SEM and their results are discussed.

2. Experimental

2.1. Materials

Poly (vinylidene fluoride-co-hexafluoropropylene) (PVdF-HFP, Mw ~ 455, 000,) in pellet form, Poly (ethyl methacrylate) (PEMA, average Mw ~ 515,000) in powder form, lithium perchlorate ($LiClO_4$, Mw = 106.39, battery grade, purity 99.99 %,) are received from Sigma Aldrich and Propylene Carbonate (PC), Diethyl Carbonate (DEC), acetone are procured (Alfa Aesar) and used after laboratory purifications.

2.2. Preparation of PVdF - HFP / PEMA Based PEs

PVdF – HFP and PEMA blend based polymer electrolytes are prepared by solution casting technique. To enhance the ionic conductivity, PVdF-HFP/PEMA (as 18:12 (wt. %)), is dissolved in volatile solvent such as acetone separately. The $LiClO_4$ (8) was dissolved in the mixture of PC and DEC at various ratio (1:1, 1:2, 1:3 and 1:4) in an appropriate amount of acetone and then the polymer mixture was stirred continuously until obtained complete homogeneous solution. Further, polymer and salt were mixed together and the solution was stirred about 24 hours until to get the transparent resultant solution. The obtained homogeneous mixture was cast onto a cleaned glass plate and dehydrated at 45°C in an oven for 2-3 hours to evaporate the residual solvent. Upon cooling at room temperature, the mechanically stable and transparent membranes were carefully peeled from the glass plates and stored in the vacuum desiccators for further characterizations.

2.3. Characterization Techniques

To analyze the complexation behavior and structural modification of the electrolyte, FTIR spectrum was recorded using Perkin-Elmer-1600 in the range of 400 – 4000 cm^{-1},

and X-Ray diffraction pattern was obtained using a computer controlled X'PERT PROPANalytical diffractometer with Cu-K_α radiation as the source at 40 kV with a scanning range between 10° to 80°. Impedance of the each sample was determined using electrochemical work station of Bio-Logic SAS instrument (SP-150 model). The measurement was carried out in the frequency range from 1 Hz to 1 MHz at various temperatures. The impedance studies were carried out by sandwich the polymer electrolyte membrane between two stainless steel (SS) electrodes under spring pressure. The thickness of each sample was measured by a Digital Caliper. The surface morphology of the films was examined using VEGA3 TESCAN Scanning Electron Microscope (SEM).

3. Result and Discussion

3.1. FTIR Analysis

"Figure 1." shows the FTIR spectrum of pure PVdF-HFP, PEMA, $LiClO_4$ and (a) PVdF-HFP/PEMA (18/12) - $LiClO_4$ (8) – PC/DEC (1:1), (b) PVdF-HFP/PEMA (18/12) - $LiClO_4$ (8) - PC/DEC (1:2), (c) PVdF-HFP/PEMA (18/12) - $LiClO_4$ (8) - PC/DEC (1:3) and (d) PVdF-HFP/PEMA (18/12) - $LiClO_4$ (8) - PC/DEC (1:4) complexes. The vibrational peaks at 485, 487, 488 cm-1 and 436, 437, 438 cm-1 are assigned to the bending and wagging vibration of $– CF_2 –$ in complexes. The crystalline phase of the PVdF – HFP polymer is identified at 969, 775 cm^{-1} and 623, 624, 625 cm^{-1} in the complexes. The peak at 1132 cm-1 is shifted from 1175, 1176 cm^{-1} in film (a and c), 1389, 1394 and 1398 cm^{-1} are assigned to the symmetrical stretching of $– CF_2 –$ and $– CH_2$ groups, respectively [11]. The peak at 878 cm^{-1} in the complexes is assigned to the vinylidene group of the polymer. The PVdF-HFP skeletal vibration of $– CF_2 –$ stretching vibration at 1054 cm^{-1} is shifted to 1030, 1032, 1033 cm^{-1} in film (b, d and c). The vibrational peaks at 2965, 2977 cm^{-1} and 2985, 2983 cm^{-1} are attributed to the asymmetric C-H stretching vibration of the ethylene group of PEMA in the complexes. The functional groups corresponding to - CH_2 - scissoring and - CH_2 - rocking are observed at 1477 cm^{-1} and 752, 753 cm^{-1} are shifted to 1469 cm^{-1} and 777 cm^{-1} in complexes. The C=O stretching band of PEMA are located at 1730, 1735 and 1724, 1726 cm^{-1} in complexes. The vibrational peaks at 713, 721, 768 and 775 cm^{-1} belongs to the anion and cations are coordinated with the carbonyl carbon and carboxylic oxygen (C=O) presence in the polymer complexes. The skeletal vibration is identified at 1469 and 1478 cm^{-1}in the complexes which is assigned to $– CH_3$- asymmetric bending of plasticizer such as propylene carbonate. The band position of C-O-C asymmetric stretching vibration and -C-O-C-O-skeletal vibration of diethyl carbonate (DEC) molecule located at 1213 and 1229, 1234 cm^{-1} are observed in complexes. The frequencies at 835 cm^{-1} are assigned to C-Cl stretching vibrations of perchlorate. The vibrational peaks of $LiClO_4$ at 1149 cm-1is shifted to 1133 cm^{-1} in complexes. Shifting of peaks and formation of new peaks imply the polymer – salt interaction in PVdF-HFP/PEMA blend based polymer electrolytes systems.

3.2. X-ray Diffraction Analysis

"Figure 2." shows the X-ray diffraction pattern of pure PVdF-HFP, PEMA, LiClO$_4$, (a) PVdF-HFP/PEMA (18/12) - LiClO$_4$ (8) – PC/DEC (1:1) and (b) PVdF-HFP/PEMA (18/12) - LiClO$_4$ (8) - PC/DEC (1:2) complexes.

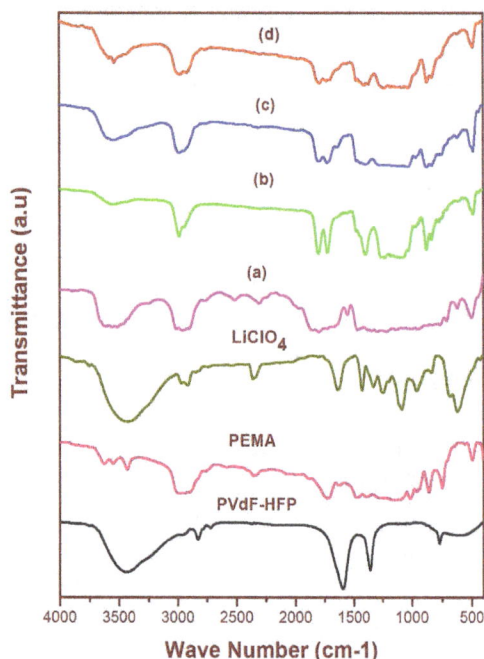

Fig. 1. *FTIR spectrum of pure PVdF-HFP, PEMA, LiClO4, (a) X - PC/DEC (1:1), (b) X - PC/DEC (1:2), (c) X - PC/DEC (1:3) and (d) X - PC/DEC (1:4). Where X = PVdF-HFP/PEMA (18/12) – LiClO4 (8).*

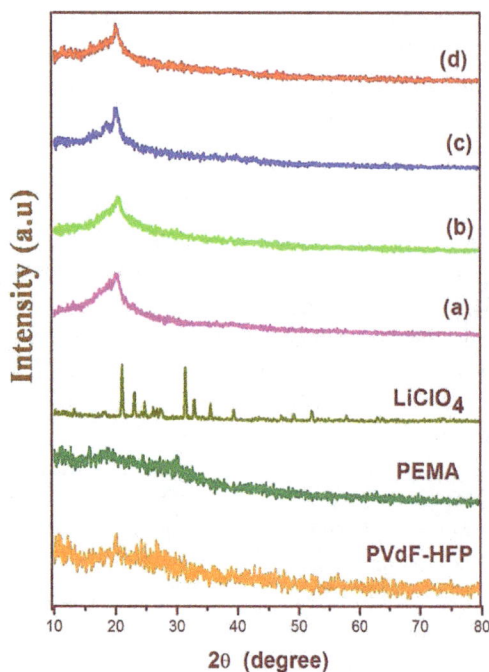

Fig. 2. *XRD pattern of pure PVdF-HFP, PEMA, LiClO4 and (a) X - PC/DEC (1:1), (b) X - PC/DEC (1:2), (c) X - PC/DEC (1:3) and (d) X - PC/DEC (1:4). Where X = PVdF-HFP / PEMA (18/12) – LiClO4 (8).*

The peaks found 2Θ = 18.05o, 20.01o, 20.19o, 20.22o, 20.45o and 43o reveal the partial crystallization of PVdF units present in the complexes, giving an overall semi crystalline morphology of PVdF-HFP [12] and the presence of broad hump in the complexes confirms the amorphous nature of electrolytes. Furthermore, no peaks are found for LiClO$_4$ reveals the completely dissolution of salt in the polymer complex and therefore from these observations no diffraction peaks are identified at any separate phase of LiClO$_4$ in complexes, which confirm that the salt dissolution was completely dissolved and trace of plasticizer is also absent in the films.

3.3. Ionic Conductivity Studies

The ionic conductivity of the each sample was calculated using the equation $\sigma = t/R_b{*}A$ (Scm^{-1}), where t is thickness of the electrolyte, R_b is the bulk resistance and A is the area of the electrolyte – electrode contact. The imaginary impedance (Z") was plotted against the real impedance (Z') and the bulk resistance was obtained from the intercept with the real-axis [13]. "Fig. 3(a- d)." shows the complex impedance spectra of P(VdF-HFP)/PEMA – PC/DEC – LiClO$_4$ for the different concentrations of polymer blend based electrolyte system at different temperature with enlarged view of each complex impedance spectra.

According to the theoretical analysis given by Watanabe and Ogata [14], two semi circles should appear in an impedance spectrum for a symmetric cell. i.e., one at higher frequency related to bulk electrolyte impedance and other at lower frequencies related to the interfacial impedance. It is also reported [15] that high frequency semi-circle does not appear in our useful impedance plots for blend polymer membrane as shown in "FIG. 3(A) AND 3(B)." This phenomenon is quite reasonable since the too unproblematic mobility in this solid electrolyte system, when compared with liquid and gel polymer electrolytes.

"Fig. 4." shows the polymer complexes obey the temperature dependant ionic conduction is good agreement with Vogel – Tamman – Fulchar (VTF) relations, which describes the transport in a viscous matrix [4]. It supports to initiative the ions movements through the plasticizer rich phase, which is the conducting medium and involved with lithium salt. The samples are having improved ionic conductivity to higher temperature from ambient temperature, which could be credited to the enhance the free volume of the polymer electrolyte membrane, hence the free volume in a solid polymer electrolyte increases as a result and the segmental motion also permits the ions to hope from one site to another site or given the pathway to ion migration.

TABLE.1 shows the a.c conductivity at different temperatures of different concentrations of plasticizer. Film (a) achieved the higher ionic conductivity as 4.77 x 10^{-3} Scm^{-1} at room temperature and 8.43 x 10^{-3} Scm^{-1} at 363 K. When the addition of plasticizer ratio as 1:1 helps to induce the amorphous region and exhibit maximum ionic conductivity at room temperature with flexible nature of electrolyte membranes. When diethyl carbonate (DEC) increases with minimum of propylene carbonates in the complex (b), (c) and (d), its overall amorphous phase becomes reduced and hence low ionic conduction with poor mechanical strength.

P.Sivakumar et al: Highly Porous Polymer Electrolytes based on PVdF-HFP/PEMA with Propylene Carbonate/Diethyl Carbonate for Lithium Battery Applications.

Table 1. *Ionic conductivity (10⁻³S/cm) at various temperatures.*

Sample ID	303K	313K	323K	333K	343K	353K	363K
a	4.77	4.95	5.21	5.59	6.12	7.09	8.43
b	3.08	4.15	4.41	4.46	4.32	4.03	4.20
c	1.50	1.60	1.72	1.94	2.25	3.15	3.67
b	1.17	1.26	1.46	1.71	2.12	2.33	2.65

Where (a) = X – PC/DEC (1:1), (b) = X - PC/DEC (1:2). (c) X - PC/DEC (1:3) and (d) X - PC/DEC (1:4). Where X = PVdF-HFP/PEMA (18/12) - LiClO₄ (8)

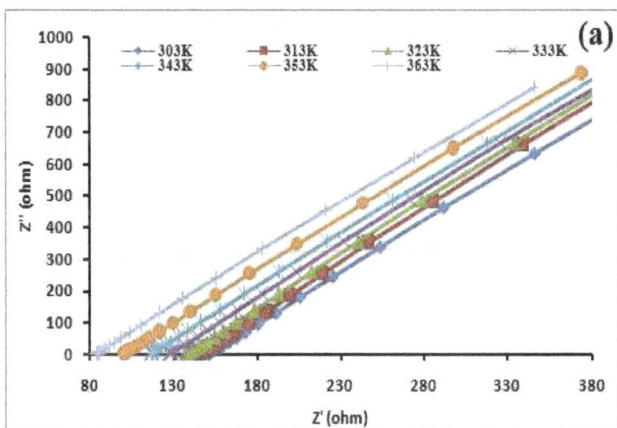

Fig. 3(a). *Impedance spectra of PVdF-HFP/PEMA (18/12) - PC/DEC (1:1/62) - LiClO₄ (8).*

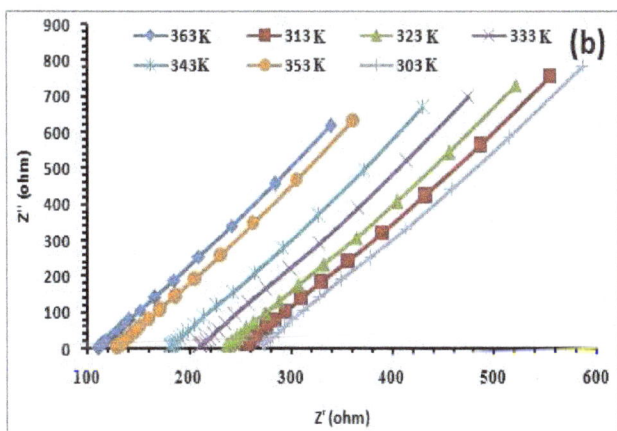

Fig. 3(b). *Impedance spectra of PVdF-HFP/PEMA (18/12) - PC/DEC (1:2/62) - LiClO₄ (8).*

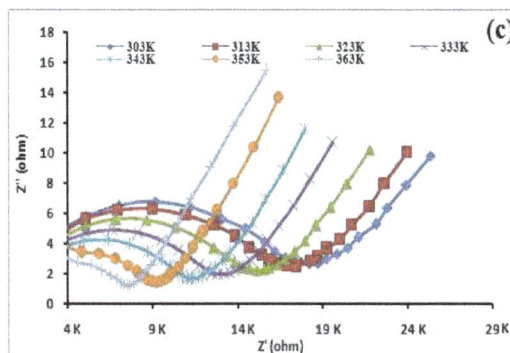

Fig. 3(c). *Impedance spectra of PVdF-HFP/PEMA (18/12) - PC/DEC (1:3/62) - LiClO₄ (8).*

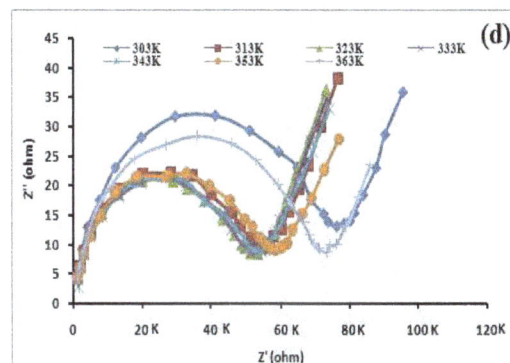

Fig. 3(d). *Impedance spectra of PVdF-HFP/PEMA (18/12) - PC/DEC (1:4/62) - LiClO₄ (8).*

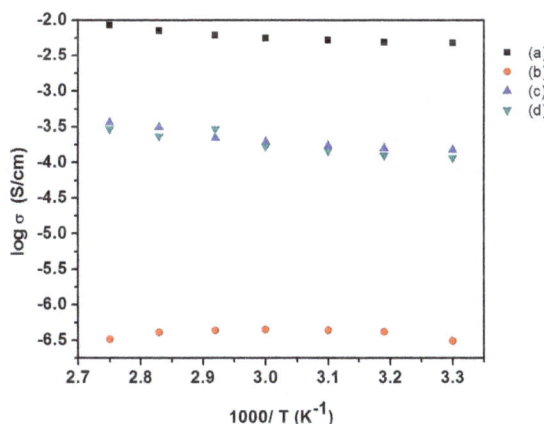

Fig. 4. *Temperature Dependance of Ionic Conductivity.*

3.4. SEM Analysis

SEM is one of the most quick-witted instruments for the examination and analysis of the microstructure characteristics of the substance. "Fig. 5 and 6." are reveals the surface morphology of the polymer electrolytes which depend upon the mixture of PVdF-HFP and PEMA polymer blend at different concentrations of propylene carbonate and diethyl carbonate (1:1, and 1:4). It is transparent that the large porosity and small pore size are fundamental for a good separator and high conductivity. When propylene carbonate and diethyl carbonate are taken as 1:1 wt %, the surface shows enormous number of pores and hence higher ion migration identified. The higher percentage of DEC shows lower conductivity which is due to fewer pores in the surface of the polymer electrolyte which is evident from SEM trace.

Figure 5. SEM image of PVdF-HFP/PEMA (18/12) -PC/DEC (1:1/62)-LiClO₄ (8).

Figure 6. SEM image of PVdF-HFP/PEMA (18/12) -PC/DEC (1:4/62)-LiClO₄ (8).

4. Conclusions

PVdF-HFP/PEMA blend based polymer electrolytes are prepared by solvent casting technique. The structural and complex formations of PVdF-HFP/PEMA-PC-DEC-LiClO$_4$ systems have been confirmed by XRD and FTIR studies. These electrolytes show appreciable ionic conductivity even at room temperature. Maximum conductivity as 4.77×10^{-3} S/cm at room temperature with good mechanical stability has been observed in PVdF-HFP (18) – PEMA (12) – PC/DEC (62) – LiClO$_4$ (8) based system when plasticizer ratio as 1:1. Higher conductivity in this composition could be attributed due to higher amorphicity. SEM analysis reveals the presence of higher porosity in the polymer complex that is also an evidence for higher conductivity.

Acknowledgment

The authors gratefully acknowledge the UGC, New Delhi for providing financial support to carry out this work (F.No: 42-807/2013(SR), Dated 14.03.2013).

References

[1] Armand, M., Endres, F., MacFarlane, D.R., Ohno, H., Scrosati, B., 2009, "Ionic-liquid materials for the electrochemical challenges of the future" Nature Materials, 8, pp. 621.

[2] Kumar, B., Rodrigues, S. J., Koka, S., 2002, "The crys-talline to amorphous transition in PEO-based composite electrolytes: Role of lithium salts", Electrochimica Acta, 47, pp. 4125-4131.

[3] Zhanliang Wang, Zhiyuan Tang, 2003, "Characterization of the polymer electrolyte based on the blend of poly(vinylidene fluoride-co-hexaflouropropylene) and poly(vinyl pyrrolidone) for lithium ion battery", Materials Chemistry and Physics 82, pp. 16–20.

[4] Rajendran, S., Sivakumar, P., Ravi Shanker Babu, 2007, "Studies on the salt concentration of a PVdF–PVC based polymer blend electrolyte", Journal of Power Sources 164, pp. 815–821.

[5] Alex H.C. Shiao, David Chua, Hsiu-ping Lin, Steven Slane, Mark Salomon, 2000, "Low temperature electrolytes for Li-ion PVDF cells", Journal of Power Sources 87, pp. 167–173.

[6] Panero, S., Scrosati, B., 2000, "Gelification of liquid–polymer systems: a valid approach for the development of various types of polymer electrolyte membranes", Journal of Power Sources 90, pp. 13–19.

[7] Rajendran, S., Shanthi Bama, V., 2010, "A study on the effect of various plasticizers in poly(vinyl acetate)-poly(methyl methacrylate) based gel electrolytes" Journal of Non-Crystalline Solids 356, pp. 2764–2768

[8] Hwang, Yun Ju., Nahm, Kee Suk., Prem Kumar, T., Manuel Stephan, A., 2008, "poly(vinylidene fluoride – hexafluoropropylene) – based membranes for lithium batteries", Journal of Membrane Science 310, pp. 349–355

[9] Isabella Nicotera, Luigi Coppola, Cesare Oliviero, Marco Castriota, Enzo Cazzanelli, 2006, "Investigation of ionic conduction and mechanical properties of PMMA–PVdF blend-based polymer electrolytes", Solid State Ionics 177, pp. 581–588.

[10] Diogo, F., Vieira, César O., Avellaneda, Agnieszka Pawlicka. 2007, "Conductivity study of a gelatin-based polymer electrolyte", Electrochimica Acta 53(4), pp. 1404-1408.

[11] Sim, L.N., Majid, S.R., Arof, A.K., 2012, "Characteristics of PEMA/PVdF-HFP blend polymeric gel films incorporated with lithium triflate salt in electrochromic device", Solid State Ionics 209–210, pp. 15-23.

[12] Silverstein, R.M., Bassler, G.C., Morill. T.C., 1991, "Spectroscopic Identification of Organic Compounds", 5th edition john willy & sons, inc., USA.

[13] Yong-Zhong Bao, Lin-Feng Cong, Zhi-Ming Huang. Zhi-Xue Weng, 2008, "Preparation and proton conductivity of poly(vinylidene fluoride)/layered double hydroxide nanocomposite gel electrolytes ", Journal of Material Science 43, pp. 390–394.

[14] Watanabe, M., Ogata, N., MacCallum, J.R., Vincent, CA. , 1987, Polymer Electrolyte Review, Volume. 1, Elsevier, New York,

[15] Saikia, D., Kumar, A., 2004, "Ionic conduction in P(VDF-HFP)/PVDF–(PC + DEC)–LiClO₄ polymer gel electrolytes", Electrochimica Acta 49,16, pp. 2581-2589.

8

CFD Analysis of Effects of Surface Fouling on Wind Turbine Airfoil Profiles

Sashank Srinivasan[1], Vikranth Kumar Surasani[2]

[1]Department of Mechanical Engineering, Birla Institute of Technology and Science-Pilani, Hyderabad Campus, India
[2]Department of Chemical Engineering, Birla Institute of Technology and Science-Pilani, Hyderabad Campus, India

Email address:
sashank.srinivasan.13@gmail.com (S. Srinivasan)

Abstract: One of the important factors that determine the long term efficiency of wind turbine blades is the extent to which the surface finish has been altered from the original state. This can happen either through corrosion or through impingement of particles. This paper aims at analyzing the effect of the later phenomenon on two specific profiles: the NREL S814 and NREL S826 profiles, at different Reynold's numbers. These are two very similar profiles in utility and shape but differ in their thickness. This fact is used to ascertain the effect that thickness of an airfoil has on preventing surface fouling based performance degradation. Surface fouling has been modeled as a roughness at the leading edge of the profile. This is assumed to cause enough flow transition so as to simulate roughness over the entire profile. CFD simulations have been used to perform the analysis and initial results have been validated with experimental data. The accuracy of turbulence models in predicting normal and surface fouled conditions has been assessed. The performance parameters that have been considered are the lift, drag, moment coefficients and the drag to lift ratio.

Keywords: Surface Fouling, Wind Turbine Blades, S814 Airfoil, S826 Airfoil

1. Introduction

To ensure smooth, unseparated flow over wind turbine blades, a great amount of care is taken to obtain a smooth surface finish. However, wind turbines generally operate in harsh environments which causes the surface finish of the blades to deteriorate. This is especially true for off shore marine wind turbines where the flow velocities are much higher than land based windmills. However, this is not the only mode of deterioration of surface finish of the blades. Over a period of use, the blades are impinged with particles from various sources. Some common sources are icing,

insects and dust. This is an equally important, if not more worying cause of surface finish deterioration. In the case of corrosion, improvement of blade materials can reduce the damage caused. However, surface fouling is a completely external factor that has to be dealt with during the design process itself. This study makes an attempt in trying to identify airfoil properties, in particular the thickness of airfoil profiles, on mitigating the performance reduction generally associated with surface fouling. Two specific airfoil shapes: the S814 and S826, both designed at NREL, USA, have been considered in this study. The characteristic features of the two profiles are shown in Table 1.

Table 1. Features of the Airfoils considered.

Airfoil	Location of use on the blade	Thickness	Design Reynold's Number	Rotor Diameter	Salient Features
S814	Root	24%	1.5×10^6	20-30m	High-Lift airfoil
S826	Tip	14%	1.5×10^6	20-40m	High-Lift Airfoil

The presence of roughness alters the aerodynamic performance of the profiles. The lift coefficients decrease while the drag increases. These are caused due to increased severity of flow separation as well as transition from laminar

to turbulent flow happening at a much earlier upstream location along the profile.

To see the difference in performance due to surface fouling, aerodynamic performance data of a smooth airfoil is required first. This, as well as other ensuing analyses have been done using a 2 dimensional model of the airfoils in ANSYS Fluent. The mesh generation was performed in ICEM-CFD.

Additionally, various turbulence models were tested out to determine their effectiveness in predicting normal and surface fouled conditions. To simulate surface fouling effects, it is assumed that a roughness specified at the leading edge of the airfoil is sufficient. No matter how much roughness is present on the surface, once the flow has transitioned to turbulent or separated from the surface, the effect of roughness becomes negligible.

2. Methodology Followed

To maintain uniformity while analyzing the airfoils, the flow domain, chord length, fluid considered, inlet velocity and meshing strategy were all held constant.

3. The Flow Domain

A square flow domain was considered to simulate the flow around the airfoils. To ensure infinity boundary conditions at the edges of the flow domain, the boundaries must be sufficiently far away from the airfoil. The norm followed in the current work is to maintain all boundaries at 40 chord lengths away from the airfoil. Since the chord length used is 1m, an 80m X 80m domain, with the airfoil at the center, was used to study the airfoil performance.

4. Grid Generation

Based on coordinates of the points on the airfoils provided by NREL, a CAD model was prepared in Pro/ ENGINEER using a single spline curve connecting all the points. The grid topology used in ICEM-CFD was H-Type for the flow

domain while an O-grid was used to mesh the region immediately near the airfoil. The O-grid helps resolve the boundary layer formed over the airfoil accurately. This was ensured by constructing meshes with y+ values less than 1. This means that the boundary layer does not remain within one single element of the mesh and hence the boundary layer features of the flow are completely resolved. 100 elements were used in the O-grid surrounding the airfoil. 65 elements were used upstream of the airfoil and 150 elements downstream of the airfoil. In total 235,000 elements were used.

Before continuing with the calculations, a mesh independency study was performed to check if the results do not vary with change in mesh element count. A coarse mesh with 150,000 elements and a fine one with 330,000 were used, apart from the eventual 235,000 element mesh, to simulate the flow around the smooth S814 airfoil at an Angle of Attack (AoA) of 0°. During the process of making the mesh coarser and finer, the y+ value was maintained at values lesser than 1 to ensure accurate results. The lift and drag coefficients were monitored to check if significant variations were predicted by the three meshes. Results for the lift and drag coefficients did not vary by more than 10^{-3} and 10^{-5} units respectively and hence it was concluded that the results were independent of the mesh used. Ultimately, the intermediate mesh size was chosen so as to ensure sufficient mesh density while ensuring computation time remained manageable.

The resulting mesh for the S814 airfoil is shown in Figure 1. The roughness that was specified at the leading edge was restricted to region on the airfoil between the two red lines. A similar mesh was created for the S826 airfoil without any changes in the mesh structure, barring those due to the change of airfoil shape.

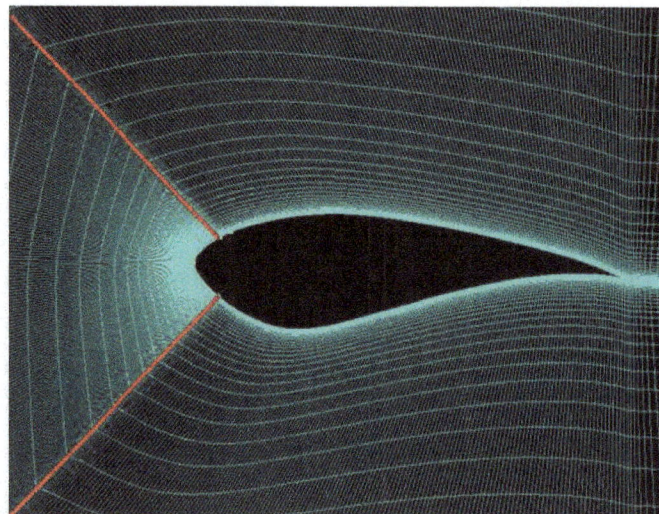

Figure 1. Grid used for the CFD runs of S814 airfoil.

5. The Solver: ANSYS Fluent

The popular commercial code: Fluent, was used to perform

the CFD calculations. The 2-D, double precision solver was chosen for this study. The fluid considered was incompressible air at atmospheric conditions. Boundary conditions specified were: No slip boundary at the surface of the airfoil and free shear flow at the upper and lower boundaries of the flow domain. The inlet velocity specified was based on the Reynold's number analyzed. A turbulence intensity of 5% was specified at the inlet. The outlet was assumed to remain at atmospheric pressure due to the far-field boundary assumption. For the present study, the steady state solver was used. The angle of attack was thus limited to the rang: -4° to 10°, after which flow separation effects play a much larger role and the entire flow field becomes unsteady in nature. Residual limits were set at 10^{-6}, although generally residuals were in the range of 10^{-10} when the simulation was deemed converged. The CFD analysis made use of two turbulence models: the 4 Equation Transition SST model and the fully turbulent SST k-ω model. To answer the question of which model predicted which case better, a separate study was made before the start of the proposed analysis using experimental data taken from literature. Details of this study are given in the next sections.

6. Specifying the Roughness

Given that the roughness was considered only at the leading edge, the height and density of the roughness must be such that the flow is immediately transitioned into the turbulent state.

Kerho and Bragg [3] advocate a Reynold's number of at least 600, calculated based on the height of the rough element. For the case of flow Reynold's number of $1.5x\ 10^6$ over the airfoil of chord length 1m, the minimum height of the rough element must be at least 0.4mm. However, if the inlet Reynold's number changes, the required minimum height would also change. To avoid this, specifying a much higher value would be prudent. The value chosen was 1.9mm for the reason that experimental data for S814 airfoil with leading edge thickness elements of the same height were reported in [2]. This data could be used to ascertain the accuracy of predictions by the CFD solver when using different turbulence models. The experimental data was for roughness elements spread over a length of 102mm over a 457mm chord length airfoil. Following the same dimensional ratio, roughness was specified over a length of 220mm, spread over the top and bottom surface of the airfoil. This norm was followed both the airfoils and at all Reynold's numbers analyzed.

7. Turbulence Modelling

The choice of turbulence model is dictated by the type of flow present in the domain and over the airfoil. For the smooth airfoil case, the flow conditions could be laminar over the initial part of airfoil and then transition to turbulent state. If the flow velocity is high enough, the flow could be turbulent right from the leading edge region. There is no definitive way of stating which conditions prevail, although chances of the former occurring are higher. For the rough airfoil, the roughness height specified was high enough to trip the boundary layer and cause immediate transition to turbulence. Here it is possible to state with more confidence that the flow over the airfoil will be fully turbulent.

A definitive check of flow transition is to plot the skin friction coefficient along the surface of the airfoil. The region where there is a sharp dip and then a sudden increase in the skin friction value denotes the location of the transition point. The reason for this is the mechanism of transition to turbulence. During the transition process, the flow momentarily separates from the surface, forming a separation bubble after which the flow reattaches to the surface of the airfoil as a fully turbulent one. The momentary loss of contact between the fluid and the surface causes the skin friction coefficient to drop to zero over a small part of the airfoil. To see this drop and increase, the transition model has to be used. The fully turbulent model assumes that the flow has already transitioned and hence will not predict this behavior.

For the current study, the 4 equation SST transition model with the k-ω-γ-Re$_θ$ correlation implemented by Menter and Langtry [6] and the SST k-ω model were used. The k-ω class of models resolve the boundary layer without the need for any wall functions and have generally been very accurate in airfoil CFD predictions. Both these modes have been used to simulate smooth and rough S814 airfoil. The predictions of lift, drag, moment coefficients along with the lift to drag ratios were compared with experimental data to see which model to use for each of the two cases. This information has then been used to predict the behavior of the S826 airfoil. The moment has been calculated at 25% of chord length.

8. Validation of Turbulence Models with S814

Experimental testing of the S814 airfoil in smooth and surface fouled conditions were conducted by Ferrer and Mandate [2]. This data was used to validate the accuracy of predictions by the SST-Transition and the SST k-ω models at the design Reynold's number. The lift, drag and drag to lift ratio were compared in this regard. Figures 2-7 show the comparison between the predictions.

From the plots shown, it was deduced that the Transition model predicted the smooth airfoil case better than the fully turbulent SST k-ω model. The rough airfoil predictions by both the models in consideration were not so conclusive.

For the rough S814, the transition model predictions of lift were better while the drag and drag to lift ratio predictions by the fully turbulent model were more accurate. However, on the account that the drag to lift ratio takes into account both the forces, the fully turbulent model was slightly better. Furthermore, the moment about 1/4th chord length was calculated. This gave a good comparison about how the models predict the distribution of forces over the airfoil.

Figure 8 shows that the fully turbulent SST k-ω model predicts the distribution of forces over the airfoil slightly better than the transition model. Additionally, the transition location was determine for the rough airfoils. The skin friction coefficient was plotted against chord length to locate the point of transition of the flow. For most Angle of Attacks, transition happens at around 7% side chord length on the upper surface and 5% on the lower side, which are both within the rough region of the leading edge.

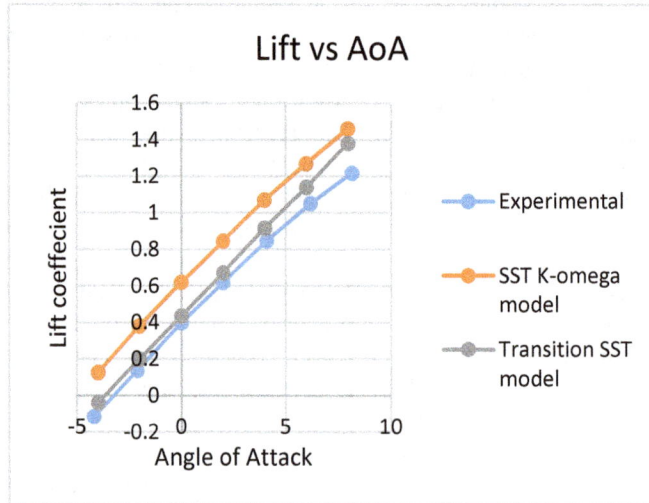

Figure 2. *Variation of lift with angle of attack for S814 (smooth case).*

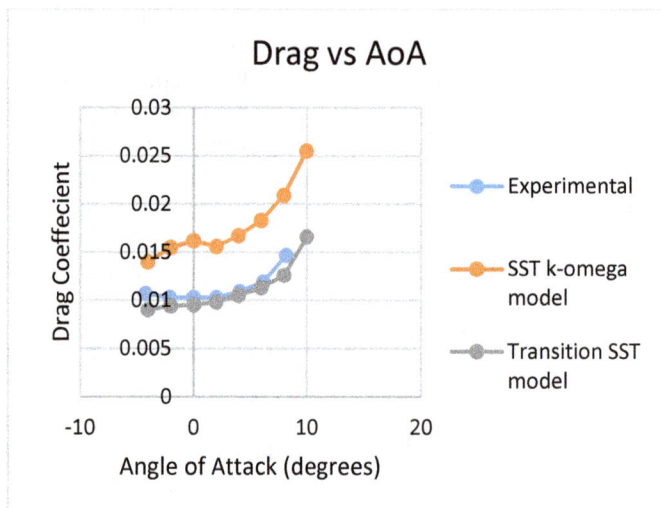

Figure 3. *Variation of lift with angle of attack for S814 (smooth case).*

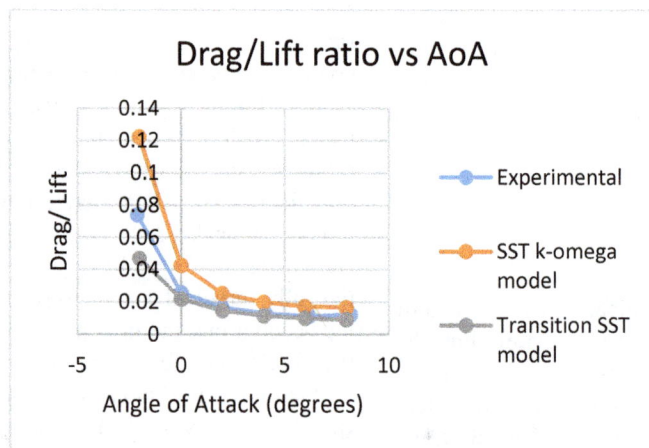

Figure 4. *Variation of drag to lift ratio with angle of attack for S814 (smooth case).*

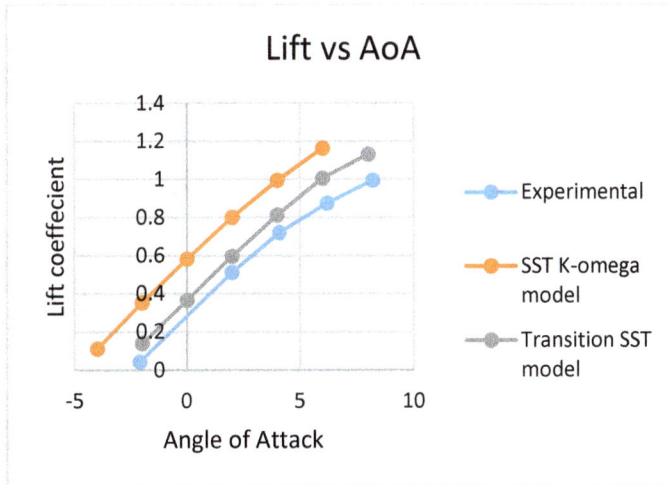

Figure 5. *Variation of lift ratio angle of attack for S814 (rough case).*

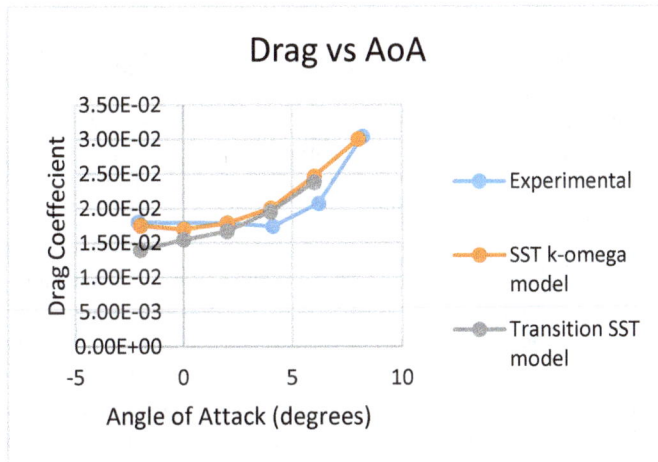

Figure 6. *Variation of drag with angle of attack for S814 (Rough case).*

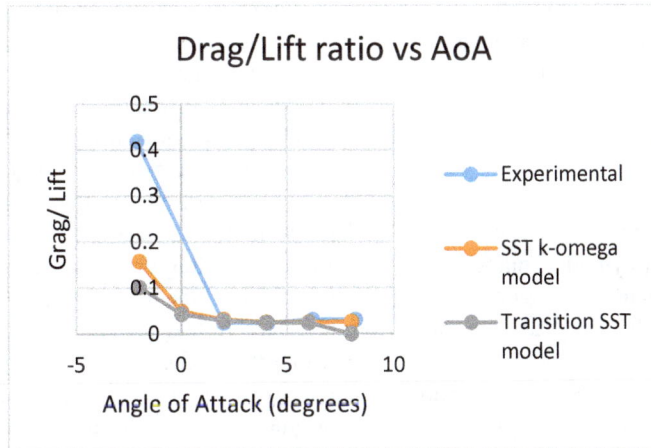

Figure 7. *Variation of drag to lift ratio with angle of attack for S814 (rough case).*

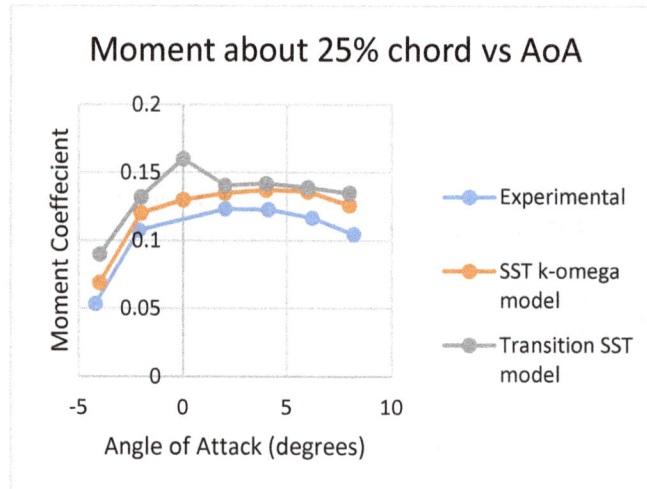

Figure 8. Variation of moment with angle of attack for S814 (Rough case).

This confirms that transition does happen within the rough region itself and the height of the rough elements are sufficient to effectively mimic surface fouled conditions. This also shows why the transition model fails, as the flow over more than 90% of the airfoil is fully turbulent.

However, it was still desirable to see if the two airfoils were resistant to the effects of surface fouling at any angle of attack and inlet velocity, even a small way. The lift, drag and moments will obviously deteriorate due to the roughness specified. However, if the transition to turbulent state happens outside the leading edge region of the airfoil, it would show that the airfoil at those operating conditions is relatively more resistant to the harm done by the fouling of the surface. Since this can be predicted only by the transition model, both the models were employed for the rough airfoil scenario. All performance data for rough airfoils were taken from the fully turbulent model only. Smooth airfoils were analyzed only by the transition model as no advantages are brought to the table by the fully turbulent model.

9. Results and Discussion

Having ascertained the turbulence models that are best suitable for the two operating scenarios, the S814 and the S826 airfoils were analyzed at Reynold's numbers of $1.5x\ 10^6$ and $7.5x\ 10^5$. The performance degradation was analyzed trough the change in lift, drag and moment coefficients. Additionally the drag to lift ratio was also monitored. The changes in these parameters were compared in terms of modulus percentage change from the smooth airfoil case. Within the data generated, two major factors that can affect the change in performance are the thickness and the Reynold's number. For ascertaining the effect of thickness in performance degradation, the S814 and S826 were contrasted against each other at both Reynold's numbers which have been considered. For determining the effect that Reynold's number has, the analysis was done by comparing the data of the same airfoil in the two operating conditions.

10. Effect of Reynold's Number

The two profiles were each tested at their design Reynold's number of $1.5x\ 10^6$ and at half that value to see if any improvement or further deterioration of performance takes place.

The change in performance characteristics of S814 airfoil are plotted in figures 9 through 12. The change in lift and moment coefficients remain the same at both Reynold's numbers. The change in drag increases slightly at the higher Reynold's number but the trends in the change is same. This increase in the drag variation from smooth to rough case can be attributed to increased turbulence at higher velocities which magnify on encountering the rough leading edge. Even this increase is at best 10-15% in the main operating range. To a large extent, the performance deterioration does not seem to be affected by the Reynold's number. This means that whatever the velocity of the flow over the turbine blade, once the surface fouling has occurred, there is no possibility of finding alternate optimum velocities through which the performance reduction can be mitigated.

The S826 airfoil was also analyzed in a similar manner. The performance changes are shown in Figures 13-16. Although the drag changes are more pronounced than in the case of S814, the variation in lift and moment remain the same. This further proves the above conclusion. This means that that the surface fouling effects can be reduced to a large extent by better airfoil designs only. The increased sensitivity to Reynold's number change in drag variation of S826 gives an opportunity for designers to optimize based on inlet velocity. In both cases, the drag to lift change is much higher than the change in drag. This is due to the effect of combining the changes in drag and lift.

Another observation from the plots is that the change in the performance parameters reduces as the angle of attack increases. This can be attributed to the already degraded state of the flow in the smooth conditions. At higher AoA, flow separation effects are more predominant in any scenario. The introduction of a rough leading edge can only do a little more   in disturbing that flow field further.

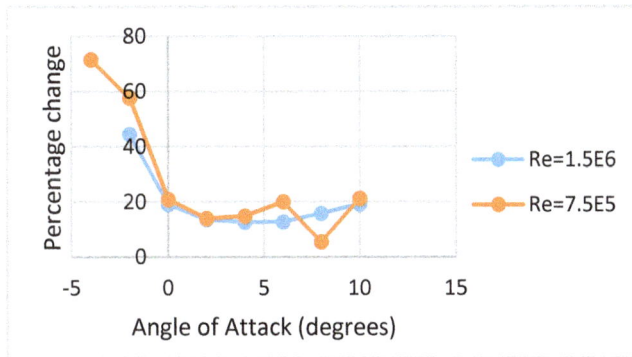

Figure 9. *S814-Change in lift coefficient.*

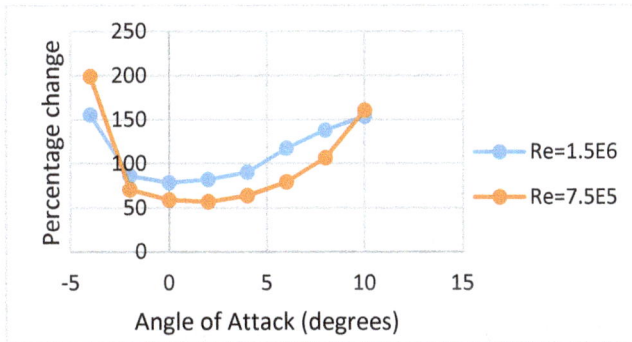

Figure 10. *S814-Change in drag coefficient.*

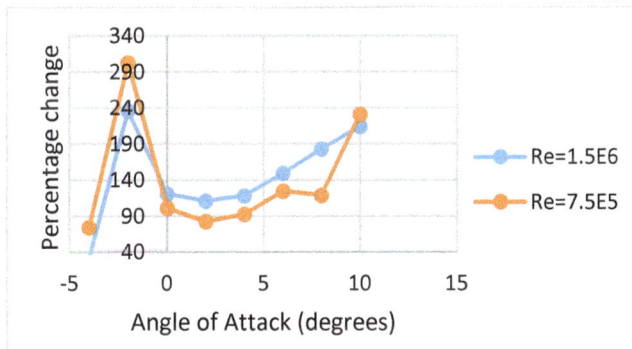

Figure 11. *S814-Change in drag to lift ratio.*

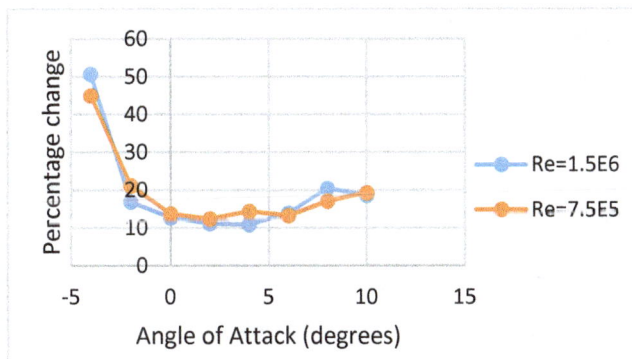

Figure 12. *S814-Change in moment coefficient.*

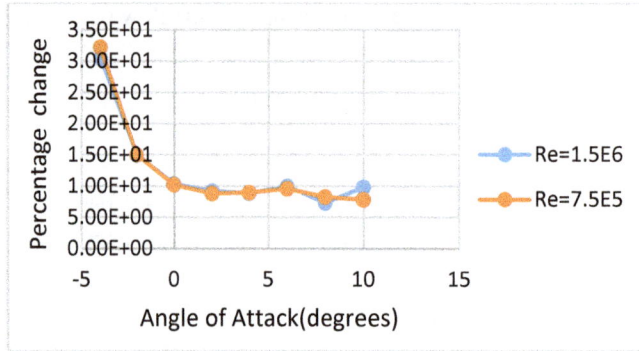

Figure 13. S826-Change in lift coefficient.

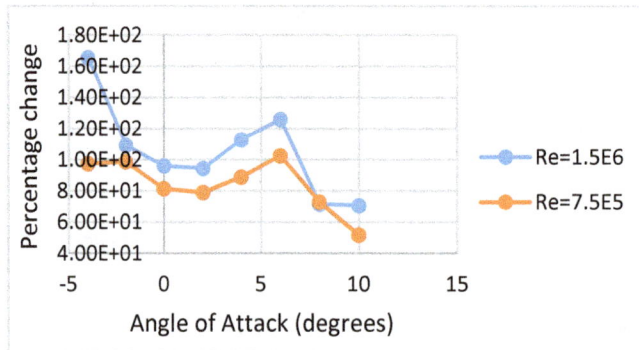

Figure 14. S826-Change in drag coefficient.

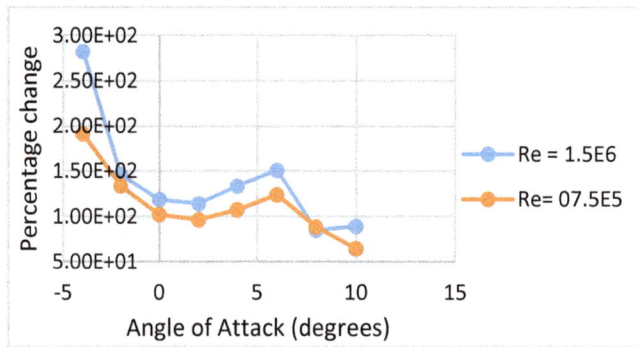

Figure 15. S826-Change in drag to lift ratio.

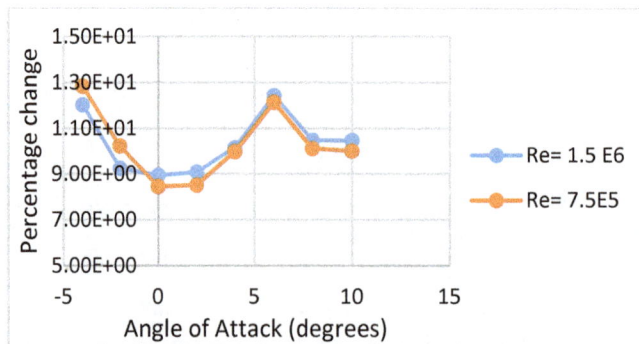

Figure 16. S826-Change in moment coefficient.

11. Effect of Thickness on Performance Degradation

The fact that the inlet velocity does not affect the variation in aerodynamic performance means that the deterioration is affected to a major extent by the airfoil properties. The thickness of an airfoil is the focus of the current study. The S814 has a thickness of 24% while the S826 is thinner with a thickness of 14%. This means that the S814 has a maximum thickness of 24% of the chord while the S826 has a maximum thickness of 14% of the chord length. Since Reynold's number does not affect the change in lift, drag and moment, the effect of thickness was studied at the design

Reynold's number only. The conclusions that arise out of this study will thus hold for analysis at any inlet velocity.

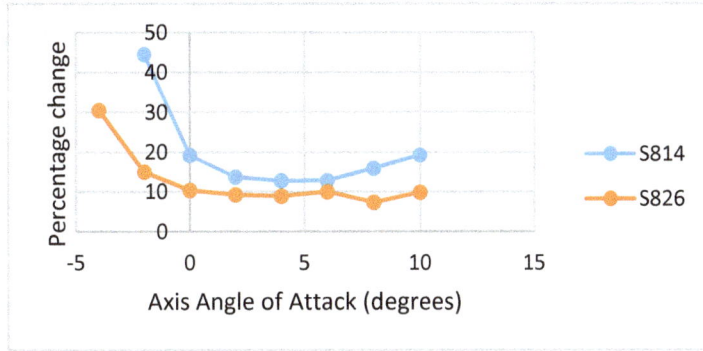

Figure 17. Comparison of change in Lift.

With regard to lift force change, the S826 is less affected when compared to the S814. In the case of drag force variation, both the airfoils remain equally affected. At higher angles of attack, the S826 is drastically less affected than the S814, meaning that the S826 is much more stable in surface fouled conditions.

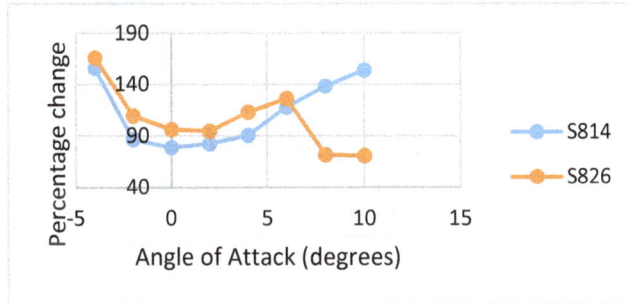

Figure 18. Comparison of change in Drag.

The moment coefficient variation is the same for the both the airfoils, except at the extremes of the AoA range considered where the S826 once again outperforms the S814. With a larger picture in mind, the moment coefficient variation is similar to lift. Considering the fact the lift and moment coefficient varied by the same order magnitude while the drag varied by higher margins, a trend is seen in the relationship between lift and moment coefficient changes.

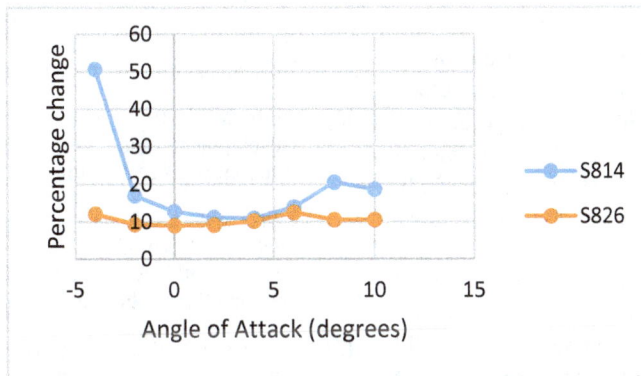

Figure 19. Moment coefficient variation comparison.

From the above comparison, the S826 seems to be better suited to tackle suraface roughness. The lift and moment coefficient variation is lesser than the S814 while the drag variation remains the same. However, it was seen that the S826 had the possibility of optimising drag based on Reynold's number and thus the thinner airfoil has an edge over the thicker one in this aspect as well.

The thicker airfoil is more prone to lift and moment based degradation. Within the range of AoA from 0° to 6°, the perfromance variation is almost the same for the moment variation. However at higher angles, the thicker airfoil becomes much more intensely afected. From this, it can be concluded that the S826 airfoil is a more resistant to surface fouling induced perfromance reduction than the S814 airfil.

12. Change in Transition Location

The height of the roughness elements has been specified large enough to cause immediate transition of flow from laminar to turbulent phase in normal flow conditions. However, it was prudent to check if the two airfoils were resistant to this transition at any angle of attack. This analysis was done using the results from the SST transition model. To determine the transition point, a plot of the skin coefficient is used. An example of this is shown in Figure 20. In this particular example, the transition happens at approximately 50% of the chord on the upper surface of the airfoil and at 25% on the lower surface of the airfoil.

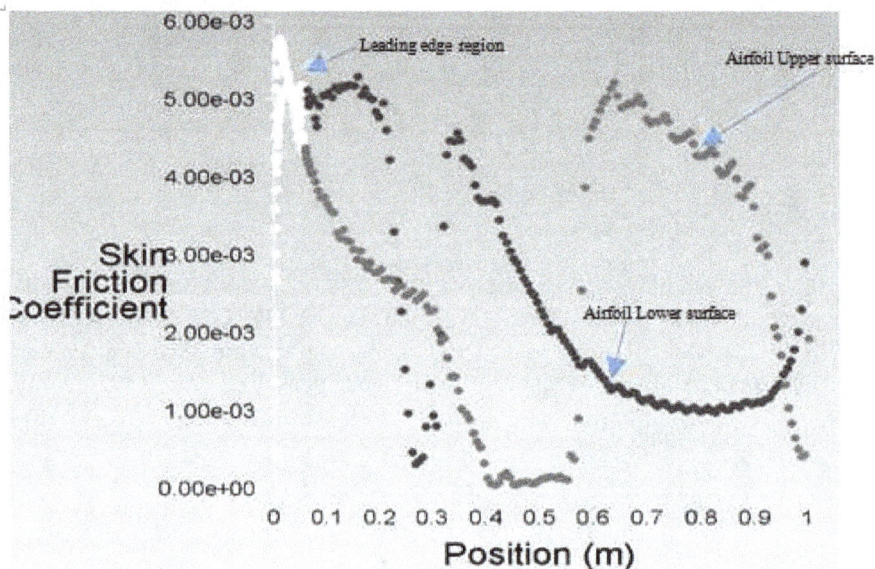

Figure 20. *Plot of skin friction.(S814, smooth, -2° AoA, Re= 1.5x 10^6).*

A similar analysis was done for the two airfoils at both the reynold's number analyzed in this study. Both the airtfoils were unable to negotiate the ill effects of the leading edge roughness element and the transition occurred within the leading edge region only.

13. Conclusions

The S814 and S826 airfoils were analyzed at two Reynold's numbers of $1.5x 10^6$ and $7.5x 10^5$. Transitional and fully turbulent models were assessed based on accuracy of predicting normal and rough airfoils using experimental data of S814. It was observed that the transition model predicted flow over smooth airfoils better while the fully turbulent model predicted surface fouled conditions more accurately.

The performance of the two airfoils were compared between normal operating conditions and surface fouled conditions. For both the airfoils, the Reynold's number did not play a major role in mitigating or worsening of aerodynamic performance during surface fouled conditions. Additionally, it was observed that the thinner S826 airfoil was more resistant to surface fouling induced performance degradation than the S814 airfoil. The location of the transition point was analyzed to see if at some angle of attack, the airfoils were able to prevent immediate transition at the leading edge. Both airfoils failed in preventing fully turbulent flow from happening over the entire airfoil.

References

[1] Dalili.A, Edrisy.A, Carriveau.R, 2009 "A review of surface engineering issues critical to wind turbine performance", Renewable and Sustainable Energy Reviews 13, pp. 428–438

[2] Ferrer.E ,Munduate.X, 2009, "CFD predictions of transition and distributed roughness over a wind turbine airfoil", 47th AIAA Aerospace Sciences Meeting Including The New Horizons Forum and Aerospace Exposition.

[3] Kerho.M.F, Bragg.M.B, 1997, "Airfoil Boundary-Layer Development and Transition with Large Leading-Edge Roughness", AIAA Journal, vol. 35.

[4] Janiszewska.J, Ramsay.R, Hoffmann.M.J, Gregorek.G.M, 1996, "Effects of Grit Roughness and Pitch Oscillations on the S814 Airfoil", Airfoil Performance Report, Revised (12/99), National Renewable Energy Laboratory, USA.

[5] van Rooij.R. P. J. O. M, Timmer. W. A, 2003, "Roughness Sensitivity Considerations for Thick Rotor Blade Airfoils" ASME Journal of Solar Energy Engineering, vol.125.

[6] Menter.FR, Langtry.RB, Likki.SR, 2006, "A Correlation-Based Transition Model Using Local Variables— Part I: Model Formulation", Journal of Turbomachinery, vol.128.

[7] Tangler J.L and Somers D.M., 1995, NREL airfoil families for hawts. Technical report, NREL.

[8] Corten G.P., 1999, "Insects cause double stall", Symposium on aerodynamics of wind turbines, Stockholm, pp.6-74.

[9] Burton. T, Sharpe. D, Jenkins. N, Bossanyi. E, 2001, " Wind Energy Handbook", John Wiley and Sons, ISBN 0-471-48997-2.

[10] Ren. N, Ou. J, 2009, "Dust effect on the performance of wind turbine airfoils", Journal of Electromagnetic Analysis & Applications, vol.1, pp 102-107

9

Magnetic Loss Inside Solid and Laminated Components under Extreme Excitations

Zhiguang Cheng[1], Behzad Forghani[2], Yang Liu[1, 3], Yana Fan[1], Tao Liu[1], Zhigang Zhao[4]

[1]Institute of Power Transmission and Transformation Technology, Baobian Electric Co., Ltd, Baoding, China
[2]Infolytica Corporation, Place du Parc, Montreal, Canada
[3]State Grid Smart Grid Research Institute, Beijing, China
[4]School of Electrical Engineering, Hebei University of Technology, Tianjin, China

Email address:

emlabzcheng@yahoo.com (Zhiguang Cheng), forghani@infolytica.com (B. Forghani), iamsam@hebut.edu.cn (Zhigang Zhao)

Abstract: The modeling and numerical analysis of the magnetic loss inside components under multi-harmonic and/or DC-biasing excitations are increasingly of concern in large and special electromagnetic devices. This paper aims to investigate efficient and reliable approaches to determine the magnetic losses inside both the solid and laminated components under such extreme excitations. All the proposed approaches presented in the paper are experimentally validated.

Keywords: Benchmark Model, Components, Device-Based Model, Experimental Validation, Extreme Excitation,
Finite Element Analysis(FEA), Magnetic Loss, Magnetic Material, Working Magnetic Properties

1. Introduction

Lately, there has been considerable focus on the modeling and prediction of the magnetic loss in electromagnetic devices, predominantly for the very high capacity and voltage level power equipment, operating under extreme excitations involving DC-biasing and/or multi-harmonic or PWM(Pulse Width Modulation) supplies [1-7].

However, the multi-scale nonlinear electromagnetic analysis and the related measurement of magnetic properties under such extreme excitations are quite challenging, additionally, making the validation of the modeling and numerical simulation very difficult, as compared to those under standard sinusoidal excitations or in the cases of lower capacity and voltage [8-9].

The purpose of this paper is to investigate efficient analysis and reliable experimental methods for evaluating the magnetic losses inside both the solid and laminated components in electromagnetic devices. The paper also addresses the working magnetic property modeling, and rigorous validation under extreme excitations, using the engineering-oriented test models and experimental setups with hybrid supply.

2. Loss Determination under Extreme Excitation

2.1. Modeling and Computation of Magnetic Loss

In the modeling and computation of the magnetic loss inside the solid and laminated components based on 3-D transient field solution under extreme excitations, including multi-harmonics and/or DC-AC hybrid supply, there are a number of key treatments and simplifications, including: 1) control of the identity of the applied excitation conditions in the experiment with that used in the transient field modeling; 2) measurement of the working magnetic property under applied excitation, at different DC-bias levels and/or with different harmonic contents; 3) non-uniformity of both the magnetic field and loss inside the solid plate due to the considerable skin effect and the laminated core-frame, caused by the lamination-joints; 4) accurate computation of the exciting coil's loss for the indirect determination of the magnetic loss in components; 5) comparison among different magnetization curves has shown that the DC magnetization curve can be practicably used in transient field analysis with hybrid excitation[10,11]; 6) for the laminated frame, if the induced eddy currents caused by the flux normal to the laminations are rather weak, then the additional magnetic loss in the frame can be neglected.

2.2. Indirect Determination of Magnetic Loss

In general, the magnetic loss P_{iron} inside components cannot be measured directly, but, it can be determined indirectly through the measured total loss, referred to as on-load loss, $P_{on-load}$, and another loss component generated in the exciting coils P_{E-coil}. If the total loss has been measured, the problem is reduced to determining the loss in the air-core exciting coils specially designed for the purpose of the validation.

In this way, three approaches for determining P_{E-coil} in the exciting coils, mainly dependent on the excitation level and the complexity of the proposed models, have been developed and realized by the authors.

1) Approach I

In the case of a weak single sinusoidal excitation and a low rated frequency, the leakage flux linked with the exciting coils does not change considerably with or without electromagnetic components. Thus the exciting coil's loss, without the electromagnetic components, can be directly measured, referred to as 'no-load' loss, P_{E-coil}, and then P_{iron} in the magnetic components can be determined, as shown in (1),

$$P_{iron} = P_{on-load} - P_{E-coil}\big|_{\text{measured without electromagnetic components}} \qquad (1)$$

2) Approach II

In the case of a strong excitation, the effect of the magnetic components on the leakage flux of the exciting coils must be considered. However, it is possible to use flux compensation to effectively determine P_{E-coil}. As an example, in a magnetic shield model, the exciting coils are located on one side of magnetic shield, which can be referred to as a high permeability plane with low eddy current reaction. In order to keep the leakage flux almost unaffected when the magnetic shield is removed, the compensatory coils, which have the completely same specifications as the exciting coils, are set up symmetrically on the other side of the high permeability plane [12].

The exciting coil's loss is measured using the leakage magnetic flux compensation coils, referred to as C-coils, and then P_{iron} in the magnetic components is determined, as shown in (2),

$$P_{iron} = P_{on-load} - \left(\frac{P_{E-coil} + P_{C-coil}}{2}\right)\Big|_{\text{measured with C-coil}} \qquad (2)$$

where P_{E-coil} and P_{C-coil} are induced in E-Coil and C-coil respectively.

3) Approach III

In the case of extreme excitation, including DC-biasing and/or multi-harmonic, and when the test model has a complicated magnetic structure, it is impossible to use Approaches I and II stated above. Fortunately, the exciting coil's loss P_{E-coil} can be accurately calculated based on a 3-D

transient field solution under complicated excitation. Therefore the magnetic loss in magnetic components P_{iron} can be determined by the following relation (3),

$$P_{iron} = P_{on-load} - \int_{\Omega_{cu}} \frac{\boldsymbol{J} \cdot \boldsymbol{J}}{\sigma}\,dv\Big|_{\text{calculated based on 3-D transient field solution}} \qquad (3)$$

Note that the second term of (3) is for the calculation of the total loss of the exciting coil, P_{E-coil}, including the eddy current loss and resistive loss caused in the exciting coil[13]. The relation (3) can now be used to indirectly determine the magnetic losses inside complex structures under extreme excitations.

3. Magnetic Loss in Solid Components under Multi-Harmonic Excitation

In this Section, the magnetic losses inside the solid components under multi-harmonics excitations are evaluated based on an upgraded benchmark model, P21⁰-B⁺ [12,13].

3.1. Magnetic Property Modeling of Magnetic Steel under Harmonic Excitation

The specific total loss of the magnetic steel plate has been measured under multi-harmonic excitations by using ring specimens [14], and two specimens prepared for comparing magnetic properties between different ring sizes. The conductivity of the magnetic steel (Q235B), $\sigma=5.7895\times10^6$ S/m, and the dimensions of the two ring specimens (RS1 and RS2) are shown in Table 1.

Table 1. Dimensions of ring specimens.

Specimens	Inner diameters (mm)	Outer Diameters (mm)	Thickness (mm)
RS1	330	400	10
RS2	430	500	10

The multi-harmonic voltage supply U is defined in (4),

$$U = U_{m1}\sum_{k=1}^{4} c_k \sin[(2k-1)\omega t] \qquad (4)$$

where U_{m1} is the peak value of the fundamental voltage, and $c_k=1, 0.3, 0.5, 0.4$ as $k=1,2,3,4$, respectively. The waveforms of multi-harmonic supply are shown in Fig.1.

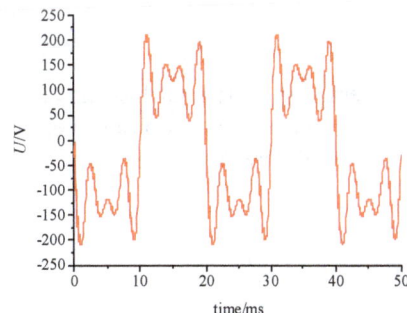

(a)

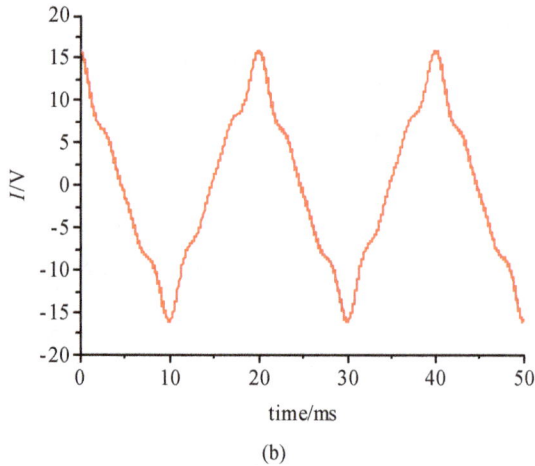

(b)

Figure 1. *Multi-harmonic supply, (a) waveform of multi-harmonic voltage, (b) waveform of multi-harmonic current.*

The block diagram of the circuit for measuring the magnetic properties, where the components include multifunction generator, WF1974, precision power amplifier, 4520/4520A, NF, power analyzer, WT3000, Yokogawa, and multi-channel temperature recorder, TP700, Toprie Electronic, are shown in Fig.2.

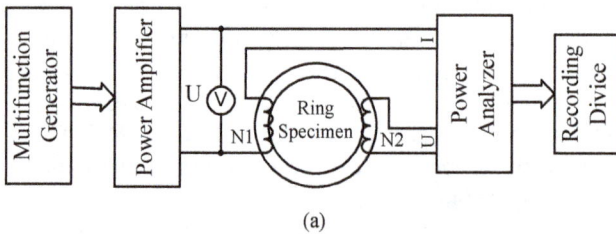

(a)

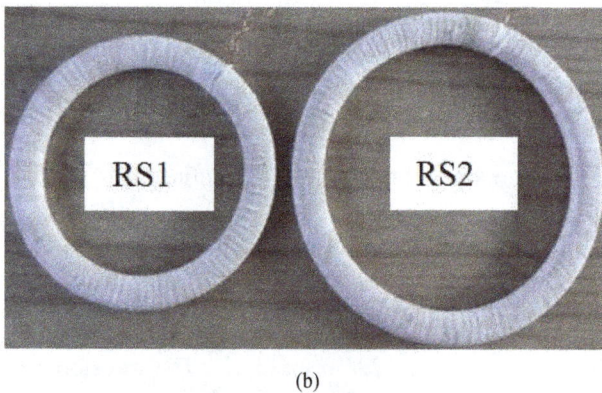

(b)

Figure 2. *Measurement of magnetic properties using ring specimens under multi-harmonics (a) block diagram of the circuit; (b) ring specimens(RS1 & RS2).*

The measured average specific total loss curves, W_{av}-B_m, with 3rd, 5th and 7th harmonics, using two ring specimens, are shown in Fig.3. A minor difference can be seen at higher flux densities.

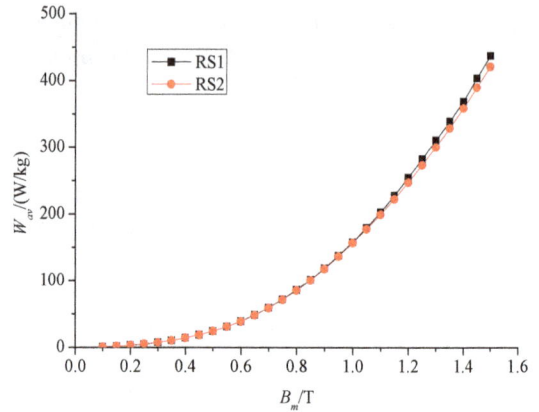

Figure 3. *Average specific total loss curve under multi-harmonic excitation(Q235B).*

The DC magnetization curve (Q235B) is used in the transient field analysis, as shown in Fig.4, provided by the WISCO, Wuhan.

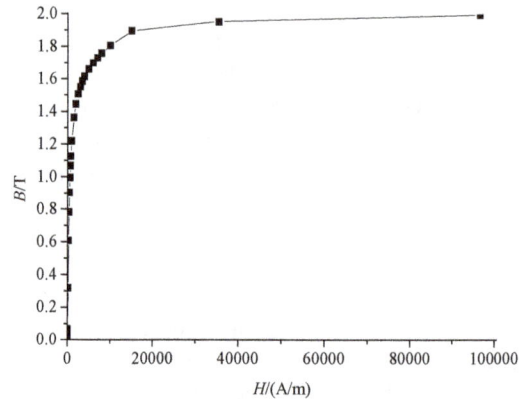

Figure 4. *DC magnetization curve (Q235B).*

3.2. Upgraded TEAM Model and Loss Evaluation

3.2.1. Upgraded TEAM Benchmark Model

Three decades ago, the TEAM benchmarking activities, now under the ICS (International Compumag Society) control, started to spread worldwide and has played a significant role in the progress of computational electromagnetic, especially for the validation of electromagnetic analysis methods. The engineering-oriented TEAM Problem 21 family, with a great deal of industrial involvement, is well established and has evolved significantly over time[13,15,16]. It is still being improved to deal with modeling and simulation under extreme excitations, making it very useful for both the development and the validation of more efficient analysis methods.

To model higher saturation levels, by increasing the excitation, or to use more complex excitation waveforms, the upgraded benchmark model P21⁰-B⁺ , based on the original member model P21⁰-B of Problem 21 family, has two leakage flux generators (E-coils 1 and 2) and two leakage flux compensators (C-coils 1 and 2) [12,13], and the magnetic plate (Q235B, size: 10×500×1000mm), as shown in Fig.5.

(a)

(b)

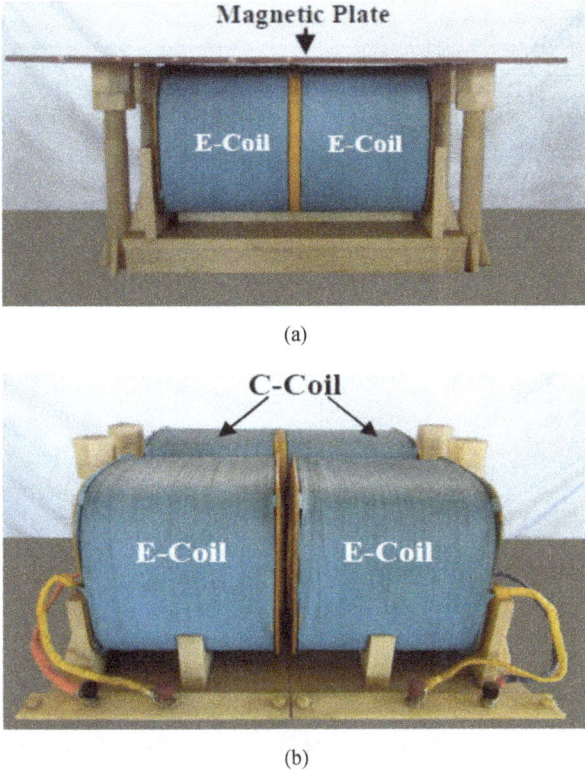

Figure 5. *Loss measurement with upgraded E-coil and moveable C-coils (P21^0-B^+), (a) on-load case with magnetic plate, (b) no-load case with leakage flux compensation but without magnetic component.*

3.2.2. Loss Evaluation

The total loss of the entire test model, $P21^0$-B^+, is measured under multi-harmonic excitations; the measurement system is shown in Fig.6.

The loss caused in the air-core exciting coils of model $P21^0$-B^+, without the magnetic plate, is calculated based on a 3-D transient field solution under either only sinusoidal or multi-harmonic excitations. The 3-D FE model of the exciting coil, using MagNet, Infolytica[13], is shown in Fig.7.

The magnetic loss inside the magnetic plate of $P21^0$-B^+ is calculated based on the 3-D transient field solution and the measured specific total loss data.

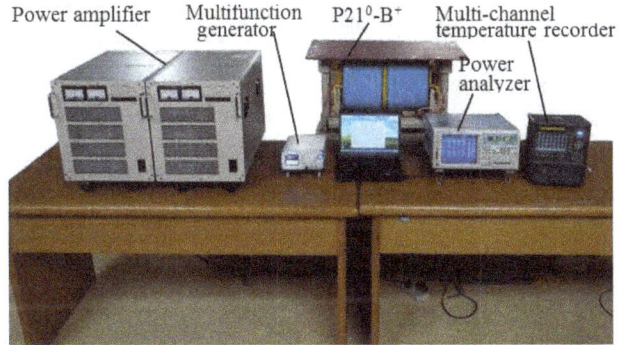

Figure 6. *Measurement of magnetic loss in magnetic components under multi-harmonic excitation.*

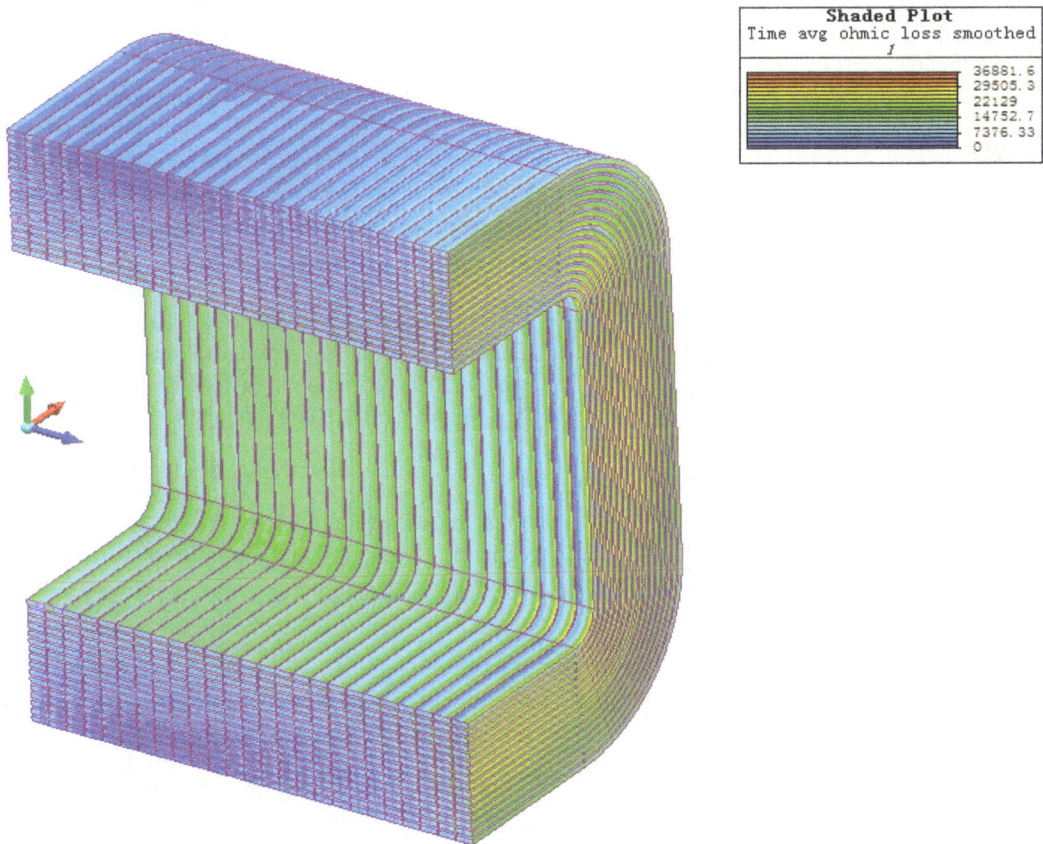

Figure 7. *3-D FE model for coil's eddy current analysis (by MagNet).*

Table 2. Loss in exciting coil under harmonics excitation.

Currents (A, rms)	5.0		7.0		9.0	
Loss(W)	Meas.	Calc.	Meas.	Calc.	Meas.	Calc.
P_{sin}	16.67	16.76	32.67	32.84	53.97	54.29
$P_{harmonic}$	18.38	19.52	36.04	38.28	59.60	63.28

The measured and calculated loss results in the air-core exciting coils of model $P21^0$-B^+, under the sinusoidal or the specified multi-harmonic excitations (as defined in (4)), are in practically good agreement. See Table 2.

The losses generated in the magnetic plate under the specified multi-harmonic excitations, as defined in (4), are obtained using different methods, i.e., the indirect determination based on the measured total loss and the calculated coil's loss (i.e., Approach III), and the numerical computation method. Both are also in practically good agreement. See Table 3.

Table 3. Loss in magnetic plate under harmonics excitation.

Currents (A, rms)	5.0		9.0	
$P_{harmonics}$ (W)	Indirectly determined by Approach III	Calc.	Indirectly determined by Approach III	Calc.
	11.07	9.60	34.83	33.12

3.2.3. Non-Uniformity of Magnetic field and Loss

According to the numerical computation results, the distributions of both the magnetic flux densities and magnetic loss in the cross-section of the ring specimens are not uniform due to skin effect. See Fig.8.

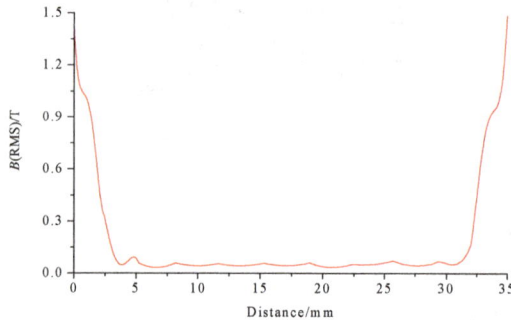

Figure 8. Distribution of magnetic flux densities (Q235B, RS1).

The heterogeneity, α, of the magnetic flux results inside the ring specimens can be expressed by the following relation (5), i.e., the standard deviation of the magnetic flux densities divided by the average flux density over the cross-section of the ring specimen,

$$\alpha = \frac{1}{B_a} \sqrt{\frac{\sum_{i=1}^{n} (B_i - B_a)^2}{n-1}} \tag{5}$$

where, B_a is the average value over all the magnetic flux densities (rms) over the cross section of the specimen; B_i is the magnetic flux density (rms) at i^{th} element; n is the total number of elements.

According to (5), the heterogeneity of the magnetic flux densities for the ring specimens RS1 under the multi-harmonic excitation (including 3rd, 5th and 7th harmonics) is around 0.48, and further examination shows that it is dependent on the average flux density and the width of the ring specimen. This fact should be considered in the accurate modeling and computation of the magnetic loss in the solid plate.

4. Magnetic Loss in Laminated Component under Hybrid Excitation

Smoothing reactors connected in series in HVDC systems are used to reduce the AC component and the transient over currents. In a magnetically-shielded smoothing reactor, there is a magnetic shielding frame outside the air-core winding to control the magnetic flux. The leakage flux enters the laminated frame from the air-core winding. This is essentially different from the usual case of a core-type power transformer.

In the stray-field loss evaluation in smoothing reactors, the measurement of the GO silicon steel material property under DC-bias with multiple harmonics, the multi-scale 3-D transient electromagnetic field analysis, due to the huge overall size of the structure and the very thin penetration depth in the magnetic parts, or the large laminated frame, are quite challenging. Nevertheless, the modeling and the prediction of the stray-field losses have become increasingly important.

This section investigates efficient numerical and reliable experimental approaches to be utilized in predicting the stray-field losses in smoothing reactors, including loss determination based on a 3-D transient field analysis, the measurement of the working magnetic properties under DC-bias and multi-harmonics, and validation based on a well-established smoothing reactor model and experimental system with DC-AC hybrid supply [17].

4.1. Loss Calculation Inside the Shielding Frame

In the smoothing reactor, the air-core exciting current $i(t)$ contains a heavy DC and multiple AC harmonics, as shown in (6),

$$I = I_{DC} + \sum_k I_{mk} \cdot \cos(k\omega t + \phi_k) \tag{6}$$

Where I_{DC} is the DC component of the total exciting current, ω, the fundamental angular frequency, and I_{mk} and ϕ_k, the amplitude and phase angle of the kth harmonic current, respectively.

Fig.9 shows the waveform of the DC-AC hybrid exciting currents, as an example, DC: 25A, and AC: 9A (rms, the rated frequency:50Hz, including 3rd, 5th and 7th harmonics).

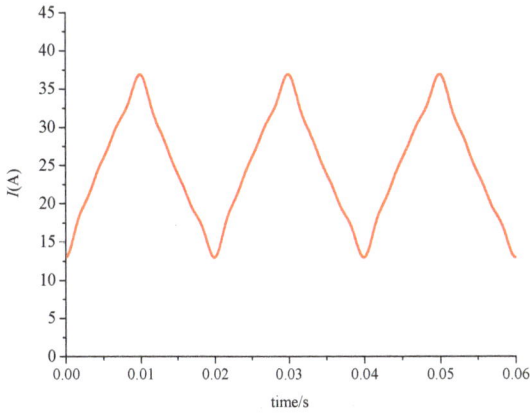

Figure 9. *Hybrid exciting currents (DC: 25A; AC:9A,rated frequency: 50Hz, rms, including 3 harmonics).*

The total magnetic loss of the laminated frame, P_{frame}, is calculated based on the 3-D transient field solution and the measured specific total magnetic loss over different zones, i.e., $P_{limb}(B_m,H_{DC})$, in the middle section of the limb, Ω_{limb}, and $P_{corner}(B_m,H_{DC})$ in the joint area of the frame, Ω_{corner}. Using (7), the magnetic loss in each element is calculated by interpolation, and then summed up for each zone,

$$P_{frame} = \begin{cases} \sum_e W_u^{(e)}(B_m^{(e)},H_{DC}^{(e)}) \cdot V^{(e)} & in\ \Omega_{limb} \\ \sum_e W_{nu}^{(e)}(B_m^{(e)},H_{DC}^{(e)}) \cdot V^{(e)} & in\ \Omega_{corner} \end{cases} \quad (7)$$

where $V^{(e)}$, $B_m^{(e)}$, the volume and magnetic flux density of the element e, respectively.

The MagNet-based scripts [13] are designed and used to compute the total loss in laminated frame.

4.2. Measurement of Working Magnetic Properties under Hybrid Excitation

In order to investigate the non-uniform specific total losses inside the shielding frame of a smoothing reactor under hybrid excitation, an efficient measurement method by means of two laminated core frames has been developed[13,18]. The two core frames, referred to as CF 1 and CF 2, are made of the same GO silicon steel, using the same joint type (multi-step lap), with the same lamination thickness and width, but different length; in fact, CF 2 is a scaled-down version of CF 1. The two core frames have different active volume(or mass), m_{a1} and m_{a2}, respectively. See Fig.10.

Two assumptions are made: 1) the magnetic field and loss distribution over the corner regions of both CF 1 and CF 2 are identical, despite the difference in the length of the limbs in the two frames; 2) the magnetic field and loss are uniform over the middle section of each limb.

In order to realize the two assumptions, it is important to keep the identity of the magnetic flux density and the magnetic field intensity by DC inside both CF 1 and CF 2 under DC-AC hybrid excitations.

The specific magnetization loss, over the indicated uniform zones of the core limbs, can be determined from the difference between the absolute power losses P_{frame1} and P_{frame2}, obtained from CF 1 and CF 2, respectively, and the corresponding total mass of the uniform zone, $(m_{a1}-m_{a2})$, implying that the frame corners have no effect on W_u as expressed by (8),

$$W_u = \frac{P_{frame1} - P_{frame2}}{m_{a1} - m_{a2}} \quad (8)$$

According to the assumptions made above, the total loss and volume (or mass) of the four corner regions of CF 1 are in fact equivalent to those of CF 2. Therefore, the average specific total loss in the non-uniform region, W_{nu}, can be determined from the absolute total power loss, measured from CF 2 (P_{frame2}), and the corresponding active mass at the 4 corners of the total non-uniform regions of CF-1, $m_{corners}$, as shown in (9),

$$W_{nu} = \frac{P_{frame2}}{m_{corners}} = \frac{P_{frame2}}{m_{a2}} \quad (9)$$

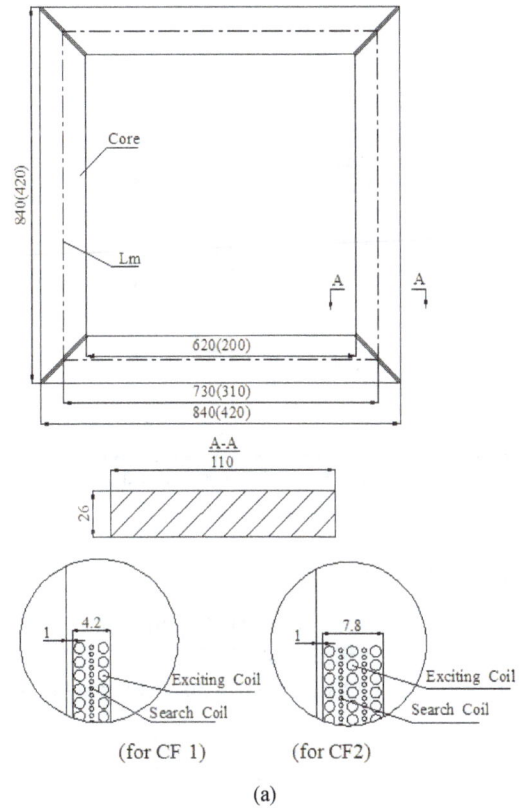

(a)

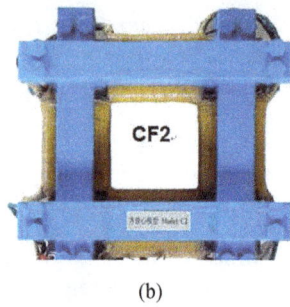

(b)

Figure 10. *Measurement of magnetic property by using two core-frames, (a) dimensions of core-frames, and the numbers within parentheses ( ) show the dimensions of CF 2; (b) tow core frames, CF1 and CF2(photos).*

The specific total loss curves measured at different DC-bias levels, described by H_{dc}, are shown in Fig.11. It can be observed that the specific total loss increases with the DC component of the total exciting currents, and the specific total losses at the corner zones (as expressed by W_{mu}) are greater than those (as expressed by W_u) at the middle of the frame's limb.

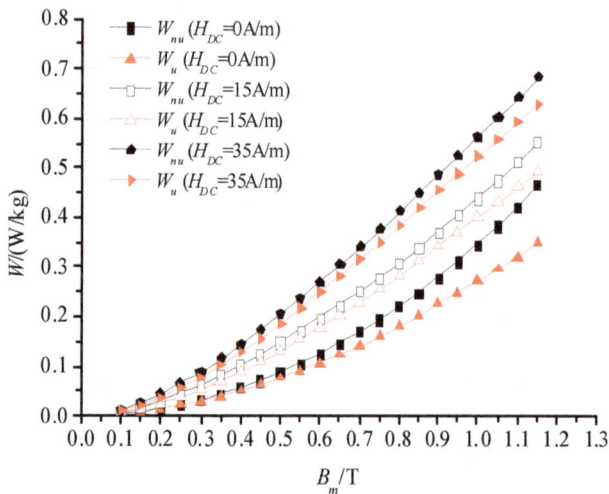

Figure 11. *Specific total loss curves measured by two core frames (30Q140).*

As mentioned above, the DC magnetization curve can be used in DC-bias transient filed analysis. Fig.12 shows the DC magnetization curve of GO silicon steel 30Q140, provided by WISCO, Wuhan.

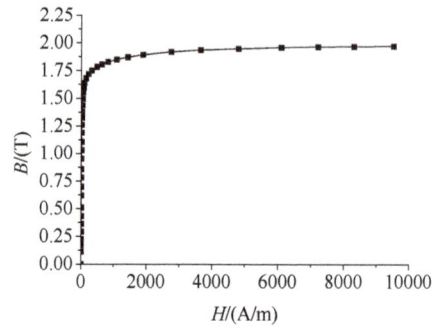

Figure 12. *DC Magnetization curve (30Q140).*

4.3. Smoothing Reactor Model and Measurements

4.3.1. Smoothing Reactor Model

The smoothing reactor model consists of an air-core elliptic exciting coil and the square laminated frame, CF 1, as used in the measurement of the magnetic property. See Fig.13. The main specification parameters of the exciting coil are: number of turns, 408; size of copper wire, 3×9mm; conductivity of wire, $\sigma=5.7143\times10^7$ S/m.

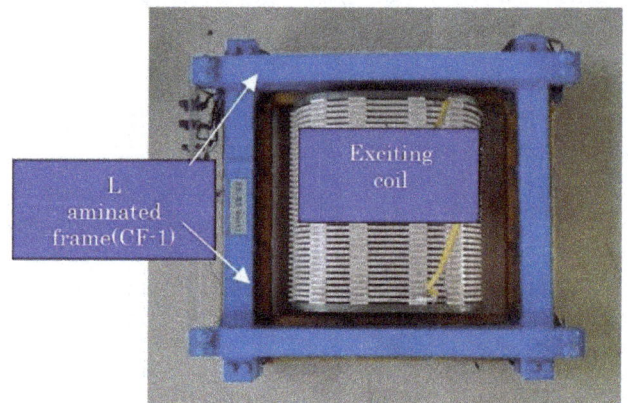

Figure 13. *Smoothing reactor model.*

4.3.2. Loss Measurements under Hybrid Excitations

The block diagram of the circuit used for measuring the stray-field loss, shown in Fig.14(a), includes the components: smoothing reactor model, multifunction generator (WF1974), precision power amplifier (4520/4520A), DC power supply (DH 400-37), NF Co., power analyzer (WT-3000, Yokogawa), Gauss/Teslameter (Model 7010, F.W.Bell), and multi-channel temperature recorder (TP700, Toprie Electronic). See Fig.14(b).

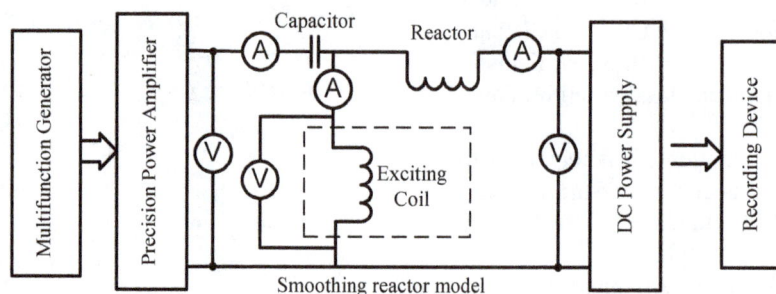

(a)

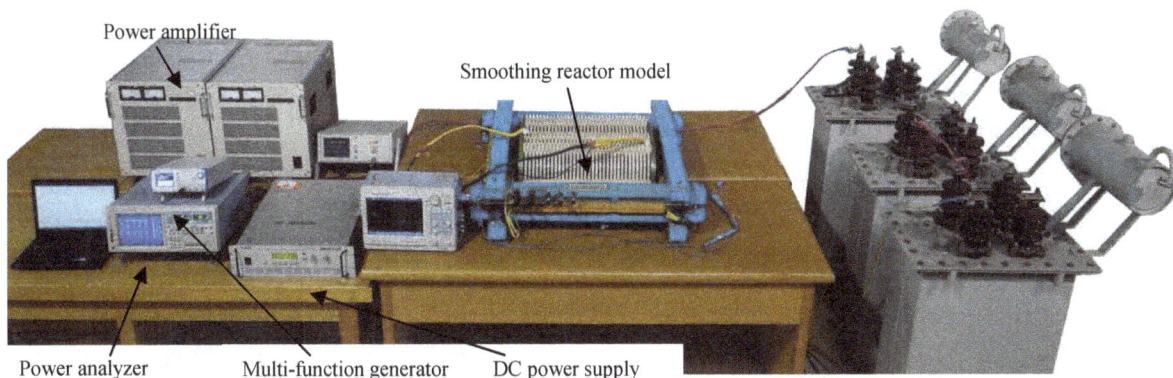

Figure 14. Measurement of stray-field loss in frame under hybrid excitations.(a) block diagram of the circuit; (b) experimental apparatus.

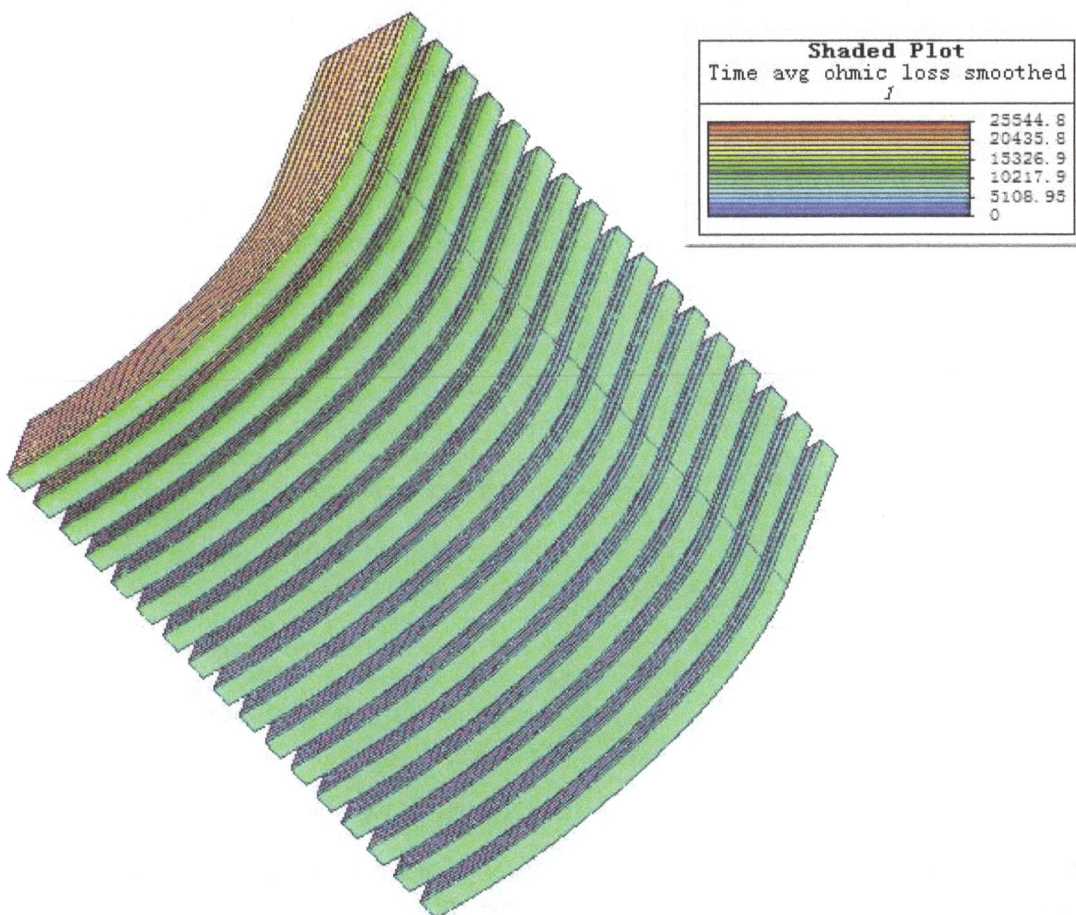

Figure 15. 3D FE model of air-core exciting coil.D

4.4. Results and Discussions

4.4.1. Total Loss of the Air-Core Exciting Coil

The total loss in the single air-core exciting coil, without the laminated frame, including both the eddy current and resistive loss, is measured using the power analyzer (WT3000, Yokogawa) and calculated using the FEA software (MagNet, Infolytica) under different excitations, including single and hybrid power supply. The solved finite element model, 1/8th of the whole coil, is shown in Fig.15. The corresponding coil's loss results are shown in Table 4.

The agreement between the calculated and measured loss results shows that the total loss of the air-core exciting coil can be accurately and numerically calculated under either single or hybrid excitations.

In order to further verify the coil's loss computation based on 3-D transient field solution, the leakage flux densities at the specified positions of the exciting coil, surrounded by the laminated frame and excited by DC-AC hybrid supply with 3 harmonics, i.e., DC:15A, AC: 9A(rms), rated frequency 50Hz, with 3rd, 5th and 7th harmonics, have been measured (by Gauss/Teslameter 7010, F.W.Bell, AC/DC can shift) and calculated (by MagNet, Infolytica), as shown in Fig.16. The agreement between the calculated and measured results demonstrates that the leakage flux linked with the exciting coil and then the induced eddy current loss can be accurately calculated under such complex excitation.

Table 4. Loss in air-core elliptic exciting coil.

Cases	Excitations	Calc.(W)	Meas. (W)
I	20.0A (DC)	119.03	118.57
II	20.0A(AC)	127.06	126.99
III	20.0(DC), 6.9A(AC)	126.88	125.29
* IV	15A (DC) 9A(AC, multiple harmonics, rms)	94.65	94.32
* V	20A (DC) 9A(AC, multiple harmonics, rms)	147.96	147.28
* VI	25A (DC) 9A(AC, multiple harmonics, rms)	217.22	215.02

Note: *Cases IV-VI, under DC-AC hybrid excitation with multiple harmonics (AC: fundamental with 3rd, 5th and 7th harmonic current components, rms).

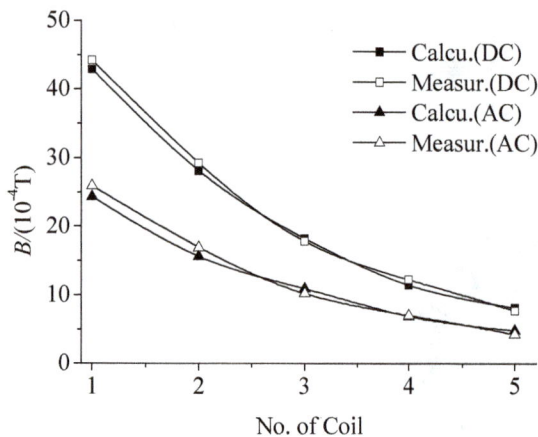

(a)

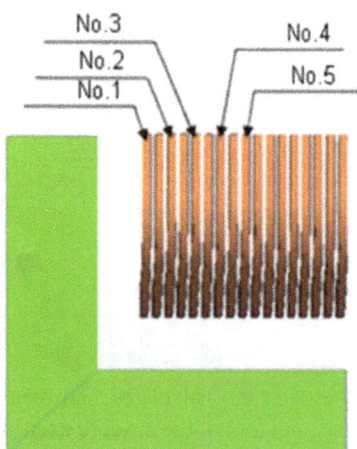

(b)

Figure 16. *Leakage flux densities under hybrid excitations. (a) measured and calculated leakage flux densities; (b) specified measurement positions.*

All the results presented herein provide a basic guarantee for the exciting coil's loss analysis. This is crucially important for indirect determination of component loss.

4.4.2. Stray-Field Loss inside the Laminated Frame

As described in Section II, the magnetic loss inside the laminated frame can be obtained using the two methods, i.e., using Approaches III and a numerical solver. The loss results involving 6 cases of DC-AC hybrid excitations are shown in Table 5, which are in practically good agreement.

Table 5. Loss in smoothing reactor model.

Exciting currents (A)			Coil's loss (W)	Total loss (W)	Loss in frame (W)	
Cases	AC	DC	$P_{excitation}$	P_{total}	Calcu.	Approach III
I	15	15	142.94	143.82	0.77	0.88
II	15	18	178.29	179.24	0.87	0.95
III	20	15	200.19	201.64	1.33	1.45
IV	13	15	123.94	124.96	0.99	1.02
*V	9	20	147.10	148.02	0.85	0.92
*VI	13	20	177.50	178.63	1.07	1.13

Notes: 1) Cases I-IV, DC+AC (only fundamental current without harmonics); 2)*Cases V-VI, under DC-AC hybrid excitation with multiple harmonics (AC: fundamental with 3rd, 5th and 7th harmonic current components, rms).

The leakage flux densities normal to the laminations of the frame of smoothing reactor model have been measured and calculated at specified positions by the authors and reported elsewhere. The results suggest that the additional magnetic loss caused by such lower normal leakage flux can be neglected in the smoothing reactor model used in this paper.

4.4.3. Further Research Works

Until now the magnetic losses in both the solid and laminated components are evaluated using DC-AC hybrid supply at the lower excitation level.

Further examinations of the loss behavior inside both the solid and laminated components under much stronger hybrid excitations, and the effects of various factors on the total and local losses under complex excitations will be undertaken by the authors, including the magnetic anisotropy, the minor loops, the harmonic phase angles in the laminated components, and the multi-steel configuration in solid components.

5. Conclusion

The multi-scale 3-D transient modeling and analysis, the measurement of working magnetic properties of both material and component, the determination of magnetic loss under multi-harmonic and/or DC-basing excitation, and the experimental validation have been carried out, which can be briefly summarized as follows:

1) The magnetic loss inside both the solid and laminated components are calculated based on the 3-D transient field solution, using the developed scripts, and the specially measured specific total loss data obtained from the non-standard excitations.

2) The experimental and numerical method used for determining the exciting coil's loss under extreme excitations are realized. This provides the confident verification for finally determining the loss inside magnetic components.

3) All the developed numerical and experimental approaches are validated, which can be effectively used in magnetic loss evaluation under extreme excitations.

4) The non-uniformity of the magnetic field and loss caused in solid plate configurations due to considerable skin effect, or in the laminated frame due to lamination-joints, must be taken into account for accurately determining the magnetic loss, under extreme excitation.

As a further research project, the magnetic property modeling under strong hybrid excitations, the detailed examinations of various effects on magnetic losses, and the efficient large-scale numerical modeling and simulation will be undertaken based on the upgraded models and enhanced experimental systems.

Acknowledgement

This project was supported in part by the National Natural Science Foundation of China under Grant 51107026, and by the State Grid Corporation of China under Grants sgri-wd-71-13-002, sgri-wd-71-14-002, and sgri-wd-71-14 -009.

References

[1]   F. Fiorillo and A. Novikov, "An improved approach to power losses in magnetic laminations under nonsinusoidal induction waveform," *IEEE Trans. Magn.*, vol. 26, no.5, pp. 2904-2910, 1990.

[2]   T. Sasaki, S. Saiki, and S. Takada, "Magnetic losses of electrical iron sheets under ac magnetization superimposed with higher harmonics", *IEEE Trans. J. Magn. Japan*, vol. 7, no. 1, pp. 64-74, 1992.

[3]   E. Barbisio, F. Fiorillo, and C. Ragusa: "Predicting loss in magnetic steels under arbitrary induction waveform and with minor hysteresis loops", *IEEE Trans. Magn.*, vol. 40, no.4, pp.1810-1819, 2004.

[4]   X. Zhao, J. Lu, L. Li, Z. Cheng, et al, "Analysis of the DC bias phenomenon by the harmonic balance finite element method," *IEEE Trans. Power Delivery*, vol.26. no.1, pp.475-485, 2011.

[5]   Y.Yao, C.Koh, G.Ni, and D.Xie, "3-D nonlinear transient eddy current, calculation of online power transformer under DC bias," *IEEE Trans. Magn.*, vol.41, no.5, pp.1840-1843, 2005.

[6]   T. Moses and J. Leicht, "Prediction and Measurement of losses under PWM Magnetisation conditions in electrical steels with different silicon content," Journal of Applied Physics 97, 10R507 (2005).

[7]   IEC TR 62383, First edition 2006-01, Determination of

magnetic loss under magnetic polarization waveforms including higher harmonic components. Measurement, modelling and calculation methods.

[8]   Z. Cheng, N. Takahashi, B. Forghani, G. Gilbert, J. Zhang, L. Liu, Y. Fan, X. Zhang, Y. Du, J. Wang, and C. Jiao, "Analysis and measurements of iron loss and flux inside silicon steel laminations," *IEEE Trans. Magn.*,vo.45, no.3, pp.1222-1225, 2009.

[9]   Z. Cheng, N. Takahashi, B. Forghani, et al, "Effect of excitation patterns on both iron loss and flux in solid and laminated steel configurations," *IEEE Trans. Magn.*, vol.46, no.8, pp.3185-3188, 2010.

[10]   Y. Zhang, Z. Peng, D. Xie, and B. Bai, "Effect of different magnetization curves on simulation for transformer core loss under DC bias," Trans., China Electrotechnical Society, vo.29, no.5, pp.43-48, 2014.

[11]   Z. Cheng, N. Takahashi, B. Forghani, Y. Du, Y. Fan, L. Liu, and H. Wang, "Effect of variation of B-H properties on both iron loss and flux in silicon steel lamination," *IEEE Trans. Magn.*, vol.47, no.5, pp.1346-1349, 2011.

[12]   Z. Cheng, N. Takahashi, B. Forghani, et al, "3-D finite element modeling and validation of power frequency multi-shielding effect," *IEEE Trans. Magn.*, vol.48, no.2, pp.243-246, 2012.

[13]   Z. Cheng, N. Takahashi, B. Forghani, et al, "*Electromagnetic and Thermal Field Modeling and Application in Electrical Engineering*," (in Chinese), ISBN 978-7-03-023561-9, Science Press, Beijing, 2009.

[14]   IEC 60404-6, Magnetic materials- Part 6: Methods of measurement of the magnetic properties of magnetically soft metallic and powder materials at frequencies in the range 20 Hz to 200 kHz by the use of ring specimens.

[15]   TEAM Benchmark Problems (no.1-n0.34) [on line] available: www.compumag.org/team.

[16]   Z. Cheng, N. Takahashi, B. Forghani, L. Liu, Y. Fan, T. Liu, Q.Hu, S. Gao, J. Zhang, and X. Wang, "Extended progress in TEAM Problem 21 family," COMPEL, 33, 1/2, pp.234-244, 2014.

[17]   Z. Cheng, B. Forghani, D. Lowther, Y. Liu, Z. Zhao, T. Liu, Y. Fan, and G. Han, "Stray-field loss modeling under hybrid excitation in smoothing reactors", presented at 20th international Conference on the Computation of Electromagnetic Fields(Compumag), June 28-July 2, 2015, Montreal, Canada.

[18]   Z. Cheng, N. Takahashi, B. Forghani, A. Moses, P. Anderson, Y. Fan, T. Liu, X. Wang, Z. Zhao, and L. Liu, "Modeling of magnetic properties of GO electrical steel based on Epstein combination and loss data weighted processing," *IEEE Trans. Magn.*, vol.50, no.1, 6300209, 2014.

10

Exploring Next Generation Energy Harvesters with PPE and IDE Electrodes

Sheetal Agrahari, Suresh Balpande

Department of Electronics Engineering, Shri Ramdeobaba College of Engineering and Management Nagpur, Maharashtra, India

Email address:

sheetalgagrahari@gmail.com (S. Agrahari), balpandes@rknec.edu (S. Balpande)

Abstract: Next generation harvesters promises environment friendly material suitable for medical field also. This paper reviews recent energy harvester for self-powered Microsystems and propose ZnO piezoelectric material for next generation harvesters. In addition to this, Vibration-powered generators are typically subjected to various design related issues which are addressed in this paper.Power can be generated from various environmental sources such as ambient heat, light, acoustic noise, radio waves, and vibration. Piezoelectric based harvester extract energy from the freely available ambient sources i.e. vibration or motion energy. The vibration energy can be converted to electrical energy by the use of piezoelectric PZT, ZnO, AlNGaAs cemented to micro cantilever. Piezoelectric electrodes play a vital role in energy extraction with higher efficiency .Piezoelectric-type harvesters have the highest reported energy density per volume. Furthermore, piezoelectric materials have an inherent capability of converting the mechanical energy into electrical energy, eliminating the need for external magnetic fields, complicated switching systems, and architectural design complexities. In this paper we have reviewed the work carried out by researchers during last few years. This review paper helps to new comers to decide best structure, material and approach to carry out their research work. Results obtained show a good scope for MEMS harvesters in numerous fields including medical field was far away because of poisonous piezoelectric material.

Keywords: Piezoelectric Material, Energy Harvesting, MEMS

1. Introduction

Recent improvements in the microelectronic and microelectromechanical system (MEMS) technologies enabled the fabrication of various sensors with remarkably small dimensions and low power requirements. Compounded with mass fabrication capabilities and low unit costs, this makes it possible to create wireless sensor networks that can monitor several parameters simultaneously using these omnipresent micro-fabricated sensors. Although success of such systems heavily depends on the performance of the sensors, the major limiting factor is generally the power management using batteries. Currently used batteries increase the cost and the size of these devices, and more importantly, it is not feasible to replace or manually recharge them. A possible solution to this problem is to apply a form of energy harvesting to convert the available ambient energy into electrical energy to recharge these batteries or substitute batteries while providing self-sustainability and intelligent power management for the overall system. In most of these applications, sensor data are only collected intermittently. Therefore, current interest is growing in utilizing harvested energy that is stored in on-chip capacitors and effectively eliminating the batteries. Although the ambient energy can be available in different forms, mechanical energy is widely preferred for energy harvesting applications because of the simplicity of the design and fabrication.

Alternative sources of energy are solar, magnetic field and wind. Outdoor solar energy has the capability of providing power density of 15, 000 $\_W/cm3$ which is about two orders of magnitudes higher than other sources. However, solar energy is not an attractive source of energy for indoor environments as the power density drops down to as low as 10–20 $\_W/cm^3$. Mechanical vibrations (300$\_W/cm^3$) and air flow (360 $\_W/cm^3$) are the other most attractive alternatives. In addition to mechanical vibrations, stray magnetic fields that are generated by AC devices and propagate through earth, concrete, and most metals, including lead, can be the source of electric energy[2].

In MEMS cantilever based energy harvester, mechanical energy is extracted by damping the motion of suspended proof masses within the devices. Mechanical energy harvesters have three main types: 1) piezoelectric,2) electromagnetic, and 3) electrostatic.

Piezoelectric-type harvesters have the highest reported energy density per volume. Furthermore, piezoelectric materials have an inherent capability of converting the mechanical energy into electrical energy, eliminating the need for external magnetic fields, complicated switching systems, and architectural design complexities. Power can be generated from various environmental sources such as ambient heat, light, acoustic noise, radio waves, and vibration[1]. Vibration energy harvesting is the most suitable power generation method because vibrations are readily available in almost all cases. A highly efficient way to harvest vibrational energy is to use piezoelectric materials for the energy transformation [3].

When base of structure is accelerated due to vibrating source(s) pressure (stress) is exerted to a material, it creates a strain or deformation in the material. The capability of the piezoelectric thin film in generating an electrical output in response to mechanical energy or vibration has given a significant impact in our daily lives. Piezoelectric thin film has been widely used in various MEMS applications such as surface acoustic wave (SAW) resonators, pressure sensors, biomedical and energy harvesting. In energy harvesting application, a piezoelectric energy micro-generator typically harvests mechanical energy or vibrations and converts it to electrical energy through piezoelectric effect. Different piezoelectric materials can affect the performance of the energy harvester due to different piezoelectric constants. Some examples of piezoelectric materials include lead zirconatetitanate (PZT), zinc oxide (ZnO) and aluminum nitrate (AlN)[9]. These parameters affect the mechanical and electrical parameters of the device. Mechanical energy in cantilever is generated due to stress and strains produced in beam as a result of acceleration of environmental vibrations.

Two types of electrodes are used in study as vibration sensing electrodes which are parallel plate electrode and interdigitated electrode. Cantilever structure helps in mechanical to electrical transduction[9]. EH are popular and penetrating in various applications due to diverse benefits: Long lasting operability, No chemical disposal, cost saving, Safety ,Maintenance free ,Inaccessible sites operability, Flexibility. It is observed that 90% of WSNs cannot be enabled without Energy Harvesting technologies (solar, thermal, vibrations)

2. Material Selection for Harvester

Most of the previous work has been concentrated on the material selection, coupling of electrode, figure of merit and their structural geometry. In case of interdigitated electrode the width, spacing and length of electrode fingers is also taken into consideration for optimization. Proper coupling

mode improves power harvesting.

Umi et al. [9], provided accurate information on the frequency, stress and voltage output of a ZnO piezoelectric energy harvester. They found out that ZnO piezoelectric energy harvester with the length of 150 µm, width 50 µm and thickness of 4 µm generates 9.9184 V electric potential under the resonance frequency of 0.71 MHz and 1µN/m2 mechanical force applied. This was a parallel plate electrode structure. Table 1 shows the effect of different piezoelectric material on output voltage, displacement and resonant frequency.

Table 1. *Comparative performance of different piezoelectric materials. [9].*

Piezoelectric Materials	Displacement (µm)	Rasonant Frequency (MHz)	Electric Potential (V)
ZnO	5.85×10^{-9}	0.17	9.91
PZT	1.08×10^{-10}	0.15	9.01
AlN	8.66×10^{-11}	0.20	9.62

Among different values of output potential obtained for different materials is observed to be highest for ZnO due highest direct piezoelectricity relationship. The cantilever depicted below is generalized structure with ZnO layer sandwiched between Al layers with bottom Silicon substrate.

3. Electrodes

This section of paper reveals drawback of parallel plate electrodes and superiority of interdigitated electrodes (IDE). Toprak et al.[1] worked to obtain optimized geometry of IDE and cantilever, including the piezoelectric and non-piezoelectric material for cantilever. Geometry with PZT thickness of 0.6 µm and an IDE consisting of 12 finger pairs gave Maximum output energy of 0.37 pJ for a 15-µN force. This energy is reduced to 1.5 fJ for 5 µm PZT thickness with 2 electrode finger pairs.

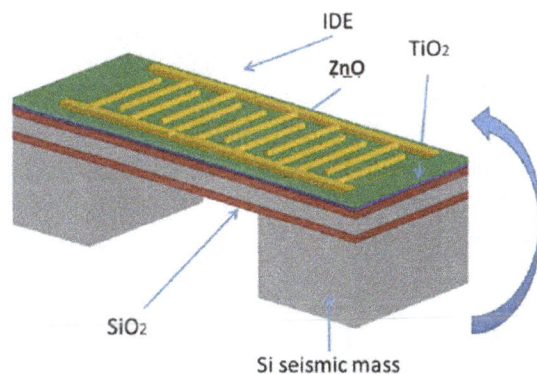

Fig. 1. *IDE electrode covering full length of cantileverbeam[11].*

Selection of IDE geometry and coupling is studied in above section of the paper. Now the one of the important parameter is mode selection which is described in next section of paper.

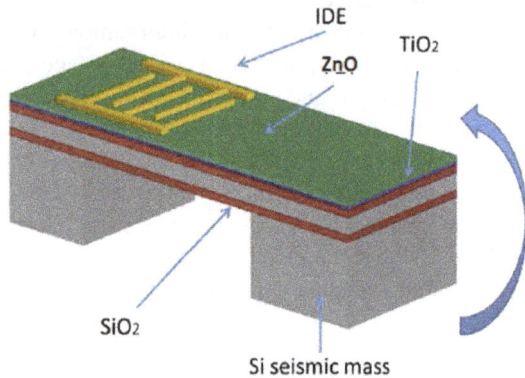

Fig. 2. IDE Covering partial length of beam[11].

Chidambaram et al.[3] The leakage current density of the IDE structure was measured to be about 4 orders of magnitude lower than that of the PPE structure. The best figure of merit (FOM) of the IDE structures was 20% superior to that of the PPE structures while also having a voltage response that was ten times higher (12.9 mV/µ strain). The IDE lower power loss inside the PZT for this kind of electrode. Some of the literature reveal study of parallel plate electrode(PPE) while some show for interdigitated electrode (IDE). Since IDE show better outputs it became part of interest due to better efficiency.

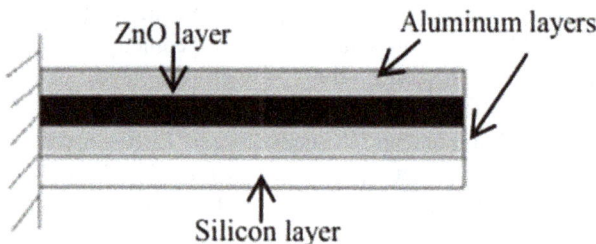

Fig. 3. Cantilever structure with parallel plates[9].

4. Mode Selection

Mode selection is equally important to maximize conversion efficiency.d33 mode provides a higher electromechanical coupling when compared with d31 mode in typical piezoelectric materials[1].

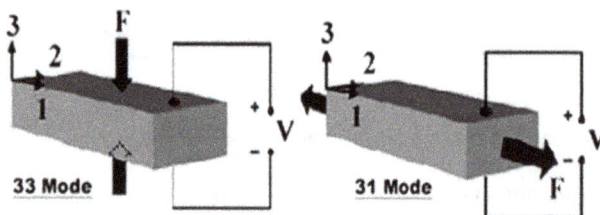

Fig. 4. d33 and d31 coupling modes[4].

Ralib et al.[5] simulated on The prototype cantilever structure consists of four layers namely: Si/ PZT/Pt interdigitated electrodes / Ni proof mass. The size of the cantilever beam is 23µm x 71 µm with thickness 37 µm operating in d33 mode. The graph shows the expected sharp change in displacement as the frequency approaches the Mode 1 value that is 55MHz which shows the highest displacement shown in Figure 5. placing 24 Pa pressure at the end of the cantilever beam again provides the highest displacement with frequency of 53.7 MHz.

Jinyu et al.[6] carried out simulations for 3x8.5x0.130 mm PPE electrode and obtained the figure of merit 59.98 for 165 Hz frequency, displacement approximately 550 µm at 1g. Output voltage is 7.70 V.g^{-1} and power 174 µWatt g^{-2}.

Ryan et al.[7] The interdigitated beam utilizes the d33 piezoelectric constant. The d33 constant for PZT is commonly known to be approximately twice as large as the d31 constant. Therefore, designing a d33 structure properly could produce more energy and larger tuning range than a d31 structure.

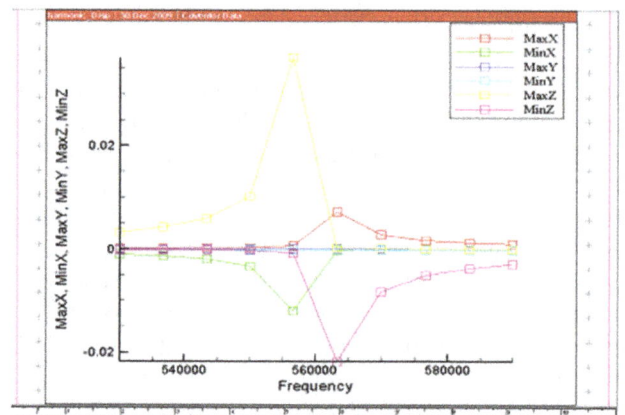

Fig. 5. Graph for displacement vs Frequency for piezoelectric harmonic analysis[5].

The coupling coefficient of piezoelectric generators depends primarily on the piezoelectric material used, although the elastic properties of the other materials used in the generator structure may also influence the values. [12]

Table 2. Device dimensions[8].

Material	Length (µm)	Width (µm)	Thickness (µm)
Silicon	71	23	30
BPSG	71	23	18
Zinc Oxide	71	23	5
Platinum (for one electrode)	0.75	18	2
Nickel	10	23	10

Hanim et al.[8] got the resonant frequency as 34.4 kHz at output voltage of 2.75V for device dimensions as shown in Table 2.

Rabbani et al.[10] used Si as proof mass (0.5592µg; dimensions:1000µm x 800µm x 300µm) at free end tip of cantilever to decrease the resonant frequency. The resonance frequencies for the cantilever having dimension 2000µm ×800µm ×20µm are 1427 Hz (first mode) and 15287 Hz (second mode). When excited with a mechanical vibration of 40 N/m2, the power generated at resonance is around 9µW.

The above discussionsshows that resonant frequency is very high as compared to frequency of vibration sources. Thus there is need of lowering this parameter. The length of cantilever is inversely proportional to resonant frequency. Dimensions should be chosen in such a way that resonant frequency of structure should be in the range of 10Hz-1Khz only. If the resonance is not taken place then vibration energy cannot be captured effectively. We propose the use of IDE for piezoelectric based energy harvester. Fabrication Process for IDE only is discussed below.

Andrea et al.[11] investigated the advantages of interdigitated electrode configurations (IDE) with respect to parallel-plate electrodes (PPE) in terms of output voltage and output power from the constitutive equations of piezoelectricity. A figure of merit for comparison has been proposed and calculated for both PPE and IDE structures. IDE has 2.12 times the figure of merit of PPE. The latter yielded higher energy densities and output voltages.

Ankita et al. [13] The cantilever of E shaped is being designed and simulated with dimensions of 6.832mm x 2.180mm. The materials play a vital role in sensitivity of sensor and PZT 5H is selected on the basis of material analysis carried out. The Z displacement of 5.3 µm to 46.14 µm has been recorded.

Ankita et al.[14] The cantilever of E shaped has been designed and simulated with dimensions of 3.332mm x 1.180mm for loading on bridges. The materials play a vital role in sensitivity of device and PZT 5H is selected on the basis of materials analysis carried out.

The voltage generation of about 3.12 mV and Z-displacement of 4.03µm has been recorded at 100 Hz of frequency.

The tuning of the resonant frequency and electrical damping force is needful to keep an optimum output power, especially when the vibration amplitude and frequency is susceptible to change over time. In this paper, we have first presented the different key issues for VEH[15].

It was found that the thinner the beam, i.e. the lower the spring constant of the cantilever, the lower the untunedresonant frequency and the larger the tuning range. For this generator, a 120 m thick beam was chosen to give a predicted untuned resonant frequency of 45.2 Hz and a tuning range from 66.4 to 108.8 Hz[16].

The EH device has a wideband and steadily increased power generation from 19.4 nW to 51.3 nW within the operation frequency bandwidth ranging from 30 Hz to 47 Hz at 1.0 g. Based on theoretical estimation, a potential output power of 0.53 µW could be harvested from low and irregular frequency vibrations by adjusting the PZT pattern and spacer thickness to achieve an optimal design[17].

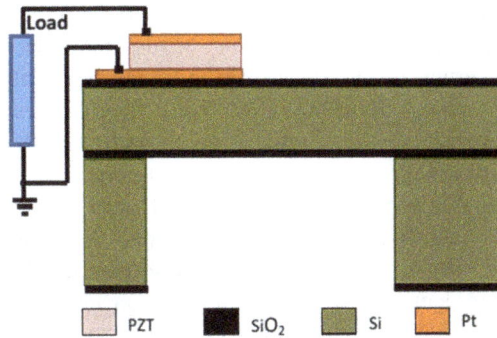

Fig. 6. *Lateral view of MEMS structure[10].*

5. Application Areas of EH

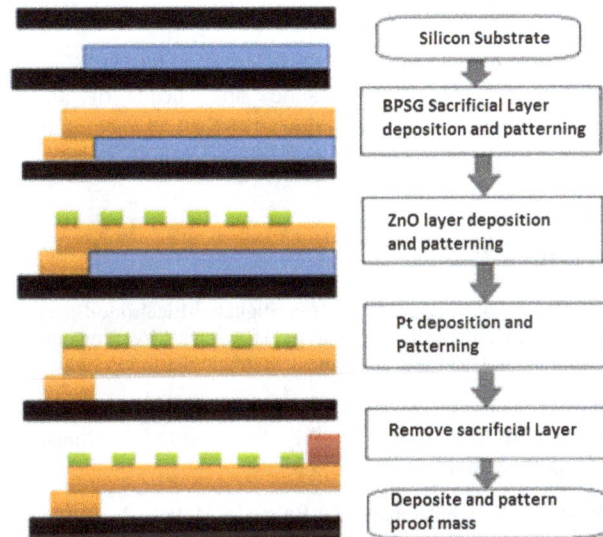

Fig. 7. *Fabrication Flow for IDE.*

- Environmental Monitoring
  - light, temperature, humidity–
- Integrated Biology
- Structural Monitoring
  - Building, Automation
- Interactive and Control
  - RFID, Real Time Locator, TAGS
  - Transport Tracking, Cars sensors
- Surveillance
  - Intrusion Detection
  - Interactive museum exhibits
- Medical remote sensing
  - Emergency medical response
  - Monitoring, pacemaker, defibrillators
- Military and Aerospaceapplications
- Not limited above all

6. Conclusion

Designs of energy harvester with ID electrodes are reviewed in this paper. Comparing IDE and PPE structures for electrodes it is observed that IDE has four times better output as compared to PPE. It is also observed that IDE has better Figure of merit. Device geometry, modes, piezoelectric materialand design has a crucial role in giving high output voltages at resonant frequency. Zinc oxide material fornext generationharvester is proposed as piezoelectric layer because of its excellence bonding to substrate material such silicon and high piezoelectric coupling coefficient. Choosing the proper interdigitated electrode layout and beam dimensions can nearly double the performance. Thus, designing a proper d33 unimorph or bimorph device will increase energy harvesting performance. These structures are being used forwireless sensor networks still there is scope for further optimization to obtain power which will be enough to drive portable devices.

Acknowledgments

Authors would like to thank Dr.S.S.Sadistap,Head, CEERI Pilani, Principal Dr. R.S. Pande and head of the department,Electronics Engineering, SRCOEM, Nagpur, Mr.Deepak Khushlani for the needful support and guidance.

References

[1] Toprak, A.,Tigli, O., 2013, "Interdigitated-Electrode-Based MEMS-Scale Piezoelectric Energy Harvester Modeling and Optimization Using Finite Element Method",IEEE.

[2] Kim, H., Tadesse, Y., Priya, S., 2009, "Energy Harvesting Technologies", S. Priya, D.J. Inman (eds.),SpringerScience+Business Media.

[3] Mazzalai, A., Chidambaram,N.,Balma,D.,Muralt, P., August 2013, "Comparison of Lead ZirconateTitanate Thin Films for Microelectromechanical Energy Harvester With Interdigitated and Parallel Plate Electrodes", IEEE.

[4] Saadon,S., Sidek,O., 2010, "A review of vibration-based MEMS piezoelectric energy harvesters", Elsevier.

[5] Ralib, A.A., Nordin,A.N. , Salleh, H., 2010, "Theoretical Modeling And Simulation Of MEMS Piezoelectric Energy Harvester", ICCCE.

[6] Ruan, J.J., Lockhart, R.A., Janphuang, P., Quintero, A., Briand,D., and Rooij,N., 2013, " An Automatic Test Bench for Complete Characterization of Vibration-Energy Harvesters", IEEE.

[7] Knight, R.R., &Changki Mo &Clark.W.W., 2011, "MEMS interdigitated electrode pattern optimization for a unimorph piezoelectric beam".

[8] Ralib,A.A., Nordin,A.N., Salleh,H., 2010, "Simulation of MEMS Piezoelectric Harvester".

[9] Jamain,U.M., Ibrahim,N.H., Rahim, R.A, IEEE-ICSE2014 , "Performance Analysis of Zinc Oxide Piezoelectric MEMS Energy Harvester".

[10] Rabbani, S., Rathore, P., Ghosh, G., and Panwar, B. S., 2010, "Mems structure for energy harvesting",COMSOL Conference 2010 India.

[11] Beeby, S.P., Tudor, M.J., and White, N.M., "Energy harvesting vibration sources for microsystems applications" Measurement Science And Technology, Institute Of Physics Publishing , Meas. Sci. Technol. 17 (2006) R175–R195

[12] Balpande, S.S. ; Lande, S.B. ; Akare, U. ; Thakre, L., 2009, " Modeling of Cantilever Based Power Harvester as an Innovative Power Source for RFID Tag" 2nd IEEE International Conference onEmerging Trends in Engineering and Technology (ICETET), pp. 13 – 18.

[13] Kumar, A., S.S.Balpande, Oct, 2014, "MEMS Based Bridge Health Monitoring System" International Journal of Advances in Science, Engineering and Technology, Volume-2,Issue-4.

[14] Kumar, A., Balpande, S.S., 2014,"Energy Scavenging From Ambient Vibrations Using MEMS Device" International Journal Of Scientific Progress And Research (IJSPR) ,05, Number-01, ISSN: 2349 – 4689

[15] Seddika,B.A., Despessea,G., Defaya,E., September 9-12, 2012 "Improved Wideband Mechanical Energy Harvester Based on Longitudinal Piezoelectric Mode".

[16] Dibin Zhu, Stephen Roberts, Michael J. Tudor , Stephen P. zeeby ,2010, "Design And Experimental Characterization Of A Tunable Vibration-BasedElectromagnetic Micro-Generator".

[17] Liu, H., Tay, C.J., Quan, C., Kobayashi, T., and Lee, C., IEEE, "Piezoelectric MEMS Energy Harvester forLow-Frequency Vibrations with Wideband OperationRange and Steadily Increased Output Power".

11

Energy Potential of Waste Derived from Some Food Crop Products in the Northern Part of Cameroon

Samomssa Inna[1], Jiokap Nono Yvette[1, *], Kamga Richard[2]

[1]University Institute of Technology (IUT) of the University of Ngaoundere, Department of Chemical Engineering and Environment, Ngaoundere, Cameroon
[2]National Advanced School of Agro-Industrial Sciences (ENSAI) of the University of Ngaoundere, Department of Applied Chemistry, Ngaoundere, Cameroon

Email address:

jiokapnonoy@yahoo.fr (J. N. Yvette)

Abstract: The purpose of this study was to quantify the agricultural crop productions in each division of the northern regions of Cameroon, to evaluate the proportion of waste derived from these and to classify them according to their potential as energy sources. To achieve these goals, statistical data from Cameroon's Ministry of Agriculture as well as standard methods of proximate analysis have been used to evaluate the proportion of each waste and its physico-chemical properties. The study reveals that agricultural activities generate an important quantity of waste (corn cobs and stalks, millet/sorghum stalks, rice hulls, cassava peelings, groundnut hulls, sweet potato peelings, Irish potato peelings and cotton hulls) of about 555 002.27 dry-bone tons per annum in the three northern regions of Cameroon. The highest waste production is found in the North region with 42.93% of the total waste, directly followed by the Far North region with 42.44%. Of the three regions, the Adamawa presents the smallest percentage (11.23%). The main agro-industrial waste of these regions includes cotton hulls, with 3.41% of the total waste. The anhydrous low heating values of the wastes derived from the selected food crop products vary between 13.51 and 29.97 MJ/(kg d-b), indicating a total biomass-energy potential in the northern part of Cameroon of 11.5 TJ per year.

Keywords: Northern Regions of Cameroon, Food Crop Products, Lignocellulosic Waste, Heating Value, Bio-energy

1. Introduction

The energy deficit observed in the world is justified by urbanization and industrialization [1]. The principal sources of energy are divided into two groups: renewable and nonrenewable energy [2]. Among the nonrenewable energies, the most used are petrol oil, nuclear energy and gas [3]. These energies are not very available, exhaustive, very expensive especially for the rural populations and are sources of pollution [4]. Whereas, there are more available and less polluting and renewable energies, namely wind energy, hydropower, biomass and solar energy [4, 5].

However, in the less developed countries like Cameroon, wind energy, solar energy and hydropower are not very accessible. They use wood for their energy need [6], which leads to a fast deforestation and consequently to an acceleration of global warming [4, 7, 8]. In this context, cheap biomass in the form of agricultural residues stands as the most attractive and promising energy source for fuel, heat and electricity [8]. The choice of this biomass type however, as modern source of energy depends on their availability. In this regards, Ackom et al. [9] assessed the biomass resource potential in Cameroon which amounts to 1.11 million bone-dry tons per year. This study did not specify the proportion of crop residues generated by each region of the country. In the Northern part of Cameroon, no systematic study on the quantification and biomass analysis of agricultural-derived wastes has been performed to date.

The aim of this study is to quantify the agricultural products in the northern parts of Cameroon, to evaluate the proportion of waste by the types of crops grown in each division, and to classify them according to their potential use for bioenergy.

2. Materials and Methods

2.1. Investigated Food Crop Products

The food crop products investigated in this study are presented in Table 1. They are the most cultivated in the northern parts of Cameroon (more than 90% of the said region total production) [10].

Table 1. *Agricultural products investigated.*

Agricultural crop products	Scientific name [11, 12]
Banana/plantain	*Musa spp*
Corn	*Zea mays*
Rice	*Oryza sativa*
Millet/Sorgho	*Pennisetum spp/Sorghum Bicolor*
Cassava	*Manihot esculenta Crantz*
Sweet Potato	*Impomoea batatas*
Yam	*Dioscorea spp*
Irish potato	*Solanum Tuberosum*
Groundnut	*Arachis hypogaea*
Cotton	*Gossypium hirsutum* *Gossypium barbadense*

Among these agricultural products, some generate agricultural residues and others house hold wastes. For the present study the agricultural residues collected from the farms are corn cobs, corn stalks, sorghum/millet stalks, cassava peelings and groundnut hulls. Plantain peelings and stocks, sweet potato peelings, Irish potato peelings, yam peelings are the most produced household wastes and are collected by the Cameroon Waste Collection Society (HYSACAM). Cotton is transformed in industry and generates an industrial waste (cotton hulls). Corn is also transformed in industry (MAISCAM-Ngaoundere) and generates corn cobs and corn stalks as wastes.

2.2. Analysis and Calculation Methods

The agricultural productions data were collected from Cameroon's Ministry of Agriculture and Rural Development [13-15]. The generated waste of each food crop product (W) was calculated by the following equation, where P is the production tonnage of the food crop and RPR the residue to product ratio obtained by dividing the mass of waste in a sample by the mass of the corresponding sample:

$$W = P \cdot RPR \qquad (1)$$

Proximate analysis of different wastes was also investigated. Analyses were carried out on a mixture of two varieties frequently present in the locality (ratio 1:1). The waste moisture content was evaluated according to ASTM E1871-82 method [16], volatile matter (%VM) by ASTM E872-82 method [17] and ash (%ash) by ASTM E1755-01 method [18]. Fixed carbon percentage (%FC) was calculated by the following equation [19]:

$$\%FC = 100 - (\%ash + \%VM) \qquad (2)$$

The lower heating value (LHV) expressed in MJ/(kg w-b) was determined using an oxygen bomb calorimeter (Parr 6100 Model A1329 DD, ID lot number M15320). Measurements were conducted on each sample previously dried in an oven at 45°C till constant weight corresponding to a moisture content MC# (g/100g w-b). The anhydrous lower heating value

(ALHV) expressed in MJ/(kg d-b) was then calculated by the following equation:

$$ALHV = LHV \cdot \frac{100}{100 - MC\#} \qquad (3)$$

ALHV was also expressed in kWh/(kg d-b) using the conversion factor:

$$\frac{1\,kWh/(kg\,d.b)}{3.6\,MJ/(kg\,d.b)} = 1 \qquad (4)$$

2.3. Presentation of the Study Area

Figure 1 shows the northern part of Cameroon, composed of fifteen divisions. From up to down the picture, we have the Far-North region with six divisions (Logone-et-Chari; Mayo-Sava; Mayo-Tsanaga ; Diamare; Mayo-Kani; Mayo-Danay) followed by the North region with four divisions (Mayo-Louti; Benoue; Faro; Mayo-Rey) and finally the Adamawa region with five divisions (Faro-et-Deo; Vina; Mayo-Banyo; Djerem; Mbere). Also presented on Table 2 are the area of each department and the area of arable land of the investigated food crops (average data 2013-2014). This table brings out that the northern part of Cameroon area is equal to 34.5% the total surface area of Cameroon [20].

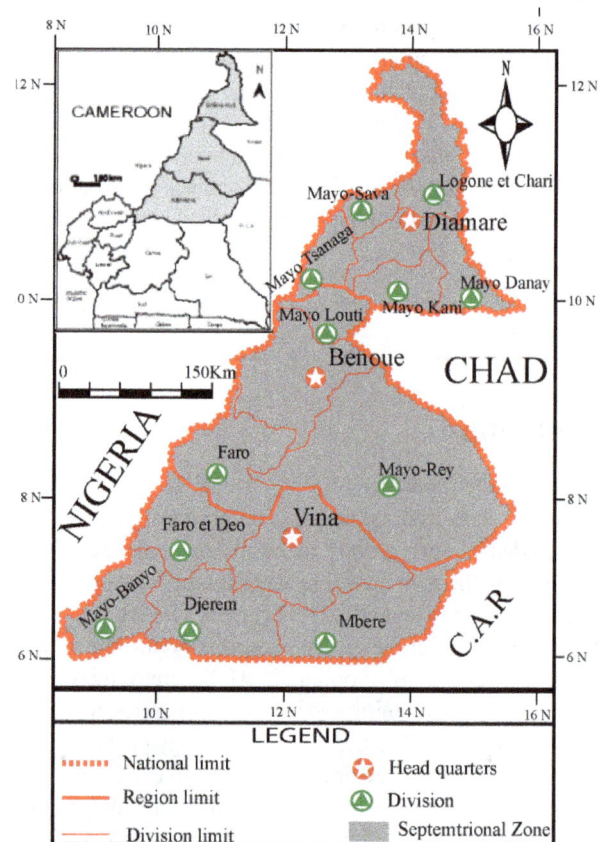

Figure 1. *Location map of the study area.*

Table 2. *Area of each department (ha) and area of arable land (ha) of the investigated food crops [13-15, 20].*

Division	Division Area	Banana/ Plantain	Corn	Cassava	Rice Area	Groundnut	Millet/ Sorghum	Potato	Yam	Irish potato	Total arable land
Vina	1 719 600.00	75.75	22 682.50	12 493.13	40.63	944.60	1 230.60	1 316.25	699.13	570.88	40 053.45
Mbere	1 426 700.00	165.13	2 439.50	7 004.75	0.00	1 146.50	27.00	550.60	327.75	80.75	11 741.98
Faro Edeo	1 043 500.00	47.61	5 585.15	1 524.85	127.08	1 873.01	7 531.81	185.08	113.50	28.40	17 016.48
Mayo Bayo	852 000.00	2 197.00	15 323.50	2 726.00	218.00	2 721.50	0.00	835.75	187.50	75.50	24 284.75
Djerem	1 328 300.00	297.25	8 058.00	5 783.00	0.00	5 760.00	0.00	412.00	479.00	35.25	20 824.50
Total area Adamawa region	6 370 100.00	2 782.74	54 088.65	29 531.73	385.70	12 445.61	8 789.41	3 299.68	1 806.88	790.78	113 921.15
Benoue	1 361 400.00	300.00	66 560.50	2 008.50	9 831.00	39 584.00	88 677.50	1 600.00	2 129.00	16.50	210 707.00
Faro	1 178 500.00	175.00	11 300.00	1 960.00	6 250.00	38 211.50	19 025.00	1 706.00	2 256.00	0.00	80 883.50
Mayo-Louti	416 200.00	325.00	32 152.00	1 850.00	601.50	38 900.50	53 086.50	1 410.50	2 730.50	0.00	131 056.50
Mayo-Rey	3 652 900.00	400.00	29 250.00	2 563.00	5 500.00	41 642.50	25 000.00	2 199.50	1 402.00	0.00	107 957.00
Total area North region	6 609 000.00	1 200.00	139 262.50	8 381.50	22 182.50	158 338.50	185 789.00	6 916.00	8 517.50	16.50	530 604.00
Diamare	466 500.00	0.00	7 061.50	167.50	2 385.00	8 262.00	145 508.00	1 065.50	0.00	0.00	164 449.50
Logone and Chary	1 213 300.00	0.00	30 445.80	442.90	10 003.50	15.00	38 394.50	600.15	0.00	0.25	79 902.10
Mayo-Danay	530 300.00	0.00	4 546.90	1 877.15	10 119.60	7 762.25	106 768.10	52.85	0.00	0.00	131 126.85
Mayo-Kani	503 300.00	0.00	7 960.00	423.50	454.50	4 705.00	95 182.50	83.50	0.00	0.00	108 809.00
Mayo Sava	273 600.00	0.00	5 162.50	13.50	470.45	8 111.30	87 352.20	11.00	0.00	0.00	101 120.95
Mayo Tsanaga	439 300.00	0.00	55 266.25	820.00	4 418.50	53 787.90	97 148.65	1 919.50	0.00	4 226.10	217 586.90
Total area Far North region	3 426 300.00	0.00	110 442.95	3 744.55	27 851.55	82 643.45	570 353.95	3 732.50	0.00	4 226.35	802 995.30
Total surface area	16 405 400.00	3 982.74	303 794.10	41 657.78	50 419.75	253 427.56	764 932.36	13948.18	10324.38	5 033.63	1447 520.45

3. Results and Discussion

3.1. Evolution of Agricultural Crop Production in the Northern Part of Cameroon

Figures 2, 3 and 4 present the evolution of the selected agricultural products in Adamawa, North and Far North regions respectively.

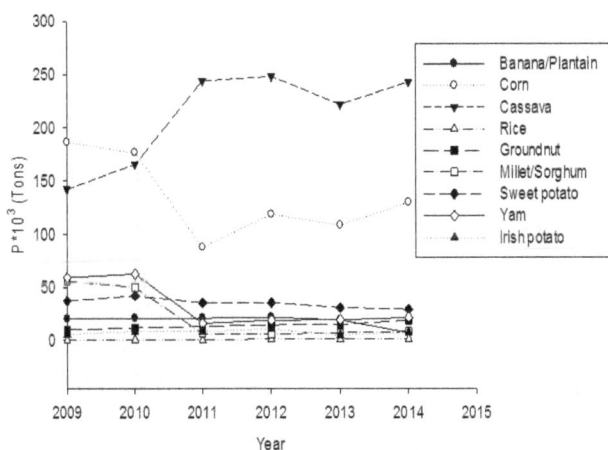

Figure 2. *Evolution of selected agricultural crop products in the Adamawa region. P is the production tonnage of the food crop.*

The Figure 2 reveals that amongst the selected agricultural products, cassava presents a global increasing production over years and a highest production level followed by corn. Over the three last years (2012, 2013 and 2014), cassava average production equals to $(238 \pm 14).10^3$ tons. Between 2010 and 2011, corn production as well as yam and millet/sorghum productions significantly decreased due to poor technical know-how of cultivation techniques, non-readily accessible improved species and fertilizers, the non-use of fallowing techniques and the presence of food scavengers [21]. Since 2014, there is a slight increase in corn production. The average corn production over the three years equals to $(119 \pm 11).10^3$ tons. Rice production is the smallest amongst all with an average production of $1\ 227 \pm 115$ tons over the three last years. All the other productions (banana/plantain, groundnut, sweet potatoes and Irish potatoes) have a substantial constant production since 2009.

It appears on the Figure 3 that corn, millet/sorghum and groundnut are the most cultivated in the North region with an average production over the three last years of $(291 \pm 7).10^3$; $(281 \pm 49).10^3$ and $(241 \pm 5).10^3$ tons respectively. Between 2012 and 2013, millet/sorghum production declined sharply due to seasonal changes and inefficient cultivation techniques [21]. For the other food crop products (rice, cassava, sweet potatoes, yam, banana/plantain and Irish potatoes), productions are smaller and vary between $(52 \pm 5).10^3$ tons for rice to 57 ± 6 tons for Irish potatoes. The smallest production of the latter has been decreasing over the years due to plant pest diseases like *Phytophthora infestans* and *Ralstonia solanacearum* [21] as well as the lack of improved varieties of the local species. Most of the selected products have increasingly evolved since 2009 with the advent of agricultural research and development by scientific institutes such as the Institute of Research and Agricultural Development (IRAD-Cameroon).

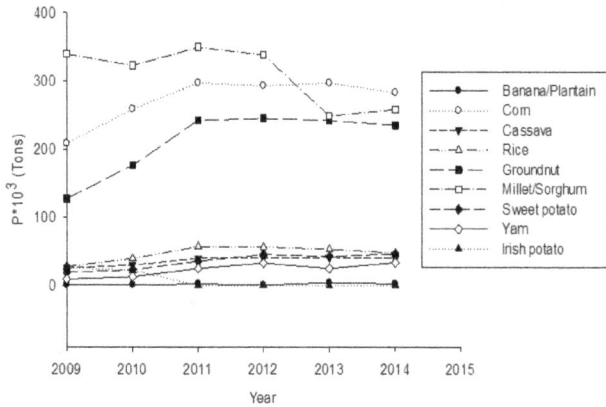

Figure 3. *Evolution of selected agricultural crop products in the North region.*

The Figure 4 shows that sorghum/millet has the highest production in the Far-North region. The average production over the three last years is $(803\pm147).10^3$ tons. This production increased from 2009 and 2012, but a sharp decrease was observed between 2012 and 2014.

Other crops like corn, groundnut, rice, sweet potatoes, cassava and Irish potatoes are less produced. In this group, the lowest production is recorded on Irish potatoes and the highest on corn. For these two, productions increased over the three

last years, to reach an average of $(38\pm32).10^3$ tons for Irish potatoes and $(208\pm42).10^3$ tons for corn. For the remaining food crop products (groundnut, rice, sweet potatoes, cassava), the production is almost constant since 2009. However, it can be noticed that yam and banana/plantain are not produced in the Far-North region.

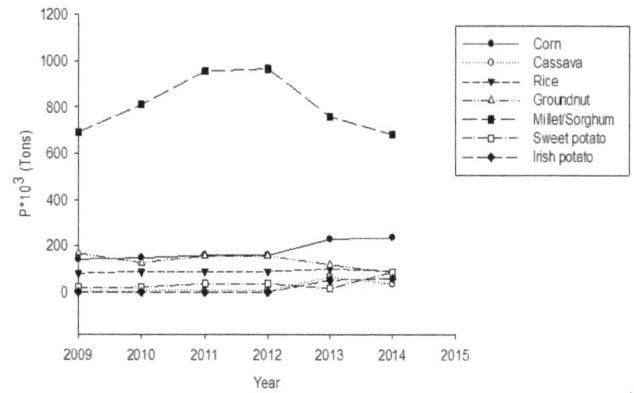

Figure 4. *Evolution of selected agricultural crop products in the Far-North region.*

3.2. Proximate Analysis of Waste Generated in the Northern Part of Cameroon

Table 3 presents the proximate analysis of some wastes generated in the northern parts of Cameroon. As indicated by Vargas-Moreno *et al.* [22] and Lee and Shah [23], the compositional information provides the science and engineering information needed to predict the feedstock behavior in a processing, and much more.

Table 3. *Proximate analysis of different wastes.*

Waste	MC (g/100g w-b)	VM (g/100g d-b)	Ash (g/100g d-b)	FC (g/100g d-b)	LHV MJ/(kg w-b)	MC# (g/100g w-b)	ALHV MJ/(kg d-b)	ALHV kWh/(kg d-b)
Corn cobs	12.00±0.23	98.67±0.28	0.92±0.14	0.41±0.02	25.33±0.03	7.33±0.29	27.33	7.59
Millet/ Sorghum stalks	10.00±0.12	94.17±1.89	2.65±0.93	3.19±0.96	15.4±0.02	8.83±0.29	16.89	4.69
Banana/plantain peelings and stocks	89.00±0.98	78.17±3.18	6.26±0.98	15.58±7.35	17.51±0.02	12.67±5.48	20.05	5.57
Cassava peelings	75.50±0.35	90±2.83	4.52±0.793	5.48±1.90	12.52±0.01	7.33±0.29	13.51	3.75
Rice hulls	9.00±0.49	70.83±4.37	11.07±2.14	18.10±2.79	16.41±0.02	8±0.50	17.84	4.95
Groundnut hulls	8.00±0.50	74.67±1.26	8.56±0.56	16.77±1.24	18.13±0.02	9.50±0.86	20.03	5.56
Sweet potato peelings	75.95±0.42	69.75±3.18	0.85±0.05	29.40±3.71	25.03±0.03	10.50±0.10	27.97	7.77
Irish potato peelings	83.00±0.92	93.00±1.50	3.29±0.60	3.71±1.01	25.77±0.03	14±0.50	29.97	8.32
Yam peelings	73.00±0.17	91.00±3.54	1.88±0.53	7.12±3.01	18.32±0.02	13.83±0.29	21.26	5.91
Cotton hulls	9.00±0.02	81.13±2.62	14.51±0.30	4.36±0.36	16.57±0.02	9.00±0.02	18.21	5.06

With MC the wet-basis moisture content of raw waste; VM the volatile mater; FC the fixed carbon; MC# the wet-basis moisture content of 45°C oven-dried raw waste; LHV the low heating value; ALHV the anhydrous low heating value.

From the Table 3, the moisture content of the different studied wastes can be categorized into 2 groups. The first group is composed of high moisture content wastes (banana/plantain peelings and stocks; cassava peelings; sweet potato peelings, Irish potato peelings and yam peelings) and the second group is composed of less humid wastes (corn cobs and stalks; millet stalks; rice hulls; groundnut hulls and cotton hulls). Most produced wastes belong to this last group. Values of volatile matter (VM), ash and fixed carbon (FC) vary respectively between 69.75% for sweet potato peelings and 98.67% for corn cobs; 0.85% for sweet potato peelings and 6.26% for banana/plantain peelings and stocks; 0.42% for corn cobs and 29.4% for sweet potato peelings. According to Vargas-Moreno et al. [22] and Debdoubi et al. [24], the higher the volatile matter and the fixed carbon contents, the higher the heating values. Also, the higher the ash content, the lower the heating value. One can therefore expect a high heating value for corn cobs, sweet potato peelings and Irish potato peelings compared to other crop wastes. This is confirmed by the heating values obtained by the bomb calorimeter and presented in the same table. Except for the cassava peelings, all the obtained heating values are greater than 4 kWh/kg d-b). ADEME [25] mentioned that the anhydrous heating value of hardwood is between 4.8 and 5.3 kWh/kg. Our results reveal that the studied wastes have heating values comparable to that of wood and can therefore be used as substituents to wood-energy.

Proximate analysis results for banana/plantain peelings show that our result for fixed carbon and volatile matter contents are in accordance with those obtained by Bianca et al. [26]. Concerning the heating values, our result is similar to those presented in the literature, ranging from 9.76 MJ/kg to 17.30 MJ/kg [26, 27]. However, from literature, the ash content data range from 9.51% to 15.3% [26-28] and are not in accordance with our result, probably due to the differences in species of varieties used.

Concerning corn cobs, data in the literature range from 1% to 2.65% for ash content [29-31]; 8.25% to 18.54% for fixed carbon; 78.70% to 90.12% for volatile matter [31-33] and 6.79 to 26.3 MJ/kg for heating values [27, 29, 34]. The literature data for corn cobs are in accordance with our results. Corn stalks moisture content obtained is (10.00±0.23) g/(100 g w-b); however, volatile matter, ash, fixed carbon contents and low heating value were not carried out in this study. Respective values in the literature are 59.83%; 11.52%; 17.00% and 16.29 MJ/kg [35].

Literature data for yam peelings proximate analysis range from 3.27% to 14.50% for fixed carbon [36]; 15.54 to 16.43 MJ/kg for heating value [27, 37] and 3.86% to 6.3% for ash content [27, 36]. Oladeji [36] obtained a volatile matter content of 82.87% for this substrate. Our heating value, volatile matter and ash content value are slightly different from those of the literature and could be due to the differences in the varieties used and the soil type.

Proximate analysis data for cassava peelings in the literature range from 3.25% to 8% for ash content [27, 38] and

12 to 16.32 MJ/kg for heating value [27]. Oladeji [36] obtained a volatile matter and fixed carbon content of 83.06% and 2.27 % respectively. These are in accordance with our results.

For groundnut hulls, literature proximate analysis data range from 0.76% to 5.70% for ash content; 6.47% to 21.60% for fixed carbon; 72.70% to 88.47% for volatile matter and 19.20 to 13.78 MJ/kg for heating value [36, 37, 39]. These are in accordance with our results.

Rice hulls proximate analysis data in the literature range from 15.5% to 20.26% for ash content [29, 40]; 63.52% to 90% for volatile matter [40, 41] and 12.1 to 16.7 MJ/kg for heating value [42]. These are in accordance with our results.

Proximate analysis of millet stalks and sorghum stalks in the literature are evaluated separately and values vary between 8.40% to 14.15 for ash content; 71.40% to 78.26% for volatile matter and 14.50 to 16.45 for fixed carbon [29, 30]. These authors also reported a heating value of 15.4 MJ/kg. Our obtained values are in these intervals.

For cotton hulls, literature proximate analysis data vary between 1.61% to 5.10% for ash content; 77.80% to 83.41% for volatile matter; 14.97% to 17.10% for fixed carbon and 18.13 to 18.28 MJ/kg for heating value [29, 40, 41]. These are in accordance with our results.

Concerning sweet potato peelings and Irish potato peelings, the literature values for volatile matter content are 75% and 90% respectively [41]. Our obtained values slightly differ from these.

3.3. Quantification of Waste Generated in the Northern Part of Cameroon

Table 4 presents the quantification (mean value over 2013 and 2014) of waste obtained in the northern part of Cameroon. Quantification is done using the total production over the three regions (Adamawa, North and Far-North) of the defined product and the residue to product ratio (RPR) as presented in section 2. Values are compared to the national production available by now (average for the year 2010).

The Table 4 shows that in the northern parts of Cameroon, millet/sorghum is the most produced (982 025.39 tons), followed by corn. Banana/plantain is the less produced (16 542. 05 tons) and this could be due to climatic conditions which are not well adapted to this crop. For the selected agricultural crop products, the annual total production is around 2 188 213.87 tons for the northern parts of Cameroon corresponding to 17.88 % of the national production.

Many estimates of the residue to product ratio (RPR) of the different studied crops are reported in the literature and there are significant differences among them. RPR values of corn cobs in the literature range from 20% to 30 % for sample moisture content between 7.53 to 15 % [43, 44]. Our obtained value is in this range. However, Lacour et al. [41] mentioned a RPR of 500% for a sample moisture content of 85%. Concerning corn stalks, the values range from 40% to 432.8% [41, 45]. Our obtained value of 13% is not in this range and could be justified by differences in moisture content and the corn variety.

Table 4. Annual quantification of waste derived from selected agricultural crop products in the northern part of Cameroon.

Food crop products production (tons)			Waste production (tons)						
Food crop products	Northern region [13-15]	National [10]	Waste generated	RPR (%)	Northern region		National	
					Wet tons	Dry-bone tons	Wet tons	Dry-bone tons
Corn	641 211.07	1 647 767.00	Corn cobs	25±1	160 302.77	141 066.44	411 941.75	362 508.74
			Corn stalks	13±1	83 357.44	75 021.70	214 209.71	192 788.74
Millet/Sorghum	982 025.39	1 166 533.50	Millet/Sorghum stalks	14±1	137 483.56	123 735.20	163 314.69	146 983.22
Banana/plantain	16 542.05	4 249 110.50	Banana/plantain peelings and stocks	40±4	6 616.82	727.85	1 699 644.20	186 960.86
Cassava	324 449.80	3 574 400.50	Cassava peelings	20±1	64 889.96	15 898.04	714 880.10	175 145.62
Rice	147 406.62	143 868.50	Rice hulls	20±1	29 481.32	26 828.00	28 773.70	26 184.07
Groundnut	357 161.20	519 681.00	Groundnut hulls	43±1	153 579.32	141 292.97	223 462.83	205 585.80
Sweet potato	147 135.11	277 524.00	Sweet potato peelings	20±1	29 427.02	7 077.20	55 504.80	13 348.90
Irish potato	62 577.75	177 612.00	Irish potato peelings	27±3	16 895.99	2 872.32	47 955.24	8 152.39
Yam	49 615.95	483 156.50	Yam peelings	12±2	5 853.91	1 580.56	57 978.78	15 654.27
Cotton	101 300.00	101 300.00	Cotton hulls*	/	20 771.43	18 902.00	20 771.48	18 902.00
Total :	2 188 213.87	12 239 653.50	/	/	708 659.53	555 002.27	3 638 437.23	1 352 214.62

*: Industrial waste.

The RPR for rice hulls in the literature ranges from 20% to 33% for 2.37% sample moisture content [44, 46] and our value is within this range. Lacour *et al.* [41] mentioned a RPR value of 500% for 50% rice hulls moisture content. Similar observations were made for millet stalks and sorghum stalks whose literature values vary between 100% to 200% and 90% to 500% respectively [41, 44]. These reports indicate the high dependency of RPR to the waste moisture content.

For cassava peelings, Koopmans and Koppejan [44] reported a RPR value of 2 to 3% for a sample moisture content of 50. Lacour *et al.* [41] reported a RPR value for cassava peelings of about 60%. Our value of 20% is different and could be explained by differences in moisture content.

The value of RPR of groundnut hulls in the literature ranges from 47.7% to 50% for 8.2% sample moisture content [44, 45, 47]. Our obtained value of 43% is slightly different from these values. Differences observed could be due to variability in crop species and soil types [48] and to the waste moisture content. Also, Lacour *et al.* [41] mentioned a RPR value of groundnut hulls of 200% for 50% moisture content.

In the literature, the residual ratios for banana/plantain, sweet potato, yam and Irish potato peelings are respectively 30%, 60%, 60% and 50% [11]. Our values are different and could be explained by moisture content variations and the peeling method used.

The Table 4 also indicates a total amount of waste generated in terms of wet basis and dry basis of about 708 659.53 and 555 002.27 tons respectively. These values correspond to waste global moisture content of 21.68%. Among the estimated waste (dry basis), corn waste (corn cobs and corn stalks), groundnut hulls and millet/sorghum stalks present larger waste proportion in the Northern region of Cameroon. Their respective productions are 216 088.14; 141 292.97 and 123 735.20 dry-bone tons respectively, corresponding globally to 86.69% of the total waste. Sweet potato, Irish potato, yam and banana/plantain peelings are the less produced and represent a respective tonnage of 7 077.20; 2 872.32; 1 580.56 and 727.85 dry-bone tons corresponding globally to 2.21% of the total waste. These four last are the essential house-hold wastes in the northern Cameroon. The remaining percentage (11.10%) is distributed among rice hulls, cotton hulls and cassava peelings with respective values of 26 828; 18 902 and 15 898 dry-bone tons. The northern waste production, dry basis, represents 41.04% of the national waste production. The same trend is observed in terms of wet mass basis, with the northern waste production representing 19.48% of the national waste production (Table 5). However, on this basis, plantain/banana and yam waste generation is lower. The others such as rice hulls, sweet potato peelings, cotton hulls and Irish potato peelings have almost equal waste proportions.

Table 5. Comparison between northern and national productions.

Annual production	Total agricultural production	Total waste-derived production		Waste weighted moisture content
		Wet tons	Dry-bone tons	
Northern parts	2 188 213.87	708 659.53	555 002.27	21.68
National	12 239 653.50	3 638 437.23	1 352 214.62	62.84
Northern parts/ National (%)	17.88	19.48	41.04	/

This table shows that almost the fifth of national agricultural production is found in the northern parts of Cameroon. It can also be noticed that the global moisture content of national total waste is 2.9 times greater than the one of the northern parts. This difference is especially due to banana/plantain peelings and stocks production, which is greater in the south parts of Cameroon and which is having the higher moisture content. However, more than the two-fifth of national waste production, dry basis, is recorded in the northern part.

3.4. Energy Potential of Investigated Wastes

Table 6 presents the annual energy potential of the investigated wastes. The total calorific value of the investigated wastes is around 11.5 TJ in the northern parts of Cameroon against 27.9 TJ for the national generated wastes. The energy potential of each waste follows the corresponding dry-bone tone production presented in the previous section. Amongst the investigated wastes, the major waste contributing to the total potential energy are corn waste (corn cobs and corn stalks), groundnut hulls and millet/sorghum stalks. Their respective contributions in the northern part of Cameroon are 45.31%; 24.61% and 18.17% respectively corresponding globally to 88.09% of the total energy potential of the northern part. The national respective values are 48.05%; 14.78% and 8.91% corresponding globally to 71.74% of the national studied waste total potential energy.

Table 6. *Energy potential of wastes in terms of type in the northern region compared to national values.*

Waste	Total energy (MJ)		Total energy (kWh)	
	Northern region	National	Northern region	National
Corn cobs	3.86E+06	9.91E+06	1.07E+06	2.75E+06
Corn stalks	1.35E+06	3.48E+06	3.76E+05	9.67E+05
Millet/sorghum stalks	2.09E+06	2.48E+06	5.81E+05	6.90E+05
Banana/plantain peelings and stocks	1.46E+04	3.75E+06	4.05E+03	1.04E+06
Cassava peelings	2.15E+05	2.37E+06	5.97E+04	6.57E+05
Rice hulls	4.79E+05	4.67E+05	1.33E+05	1.30E+05
Groundnut hulls	2.83E+06	4.12E+06	7.86E+05	1.14E+06
Sweet potato peelings	1.98E+05	3.73E+05	5.50E+04	1.04E+05
Irish potato peelings	8.61E+04	2.44E+05	2.39E+04	6.79E+04
Yam peelings	3.36E+04	3.33E+05	9.33E+03	9.24E+04
Cotton hulls	3.44E+05	3.44E+05	9.56E+04	9.56E+04
TOTAL	1.15E+07	2.79E+07	3.19E+06	7.74E+06

3.5. Waste Distribution in the Northern Part of Cameroon

3.5.1. Banana/Plantain Peelings and Stocks Distribution

Figure 5 presents banana/plantain waste production in nine divisions of the northern parts of Cameroun. In the six divisions of the Far-North region, banana/plantain is not cultivated. This explains the very low aforementioned quantity of banana/plantain waste generated as compared to the other agricultural crops.

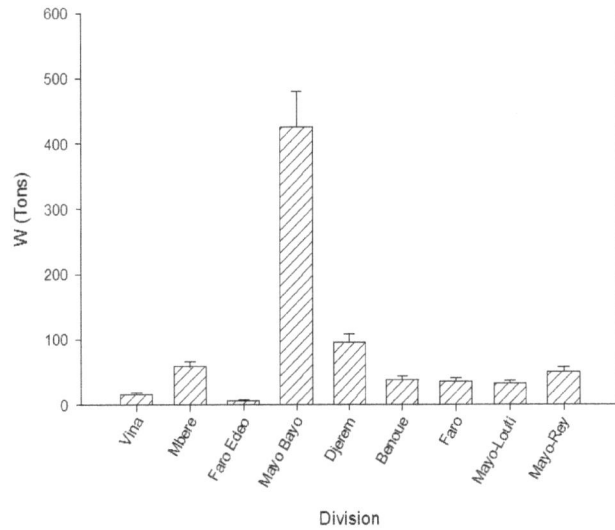

Figure 5. *Generation of banana/plantain waste.*

The Figure 5 reveals that banana/plantain waste is more generated in the Mayo Bayo division than in the others, where the generation is low with some slight variation from one division to another.

However, there is a significant quantity of plantain which enters Adamawa region by railway. This quantity was around 10 890 tons (wet basis) between 2013 and 2014. That is without counting the quantity which enters this region by road. The corresponding wet tons and dry-bone tons of the waste generated are respectively 4356 and 479 tons. This value is more than the half of the banana/plantain waste generated in the northern part of Cameroon.

3.5.2. Corn Cobs and Stalks Distribution

Corn waste production is presented on Figure 6.

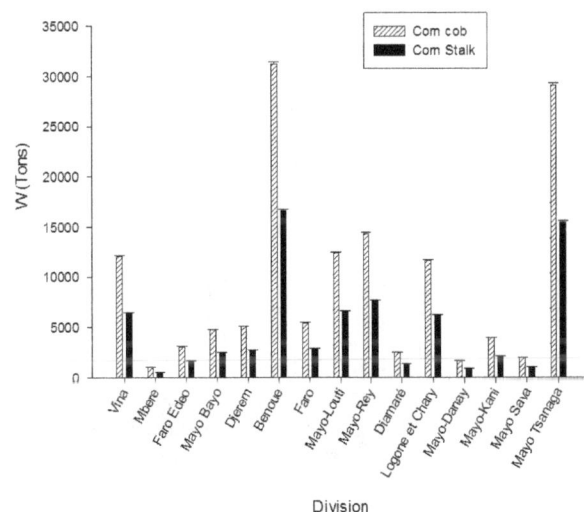

Figure 6. *Corn cobs and corn stalks dry-bone tons generation.*

From the Figure 6, it appears that the corn cobs fraction is always greater than that of the corn stalks. It is also observed that the tonnage of corn stalks and cobs is the highest in

Benoue division (North region) and Mayo Tsanaga division (Far-North region). These are followed by the Vina division (Adamawa region), Mayo-Rey, Mayo-Louti (North region) and Logone and Chary (Far-North region) with almost an equal waste production. The waste proportions in the other divisions are low and present slight variations between them.

3.5.3. Cassava Peelings Distribution

The tonnage of the cassava peelings in the various areas of the Far-North is presented on Figure 7.

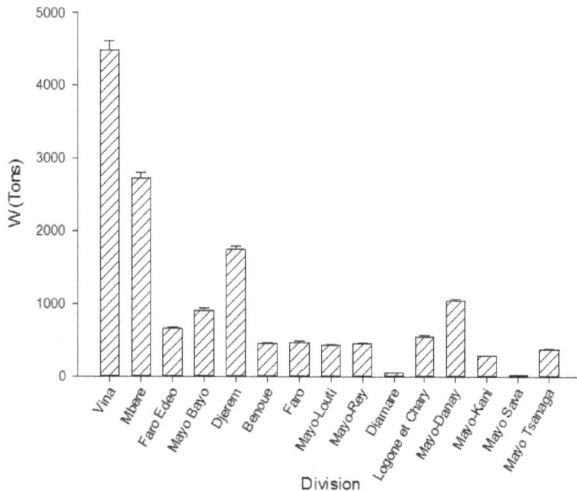

Figure 7. Cassava peelings dry-bone tons generation.

Cassava peelings in the Vina division presents the highest generation followed by Mbere and Djerem divisions. The lowest tonnage of cassava peelings is recorded in Mayo Sava and Diamare divisions. The tonnage of the other divisions is small and slightly equal.

3.5.4. Rice Hulls Distribution

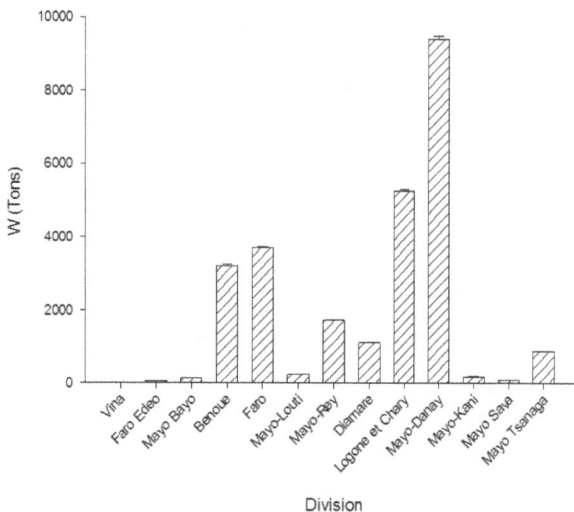

Figure 8. Rice hulls dry-bone tons generation.

Results are given on Figure 8. The Mayo-Danay division (Far-North region) stands out from other divisions with the highest rice hulls production, followed by Logone and Chary

(Far-North region), then by the Faro and Benoue division (North region). A lower generation of rice hulls is observed in the Adamawa region as compared to the North and Far-North regions. The generation in the other divisions is less with slight variation between them. It can be noticed that Djerem and Mbere divisions didn't cultivate this food crop during the year 2013 and 2014.

3.5.5. Groundnut Hulls Distribution

Groundnut hulls generation is presented on Figure 9. This figure shows that the Mayo Tsanaga division (Far-North region) has the highest tonnage of groundnut hulls, followed by those of Benoue, Faro, Mayo-Louti and Mayo-Rey division (North region). The productions recorded in Diamare, Logone and Chary, Mayo-Danay, Mayo-Kani and Mayo-Sava division (Far-North region) are low and the lowest proportion is found in the Adamawa region.

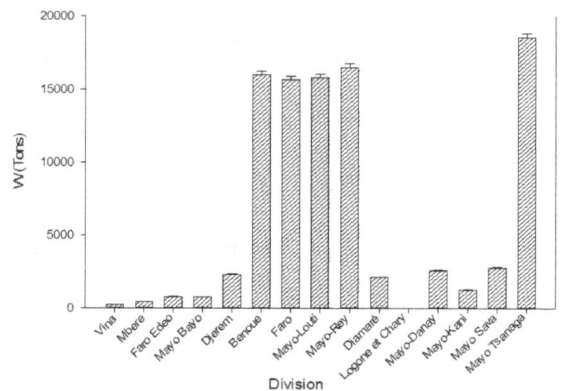

Figure 9. Groundnut hulls dry-bone tons generation.

3.5.6. Sorghum/Millet Waste Distribution

Figure 10 presents the production of the millet/sorghum stalks in thirteen divisions of the northern region of Cameroon. Djerem and Mayo Bayo do not cultivate this food crops.

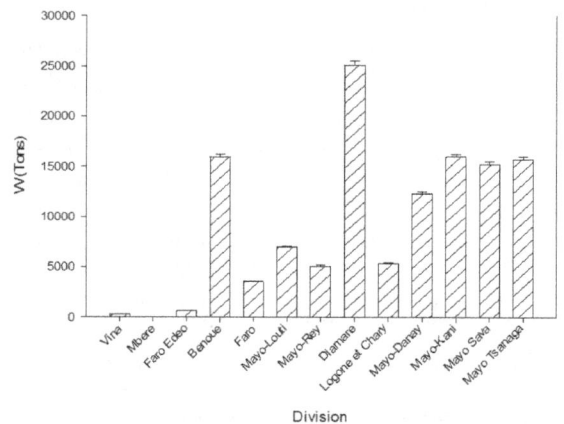

Figure 10. Sorghum/millet stalks dry-bone tons generation.

It is clear from this figure that the tonnage of millet/sorghum stalks is higher in Diamare division (Far-North region), followed by the tonnage of Mayo-Kani, Mayo-Tsanaga, Mayo-Sava Mayo-Danay division (Far-North region) and Benoue (North region). The lowest tonnage of

Millet/sorghum stalks is found in Mbere, Vina and Faro Edeo (Adamawa division).

3.5.7. Sweet Potato Peelings Distribution

Production of sweet potato peelings in the northern parts of Cameroon (Figure 11) reveals that this waste proportion is the highest in Mayo Tsanaga, followed by Logone and Chary division (Far-North region) and Vina division (Adamawa region). The smallest productions are recorded in the Far-North region, especially in Mayo- Sava, Mayo-Danay and Mayo-Kani divisions. There is an average and slightly equal tonnage of sweet potato peelings in the other divisions.

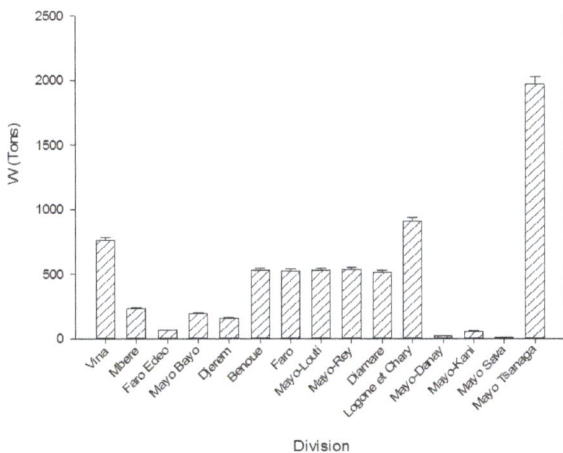

Figure 11. Sweet potatoes peelings dry-bone tons generation.

3.5.8. Yam Peelings Distribution

Results obtained in nine divisions of the Northern region are presented on Figure 12. We observe an almost equal tonnage of yam peelings in the various divisions. However, the lowest productions are recorded in Mayo Bayo and Faro Edeo divisions. This food crop is not cultivated in the Far-North region composed of six divisions (Diamare, Mayo Danay, Mayo Sava, Mayo Kani, Logone and Chary and Mayo Tsanaga).

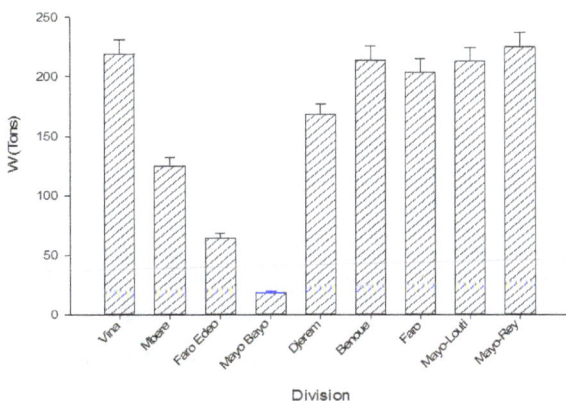

Figure 12. Yam peelings dry-bone tons generation.

3.5.9. Irish Potato Peelings Distribution

Figure 13 presents the tonnage of the potato peelings in the various divisions of Adamawa, North and Far-North regions. Amongst the fifteen divisions of this study area, a noticeable

production is only depicted in Mayo-Tsanaga division (Far-North region), where the highest production is found, and in the Vina division (Adamawa region). The proportions recorded in the other division are very low.

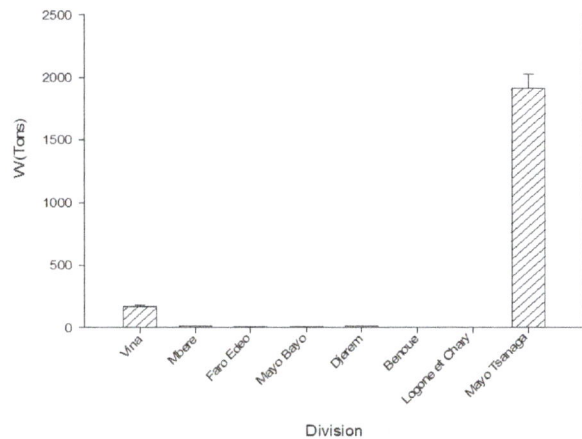

Figure 13. Irish potato peelings dry-bone tons generation.

4. Conclusions

The general objective of this work was to quantify, in terms of types, the waste generated in the northern part of Cameroon and to evaluate their associated energy potential.

Amongst the investigated food crops, there is a significant quantity of waste generated in the northern part of Cameroon of about 555 002.27 dry-bone tons per annum corresponding to a 41% of the national generated waste from the same crops. Waste generated by this food crops is classify into three groups: house-hold wastes, agricultural residues and industrial wastes.

The North region has the highest production of waste with a percentage of 42.93% followed by the Far-North region with a percentage of 42.44%. The Adamawa region has the smallest proportion of 11.23%. The cotton hulls represent 3.41%.

The evaluated potential energy of wastes generated in the northern part of Cameroon is about 11.5 TJ against 27.9 TJ for the national and for the same food crops. These values are underestimated, give that about 10% of food crops produced in the northern part were not investigated in the present study. The conversion of these wastes into energy will provide a major contribution to the Cameroon energy deficit.

To complete this work, it will be necessary to identify the collected area of house hold wastes and to propose valorization technics adapted to developing countries such as Cameroon.

Acknowledgment

The authors thank the Ministry of Higher Education of Cameroon for its financial support through the Special Fund Account for the modernization of research in state Universities. They also thank the Geomatic Laboratory of the University of Ngaoundere for the location map building of the study area, and the Agence Universitaire de la Francophonie

(AUF) for the support under the grant "Soutien aux Equipes de Recherche" (BACGL-2014-53).

References

[1] Seifried D. and Witzel W., 2007. Renewable Energy –The Facts, *Energieagentur Regio Freiburg*, 221 p.

[2] Aldo Vieira da Rosa, 2005. Fundamentals of renewable energy processes, Elsevier academic press, 446-610.

[3] Kaltschmitt M. and Wolfgang Streicher A. W., 2007. Renewable energy: Technology, Economics and Environment; *Springer-Verlag Berlin Heidelberg*, 535 p.

[4] Debra Miller A. 2011. Energy production and alternative energy, *Library of congress cataloging-in-publication data*, 10-43.

[5] Vertes A.A., Qureshi Nasib, Blaschek Hans .P, Hideaki Yukawa, 2010. Biomass to Biofuels: Strategies for Global Industries, Wiley, 547 p.

[6] Chaney J., 2010. Combustion Characteristics of Biomass Briquettes, *Thesis submitted to The University of Nottingham*, 225 p.

[7] Ningham and Singh, 2011, Production of liquid biofuels from renewable resources, Progress in Energy and Combustion Science, 37, 52-68.

[8] Sarkar N., Ghosh S.K., Bannerjee S., and Aikat K., 2012. Bioethanol from agricultural wastes. An overview. Renewable energy, 37, 19-27.

[9] Ackom E.K., Alemagi D., Ackom Nana B., Minang P.A. and Tchoundjeu Z., 2013 Modern bioenergy from agricultural and forestry residues in Cameroon: Potential, challenges and the way forward, *Energy policy*, 101-113.

[10] Agri-Stat, 2012. Annuaire des Statistiques du Secteur Agricole Campagnes 2009 et 2010, N°17, *Direction des Enquêtes et des Statistiques Agricoles*, 123 p.

[11] Devendra C. 1980, Non-conventional feed resources in Asia and Far East. Bangkok: FAO-APHCA, FAO Far East Regional Office, 104 p.

[12] FAO, 2002. Index des noms de plantes et des cultures de remplacement, améliorer la nutrition grâce aux jardins potagers. Module de formation à l'intention des agents de terrain en Afrique, Service des programmes nutritionnels, Division de l'alimentation et de la nutrition. Annexe 1, 277-280.

[13] Délégation régionale de l'Extrême-Nord, 2015. Données sur les principales réalisations Agricole de la région du Nord, *MINADER*, 33-41.

[14] Délégation régionale de l'Adamawa, 2015. Données sur les principales réalisations Agricoles de la région de l'Adamaoua, *MINADER*, 1-11.

[15] Délégation régionale du Nord, 2015. Données sur les principales réalisations Agricoles de la région du Nord, *MINADER*, 61-62.

[16] ASTM standards, 2006, Standard Test Method Moisture Analysis of Particulate Wood Fuels, E 1871-82.

[17] ASTM standards, 2006, Standard Test Method for Volatile Matter in the Analysis of Particulate Wood Fuels, E 872 – 82.

[18] ASTM standards, 2007, Standard Test Method for Ash in Biomass, E 1755-01.

[19] García R., Pizarro C., Lavín A.V., Bueno J.L., 2012. Characterization of Spanish biomass wastes for energy use, Bioresource Technol, *DOI:10.1016/j.biortech.2011.10.004*; 103, 249 -258.

[20] Institut National de la statistique Cameroun, 2015. Ministère de l'économie, de la planification et de l'aménagement du territoire, Annuaire de statistique 2015.

[21] IRAD, 2014. Cinquante ans de recherche agricole au Cameroun: Principaux résultats et acquis, Ministère de la recherche sciences et innovation, 93 p.

[22] Vargas-Moreno J.M, Callejón-Ferre A.J, Pérez-Alonso J. and Velázquez-Martí B, 2012. A review of the mathematical models for predicting the heating value of biomass Materials, Renewable and Sustainable Energy Reviews, 3065–3083.

[23] Lee Sunggyu and Shah Y. T. 2013, Biofuels and Bioenergy Processes and Technologies, CRC Press Taylor & Francis Group, 292 p.

[24] Debdoubi A., El amarti A. and Colacio E., 2004. Production of fuel briquettes from esparto partially pyrolyzed, Energy Conversion and Management, 1877–1884.

[25] ADEME, 2008. Référentiel combustible bois énergie : les connexes des industries du bois définition et exigences. Convention 0601c0005, République françaises, 62 p.

[26] Bianca G. de Oliveira Maia, Ozair Souza, Cintia Marangoni, Dachamir Hotza, Antonio Pedro N. De Oliveira, Noeli Sellin, 2014. Production and Characterization of Fuel Briquettes from Banana Leaves Waste, Chemical Engineering Transactions, VOL. 37 P 439-444.

[27] Onwuka C.F.I, Adetiloye P.O. b, Afolami C.A., 1997. Use of household wastes and crop residues in small ruminant feeding in Nigeria, *Small Ruminant Research*, 24, 233-237.

[28] Adeyi and Oladayo, 2010. Proximate composition of some agricultural wastes in Nigeria and their potential use in activated carbon production. J. Appl. Sci. Environ. Manage, 14 (1), 55-58.

[29] Sami M., Annamalai K. and Wooldridge M., 2000. Co-firing of coal and biomass fuel blends; Progress in Energy and Combustion Science; 171–214.

[30] Jigisha Parikha, S.A. Channiwalab, G.K. Ghosal, 2005. A correlation for calculating HHV from proximate analysis of solid fuels, Fuel 84, 487-494.

[31] Murali S., Shrivastavo R. and Saxena M., 2007. Quantification of agricultural residues for energy generation -A case Study, Journal of IPHE India, 27-31.

[32] Jenkins BM, Ebeling JM, 1985. Correlation of physical and chemical properties of terrestrial biomass with conversion: symposium energy from biomass and waste IX IGT, 371 p.

[33] Bhajan Dass and Pushpa Jha, 2015 Biomass Characterization For Various Thermochemical Applications, International Journal Of Current Engineering And Scientific Research, 59-63.

[34] Demirbas A., 1997. Calculation of higher heating values of biomass fuels, Fuel 76 (5), 431 p.

[35] Slavko N., Djuric, Saša D. Brankov, Tijana R. Kosanic, Mirjana B.C., and Branka B.N.S., 2014. The composition of gaseous products from corn stalk pyrolysis process, *Thermal Science*, 18, 533-542.

[36] Oladeji J.T, 2012. Comparative Study of Briquetting of Few Selected Agro-Residues Commonly Found in Nigeria "The Pacific Journal of Science and Technology, 13, 80-86.

[37] Jekayinfa S. and O. Omisakin., 2005. The Energy Potentials of some Agricultural Wastes as Local Fuel Materials in Nigeria; Agricultural Engineering International: *The CIGR Ejournal.*, Vol.VII. Manuscript EE 05 003, 10 p.

[38] Olufunke O. Ezekiel1, Ogugua C. Aworh, Hans P. Blaschek and Thaddeus C. Ezeji, 2010. Protein enrichment of cassava peel by submerged fermentation with *Trichoderma viride* (ATCC 36316), African Journal of Biotechnology Vol. 9 (2), 187-194.

[39] Jianfeng Shen, Shuguang Zhu, Xinzhi Liu, Houlei Zhang and Junjie Tan, 2009. The prediction of elemental composition of biomass based on proximate analysis, *Energy Conversion and Management*, 984-987.

[40] Jenkins B.M., Baxter L.L., Miles Jr. T.R. and Miles T.R., 1998. Combustion properties of biomass, *Fuel Processing Technology*, 17-46.

[41] Lacour J.R., Bayard E. and Gourdon E.R., 2011. Evaluation du potentiel de valorisation par digestion anaérobie des gisements de déchets organiques d'origine agricole et assimilés en Haïti, *Déchets - revue francophone d'écologie industrielle - n° 60*, 32-36.

[42] Onesias Gup-pens and Gerin P., 2009. Valorisation de la biomasse-énergie en Haiti : Analyse de la situation et perspectives d'amélioration, *Mémoire de Master*, Université Quisqueya, 50 p.

[43] Freitas Komlanvi. I, 1976. Etude des produits et sous-produits agro-industriels du Togo possibilités de leurs utilisations en élevage, Thèse, université de Dakar-Togo, 131 p.

[44] Koopmans A. and Koppejan J., 1998. Agricultural And Forest Residues Generation, Utilization And Availability Wood Energy Conservation Specialists Regional Wood Energy Development Programme in Asia, 23 p.

[45] Kimutai S.K., Muumbo A.M., SIAGI Z.O. and Kiprop A.K., 2014. A study on agricultural residues as a substitute to fire wood in Kenya: a review on major crops. Journal of Energy Technologies and Policy, 4(9), 45-51.

[46] OECD/IEA Suitainable production of second generation biofuel, potential and perspective in major economies and developing countries information paper accessed 7th (http:iea.org/papers/2010/second), 12 p.

[47] Ryan P. and Openshaw K. (1991), Assessment of Biomass Energy Resources: A discussion on its needs and methodology, The World Bank Industry and Energy Department, 77 p.

[48] UNEP, 2010. Project on converting waste agricultural biomass to an energy / material resource. Report II, Waste biomass quantification and characterization. National Cleaner Production Center, Sri Lanka, 121 p.

Design of Fast Real Time Controller for the Dynamic Voltage Restorer Based on Instantaneous Power Theory

Mohammed Y. Suliman[1], Sameer Sadoon Al-Juboori[2, *]

[1]Department of Electrical Engineering, Technical College, Mosul, Iraq
[2]Department of Electronic and Control Engineering, Technical College, Kirkuk, Iraq

Email address:

m_yahya1973@yahoo.com (M. Y. Salman), sameer.al-juboori@gis.lu.se. (S. S. Al-Juboori)

Abstract: The fast variations in the source voltage can affect the performance of the loads such as (a) semiconductor fabrication plants (b) paper mills (c) food processing plants and (d) automotive assembly plants. The common disturbances in the source voltages are the voltage sags or voltage swells this can be due to (i) disturbances arising in the transmission system, (ii) adjacent feeder faults and (iii) fuse or breaker operation. Voltage sags of 10% lasting for 5-10 cycles can result in costly damage in the loads. To mitigate the problems of poor quality power supply, voltage source converters can be connected in series with transmission lines as compensators. These are known as Dynamic Voltage Restorer (DVR) or Static Voltage Restorer. In this paper, a new scheme to control DVR using adaptive neuro fuzzy logic is proposed. In this controller, Takagi-Sugeno fuzzy rules are trained using off-line neuro fuzzy system. Also, instantaneous power theory is used to calculate the phase voltage due to its high accuracy and less computation. The simulation and practical results show that real time application of the proposed controller is possible and robust compared to conventional controllers previously investigated. The experiment results obtained using the dSPACE data acquisition system and Matlab real time toolbox.

Keywords: DVR, SSC Instantaneous Power Theory, Fuzzy Logic, ANFIS, TS controller

1. Introduction

The SSC provides three phase controllable voltage, whose vector (magnitude and angle) adds to the source voltage to restore the load voltage to pre-disturbance (sag and swell), Static Series Compensator also named Dynamic Voltage Restorer (DVR), where connected between the source and load as shown in figure (1) [1-2]. Voltage sag, which is a momentary decrease in r.m.s voltage magnitude in the range of 0.1 to 0.9pu, is considered as the most serious problem of power quality. It is often caused by faults in power systems or by starting of large induction motors. It occurs more frequently than any other power quality phenomenon does.

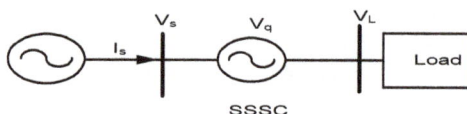

Figure 1. The Schematic diagram of SSSC or (DVR) System.

Therefore, the loss resulted due to voltage sag problem for a

customer at the load-end is huge. Swell is defined as an increase in r.m.s voltage or current at the power frequency, typical magnitudes is between 1.1 and 1.8 pu. Swell magnitude is also described by its remaining voltage for durations from 0.5 cycle to 1 min [3]. The definition of sag and swell are shown in Figure (2).

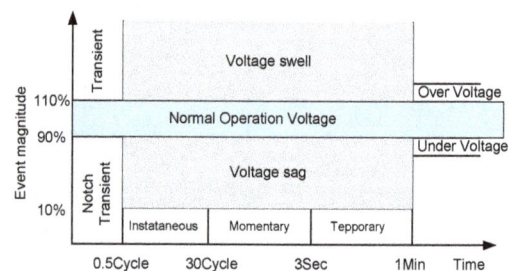

Figure 2. The Range of Different Events Magnitude.

There are three basic control strategies as follows [5]:
- Pre-Sag Compensation
- In-phase Compensation

- Minimum Energy Compensation.

These strategies control are shown in Figure3.

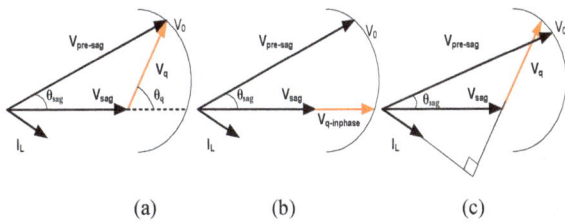

(a)   (b)   (c)

Figure 3. *Control Strategies of DVR (a) Pre-Sag (b) In-Phase (c) Minimum Energy.*

Artificial neural networks (ANN) methodology has captured the interest in a large number of applications as well as electrical power engineering. The applications include, but not limited to, economical load dispatching and power system stabilizers and controllers. The results have shown that ANNs have great potential in improving power system on-line and off-line applications [4].

The fuzzy logic control has been applied successfully in power system computer simulation [5], it does not require a detailed mathematical model of the controlled system. The intensive computation time and the huge data memory required are the limit factors of implementing real time fuzzy controllers with large number of control rules. The first type fuzzy logic controllers (using Mamdani membership functions) may not be able to provide a wide variation of control gains for the SSSC to perform robustly at different compensation levels. Alternatively, a second type, based on Takagi-Sugeno (TS) fuzzy controller can provide a wide range of control gain variation and also can use either linear / non-linear consequent expressions of the fuzzy rule base, the coefficients of the consequents are system dependent and have not been systematically chosen [6].

The purpose of this paper is to design DVR based on TS fuzzy controller in order to regulate the power flow in a transmission line. The controller's rules are optimised-using Adaptive Neuro-Fuzzy Inference System (ANFIS). The fuzzy rules are trained using gradient descent and least squares estimate for tuning every rule antecedent and consequent, respectively. The proposed controller has small computation time compared to classical fuzzy controllers; thereby it is implemented in real time using SIMULINK and dSPACE ds1103 data acquisition system. The simulation and experimental results highlight the effectiveness of the adaptive TS fuzzy controller in optimising the SSSC performance. In order to obtain fast responses, the instantaneous power theory for active and reactive measurements was used to calculate the active and reactive power flow in the system that needed to be controlled.

2. DVR Model and Control

Static synchronous series compensator is a series compensator of the FACTS family. It injects an almost sinusoidal voltage (based on switching frequency and inverter configuration) with variable amplitude. It is equivalent to an inductive or a capacitive reactance in series with the transmission line. The heart of DVR is a Voltage Source Inverter (VSI) that is supplied from a DC storage capacitor. With no external DC link, the injected voltage has two parts: the main part is in quadrature with the line current and emulates an inductive or capacitive reactance in series with the transmission line. The less significant part is in phase with the line current to supply the inverter losses. When the injected voltage is lagging the line current, it will emulate a capacitive reactance in series with the line, causing the line current as well as power flow through the line to increase. When the injected voltage is leading the line current, it will emulate an inductive reactance in series with the line, causing the line current as well as power flow through the line to decrease. DVR is superior to other FACTS series-connected devices and the benefits of using DVR are:

- Elimination of bulky passive components - capacitors and reactors.
- Symmetric capability in both inductive and capacitive operating modes.
- Possibility of connecting an energy source on the DC side to exchange real power with the AC network.

DVR comprises a voltage source inverter and a coupling transformer that is used to insert the AC output voltage of the inverter in series with the transmission line as shown in Figure4. The magnitude and phase of this inserted AC compensating voltage can be rapidly adjusted by the SSSC controls.

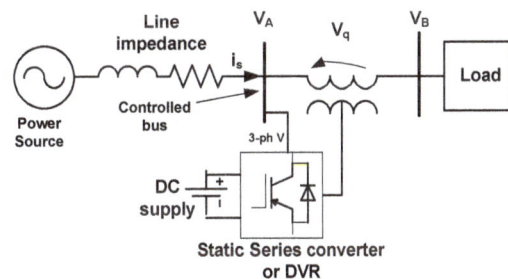

Figure 4. *Elementary system with an DVR.*

The SSSC injects the compensating voltage in series with the line irrespective of the line current magnitude. The transmitted power (p_q), therefore becomes a parametric function of the injected voltage, and can be expressed as follows:

$$V_B = V_A \pm V_q \qquad (1)$$

The DVR, therefore can increase the phase voltage, and also decrease it, simply by reversing the polarity of the injected ac voltage. The reversed (180° phase-shifted) voltage adds directly to the reactive voltage drop of the line as if the reactive line impedance was increased. Furthermore, if the injected voltage is made larger than the voltage impressed across the uncompensated line by the sending- and receiving-end systems, that is, if $|Vq| > |Vs - Vr|$, then the power flow can reverse. Apart from the stable operation of the system with both positive and negative power flows .It can also be observed that the DVR has an excellent (sub-cycle) response time and that the transition from positive to negative power flow through zero voltage injection is perfectly smooth and continuous [6].

3. Application of Instantaneous Power Theory

For fast measuring active and reactive power, instantaneous power theory was used. The p-q theory, or "Instantaneous Power Theory", was developed by Akagi et al in 1983, with the objective to apply for controlling active power filter. This theory is based on time-Domain, which makes it valid for operation in steady-state or transitory regime, as well as for generic voltage and current power system waveforms, allowing to control the active power filters in real-time. Another important characteristic of this theory is the simplicity of calculations, which involves only algebraic calculation exception to the need of separating the mean and alternated values of the calculated power component. The p-q theory performs a transformation known as "Clarke Transformation" of a stationary reference system of coordinates a-b-c to α-β-0 coordinates [7].

$$\begin{bmatrix} u_0 \\ u_\alpha \\ u_\beta \end{bmatrix} = \sqrt{\frac{2}{3}} \begin{bmatrix} \frac{1}{\sqrt{2}} & \frac{1}{\sqrt{2}} & \frac{1}{\sqrt{2}} \\ 1 & -\frac{1}{2} & -\frac{1}{2} \\ 0 & \frac{\sqrt{3}}{2} & -\frac{\sqrt{3}}{2} \end{bmatrix} \begin{bmatrix} u_1 \\ u_2 \\ u_3 \end{bmatrix} \quad (2)$$

$$\begin{bmatrix} i_0 \\ i_\alpha \\ i_\beta \end{bmatrix} = \sqrt{\frac{2}{3}} \begin{bmatrix} \frac{1}{\sqrt{2}} & \frac{1}{\sqrt{2}} & \frac{1}{\sqrt{2}} \\ 1 & -\frac{1}{2} & -\frac{1}{2} \\ 0 & \frac{\sqrt{3}}{2} & -\frac{\sqrt{3}}{2} \end{bmatrix} \begin{bmatrix} i_1 \\ i_2 \\ i_3 \end{bmatrix} \quad (3)$$

Then the active and reactive power compensated calculated by:

$$\begin{bmatrix} p \\ q \end{bmatrix} = \begin{bmatrix} v_\alpha & v_\beta \\ v_\beta & -v_\alpha \end{bmatrix} \begin{bmatrix} i_{S\alpha} \\ i_{S\beta} \end{bmatrix} \quad (4)$$

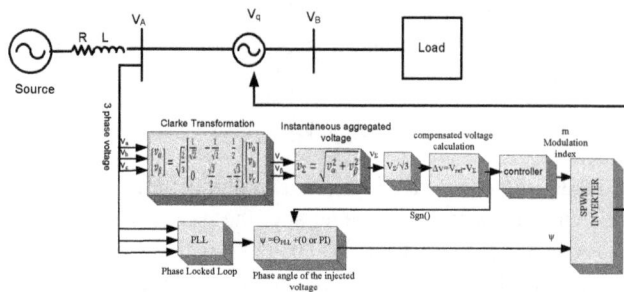

Figure 5. Block diagram for the DVR control system.

Where i_α and i_β are the two-orthogonal components of the current and v_α and v_β are the two-orthogonal components of the voltage. The compensated voltage is [8]:

$$v_\Sigma = \sqrt{v_\alpha^2 + v_\beta^2} \quad (5)$$

To get phase voltage:

$$v_\Sigma = v_\Sigma / \sqrt{3} \quad (6)$$

4. Control Scheme of SSSC

The basic control system of the DVR is shown in Figure 5. The system consists of generating machine with transmission line and load. The compensator is provided with a DC voltage source which helps in feeding or absorbing the active and reactive power from the system. For the control circuit as shown, the phase voltage is sensed; v_α and v_β are the quadrature components voltage are calculated using Clark's transformations. Then the compensated voltage is calculated using eq 5. The compensated voltage is compared with The desired voltage to generate error signals ΔV. This error signal is processed in the controller where:

$$E_{\Delta V} = (V_{ref} - V_\Sigma)(K_{p1} + \frac{K_{i1}}{S}) \quad (7)$$

5. ANFIS Based Control System

Fuzzy systems are suitable for uncertain or approximate reasoning, especially for the system with mathematical model that is difficult to derive. Fuzzy logic controllers play an important role in many practical applications. There are many fuzzy inference mechanisms in fuzzy logic control system from which Takagi-Sugeno is chosen in this study. The Artificial Neural Network (ANN) will be used in this study to tune the membership functions of the TS fuzzy-like-PI controller.

The general TS rule structure for two inputs single output system is given as:

Rule i: if x is A and y is B then $f_i = f(x, y)$ (8)

It provides a simple structure defuzzification process, reduces the overall computation time and offers a wide range of control gain variation based on its variable rule consequent. However, there are no standard methods for transforming human knowledge into the rule-base of the fuzzy inference system. Hence, the selection of the size, type and parameters of the input and output membership functions are often determined depending on the designer experience or by trial and error. There is a need for effective methods of tuning the membership functions and reducing the rule base to the minimum essential rules. Where, *Ai* and *Bi* represent the linguistic variables of the corresponding input membership functions (MF).

fi is the output represented as a function of the system variables, it could have different structure e.g.

$$f i = c \ (\text{zero-order TS model}) \quad (9)$$

$$f_i = g_i x + h_i y + r i \quad (\text{first-order TS model}) \quad (10)$$

The coefficients (c, g_i h_i, and r_i) represent the milestone of the TS fuzzy control system design, that is shown in Figure 6. It produces wide variations of the controller gain. Arbitrary selection of these parameters may lead to an adequate system response or instability.

A better system response may be achieved by using Neuro-Fuzzy system to adapt the fuzzy system parameters and rules by employing ANN learning algorithm. The adaptive TS fuzzy controller used in this work consists of seven triangle membership functions as shown in Figure 7. Adaptive Neuro-Fuzzy Inference System (ANFIS) was proposed to overcome the above difficulties. Since it combines the fuzzy qualitative approach with the adaptive learning capabilities of the neural network, such a system can be trained without a great amount of expert knowledge usually required for the standard fuzzy logic. As a result, the rule-base can be reduced. A typical architecture of ANFIS based on the first order Takagi-Sugeno model is shown in Figure(8), with two-inputs (x, y) and one-output (f). The architecture is expanded as follows:

Rule ij:if x is A_i and y is B_j then $f_{ij} = g_{ij} x + h_{ij} y + r_{ij}$ (11)

Where, A_i and B_j represent the input membership functions (MF). g_{ij}, h_{ij} and r_{ij} are the parameters of the output membership functions. The parameters of the input and output membership functions are to be determined during the training stage. ANFIS consists of five layers, each layer has either fixed nodes (that have no parameters to be tuned) represented by a circle or adaptive nodes (that have parameters to be tuned during training) represented by a square, as shown in Figure8a. The output of the five layers which emulate the fuzzy system design steps is given as follows, referring to [9] for more details.

$$O_{1i} = \mu_{Ai}(x) \text{ or } O_j = \mu_{Bj}(y) \quad (12)$$

$$O_{2ij} = w_{ij} = \mu_{Ai}(x)\mu_{Bj}(y) \quad (13)$$

$$i = 1, 2, \ldots N, j = 1, 2 \ldots M$$

$$O_{3ij} = \overline{w}_{ij} = \frac{w_{ij}}{\sum_{\forall i,j} w_{ij}} \quad (14)$$

$$O_{4ij} = \overline{w}_{ij} f_{ij} = \overline{w}_{ij}(g_{ij}x + h_{ij}y + r_{ij}), \quad (15)$$

$$i = 1, 2, \ldots N, j = 1, 2, \ldots M$$

$$O_5 = f = \sum_{\forall ij} \overline{w}_{ij} f_{ij} = \sum_{\forall i,j} \frac{w_{ij} f_{ij}}{\sum_{\forall i,j} w_{ij}} = \frac{\sum_{\forall i,j} w_{ij}}{\sum_{\forall i,j} w_{ij}} \quad (16)$$

The objective of the learning algorithm is to adjust the parameters of the input and output membership functions so that the ANFIS output best matching the training data. A hybrid learning strategy (Gradient Descent-GD and Least Squares Estimate-LSE) is applied to identify the network parameters. The GD method updates the antecedent membership function parameters (A_i, B_i) while LSE identifies the consequent parameters (g_{ij}, h_{ij}, r_{ij}). To tune the TS rules using ANFIS, two sets of data are to be generated [10].

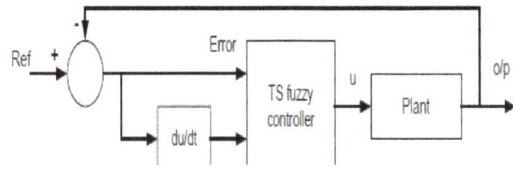

Figure 6. *TS fuzzy control scheme*

The input data is a vector of the error and change of error of the active and reactive power flow of the transmission system controlled parameters. The input universe of discourse is split into 7 triangular membership function with 50% overlapping. Therefore, for two inputs, 49-control rule consequent linear functions need to be determined. To initialise the coefficients of the consequents, the data extracted from the standard Mamdani fuzzy like PI controller as described in Table 1 is used to start the training procedure and the error and change of error relation with the output surface shown in Figure 9. This procedure is performed using the ANFIS included in the MATLAB/FUZZY Logic Toolbox.

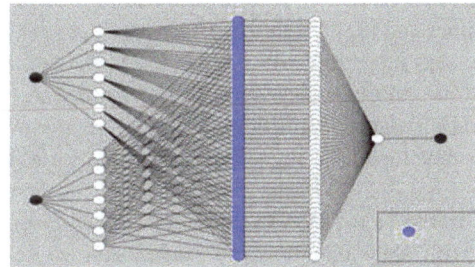

Figure 7. *The ANFIS model structure*

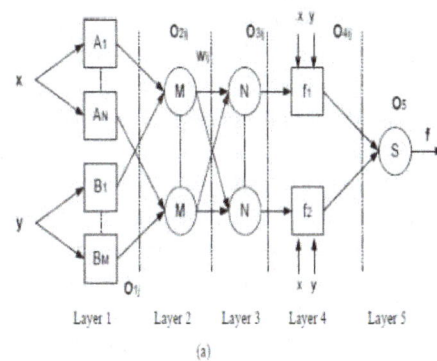

(a)

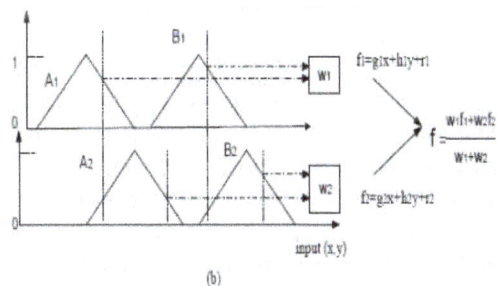

(b)

Figure 8. *a-ANFIS structure b-Takagi-Sugeno fuzzy inference*

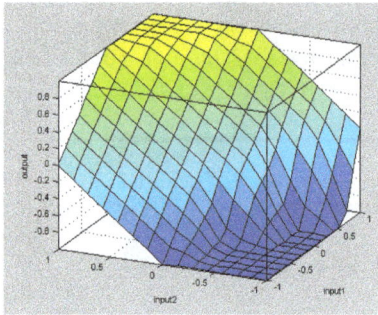

Figure 9. Control surface of SSSC-based neuro-fuzzy controller

6. Simulation Study

The system shown in Figure (5) is simulated to investigate the performance of the proposed intelligent controller under the step change of the power flow condition. Sag disturbance has been generated to validate the DVR in voltage regulation mode, these disturbances were made in the control bus position (V_A) in the Matlab model, the phase voltages were measured before and after DVR (before and after injection voltage), the range of the SSSC to compensate voltage was ±0.1 pu of the phase voltage of the controlled bus as shown in Figure (10).

Table 1. *The Error and Change of Error of The Voltage.*

e\Δe	-0.1	-0.0667	-0.033	0	0.033	0.0667	0.1
-0.1	-0.1	-0.1	-0.1	-0.1	-0.066	-0.033	0
-0.066	-0.1	-0.1	-0.1	-0.066	-0.033	0	0.033
-0.033	-0.1	-0.1	-0.0667	-0.033	0	0.033	0.066
0	-0.1	-0.0667	-0.033	0	0.033	0.0667	0.1
0.033	-0.066	-0.033	0	0.033	0.066	0.1	0.1
0.066	-0.033	0	0.033	0.066	0.1	0.1	0.1
0.1	0	0.033	0.066	0.1	0.1	0.1	0.1

7. Experimental Results

The laboratory model used in this study includes, the host computer that was interfaced with the DVR and transmission system hardware through the Control Desk software and dSPACE ds1103 data acquisition board. The controller algorithm is developed in the SIMULINK platform then downloaded to the ds1103 board. A 6-pulse PWM converter connected to the a.c. system through an appropriate transformer with the suitable turns' ratio. The switching frequency of the converter is set to 9 times the system frequency in order to eliminate both the evens and tripled harmonics. The DVR operates as a voltage regulation by injection voltage in-phase or anti-phase with the phase voltage of controlled bus. To validate the DVR as a voltage regulation, a variable load was added to the controlled bus V1 and a step change in the load suddenly to check the DVR response for both disturbance cases in sag and swell conditions. Figure 11 shows the experimental setup for voltage regulation. The limitation of the DVR injected voltage was ±10% of the magnitude of the phase voltage (reference signal), and the compensator will fixed to this amplitude even though the phase voltage dropped to less than this level. Figure 12 shows two duration of DVR injected voltage, For the periods from 0.16 to 0.24 second the injected voltage was 10% of the phase voltage and restore the phase voltage to its nominal operation, and from 0.26 to 0.4 seconds the increased in the load led to further decrease in the phase voltage to less than 0.7 pu the action of the DVR starts at t=0.29 to 0.4 and inject 10% of the phase voltage to 0.8 pu. In this case the performance of SSSC was to mitigate the sag condition. The active power flow response for two types of controllers PI and Fuzzy-TS is shown in Figure13a &b It is clear that the system more stable and faster to reach the steady state. Figure14 shows the

prototype in the laboratory.

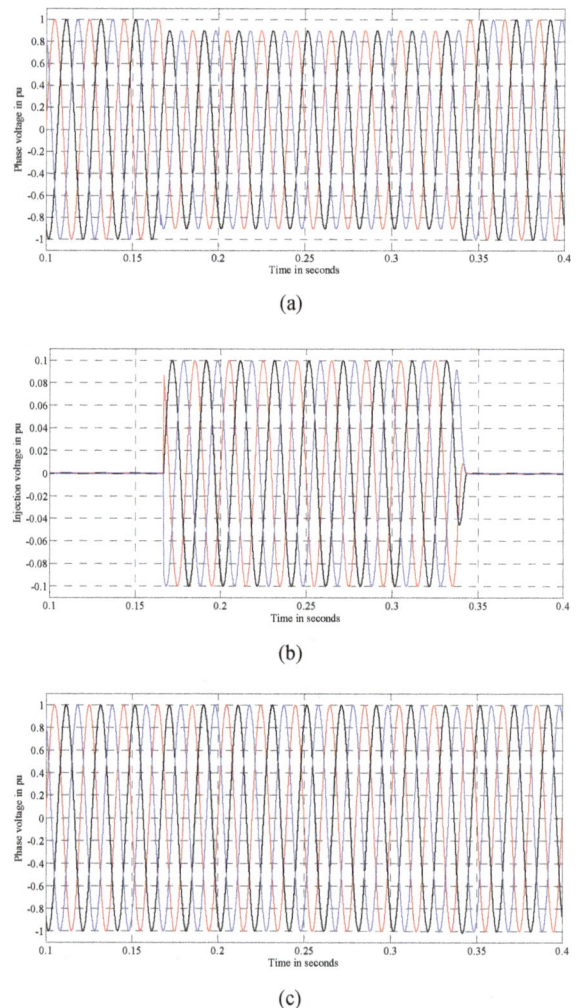

(a)

(b)

(c)

Figure 11. *Three Phase Voltages of Controlled Bus in Sag Condition (a) Before Compensation (b) The Injected Voltage (c) After Compensation.*

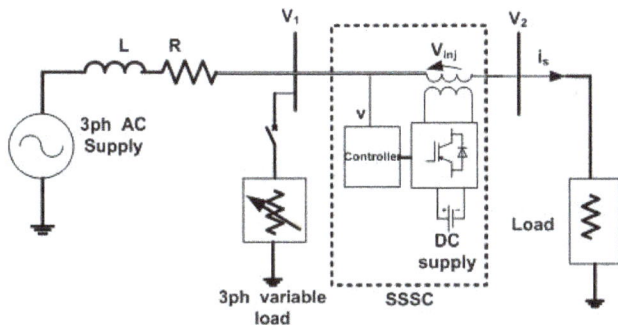

Figure 10. Experimental setup for DVR.

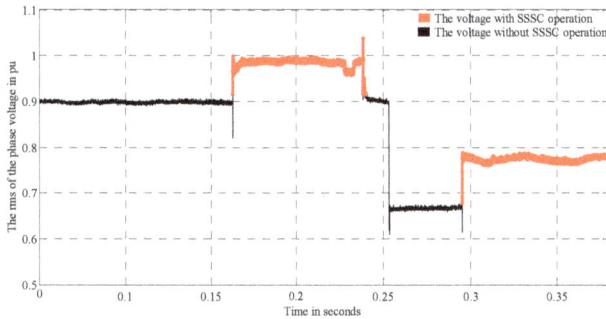

Figure 12. Two Step Changes in the Load Voltages.

8. Conclusion

In this paper, Takagi-Sugeno fuzzy control system algorithm is used to control the SSSC. The tuning algorithm is performed off-line employing the concept of Adaptive Neuro-Fuzzy Inference System (ANFIS). The rules defined by Mamdani fuzzy-like-PI controller are used to initiate the tuning process. The small computation time of the controller has the potential of implementation in real time. The proposed controller has been applied successfully to control the power flow in transmission system. The simulation and experimental results show that the proposed controller can provide an adequate performance for the SSSC operation. Also the use of instantaneous power theory gave a fast measurement tool for measuring the active and reactive power.

References

[1]   synchronous series compensator: a solid state approach to the series compensation of transmission lines", IEEE Trans. Power Delv., vol. 12, pp. 406-417, 1997.

[2]   P. Kapil, C. Vibhakar, S. Rajani and K. Bhayani," Voltage Sag/Swell Compensation Using Dynamic Voltage Restorer (DVR)", International Journal of Application or Innovation in Engineering & Management (IJAIEM), vol 4, no 3, pp 1-8, 2013.

[3]   A. Damor and V. Babaria, " Voltage Sag Control Using DVR", National Conference on Recent Trends in Engineering & Technology India, pp 1-4, 13-14 May 2011S.

[4]   M. Sharanya, B. Basavaraja and M. Sasikala," Dynamic Voltage Restorer (DVR) for Voltage Sag Mitigation", International Journal on Electrical Engineering and Informatics, vol 3, no 1, pp1-11, 2011.

[5]   S. Panda, "Multi-objective evolutionary algorithm for SSSC-based controller design", *Electr. Power Syst. Res.*, vol.79, no. 6, pp. 937-944, 2009.

[6]   E H Watanabe and Akagi H, "Instantaneous p–q power theory for control of compensators in micro-grids" IEEE No sinusoidal Currents and Compensation (ISNCC), 2010 Pages 17-26.

[7]   H. Akagi, Y. Kanazawa and A. Nabae," Instantaneous Reactive Power Compensators Comprising Switching Devices without Energy Storage Components", IEEE Transactions On Industry Applications, vol 20, no. 3, pp 625-630,1984.

[8]   Mohammed Y. Suliman and S. M. Bashi," Instantaneous Active and Reactive Power Measuring in Three Phase Power System", 3rd International Scientific Conference of F.T.E,Najaf,Iraq,20-21 Feb 2013, Page(s): 926-936.

[9]   S. Mishra, P.K. Dash, and G. Panda, "TS –fuzzy controller for UPFC in a multimachine power system," IEE Proc. Gener. Transm. Distrib, vol 147, no 1, pp. 15-22, 2000.

[10] Farrag M. E. A, G. A. Putrus," Design of adaptive neuro-fuzzy inference controller for a transmission system incorporating UPFC", IEEE, Transaction on Power Delivery, Jan. 2012, Volume:27,Issues:1, pp53-61.

13

FEMAG: A High Performance Parallel Finite Element Toolbox for Electromagnetic Computations

Tao Cui[1], Xue Jiang[2], Weiying Zheng[1, *]

[1]National Center for Mathematics and Interdisciplinary Sciences, State Key Laboratory of Scientific and Engineering Computing, Institute of Computational Mathematics and Scientific/Engineering Computing, Academy of Mathematics and Systems Science, Chinese Academy of Sciences, Beijing, China
[2]Department of Mathematics, Beijing University of Posts and Telecommunications, Beijing, China

Email address:

tcui@lsec.cc.ac.cn (Tao Cui), jxue@lsec.cc.ac.cn (Xue Jiang), zwy@lsec.cc.ac.cn (Weiying Zheng)

Abstract: This paper presents a parallel finite element toolbox for computing large electromagnetic devices on unstructured tetrahedral meshes, FEMAG—Fem for ElectroMagnetics on Adaptive Grids. The finite element toolbox deals with unstructured tetrahedral meshes and can solve electromagnetic eddy current problems in both frequency domain and time domain. It adopts high-order edge element methods and refines the mesh adaptively based on reliable and efficient finite element a posteriori error estimates. We demonstrate the competitive performance of FEMAG by extensive numerical experiments, including TEAM (Testing Electromagnetic Analysis Methods) Problem 21 and the simulation for a single-phase power transformer.

Keywords: FEMAG, Eddy Current Problem, Adaptive Finite Element Method, Parallel Computation, Large Electromagnetic Device

1. Introduction

Large devices in electric engineering usually have very complicated structures and are made of anisotropic and nonlinear materials. A large power transformer, for example, consists of nonmagnetic plates, magnetic oil tank, grain-oriented steel laminations, complex exciting coils and so on. It is extremely difficult to simulate the whole structure of a large power transformer. Commercial software is usually not competent for such a task because of memory limitation and low scalability for large computers. Moreover, the inefficiency of algebraic system solver also limits their applications to large-scale computing. The purpose of this paper is to propose a parallel finite element toolbox, FEMAG (Fem for ElectroMagnetics on Adaptive Grids), and to demonstrate the competitive performance of our finite element algorithms and the FEMAG. We propose to study the following eddy current problem:

$$\frac{\partial \mathbf{B}}{\partial t} + \nabla \times \mathbf{E} = 0 \quad \text{in } \Re^3, \quad \text{(Farady's law)}$$

$$\nabla \times \mathbf{H} = \mathbf{J} \quad \text{in } \Re^3, \quad \text{(Ampere's law)}$$

(1)

where $\mathbf{E}$ is the electric field, $\mathbf{B} = \mu \mathbf{H}$ is the magnetic flux, $\mathbf{H}$ is the magnetic field, and $\mathbf{J}$ is the current density defined by

$$\mathbf{J} = \begin{cases} \sigma \mathbf{E} & \text{in } \Omega_c, \quad \text{(conducting region)} \\ \mathbf{J}_s & \text{in } \Re^3 \setminus \Omega_c, \quad \text{(nonconducting region)} \end{cases}$$

(2)

Here σ is the electric conductivity, $\mathbf{J}_s$ is the source current density carried by some coils, and Ω_c denotes the conducting region. The eddy current problem is a quasi-static approximation of Maxwell's equations at very low frequency by neglecting the displacement currents in Ampere's law (see [1]). The material parameters could be very complicated for electric engineering applications. For example, grain-oriented steel laminations are widely used in iron cores and magnetic shields of large power transformers (see Figure 1 for a TEAM benchmark model) [2]. They are very anisotropic and have multiple scales. The transformer size could be 10^6 times the thickness of coating films over laminations. The magnetic permeability differs a lot in the rolling direction and the other two orthogonal directions.

Furthermore, the energy loss in coils should also be considered for complex exciting source. This necessitates huge number of elements in partitioning the domain. Full three-dimensional (3D) finite element simulation is very difficult due to extensive unknowns. With the evolution of the computer technology, the massively parallel computing with ten thousands of even hundred thousands of CPU cores becomes one of the trends of the scientific computing. The main challenge for simulating large transformers is focused on developing efficient algorithms for solving the discrete problem and scalable parallel finite element codes for large supercomputers.

The second issue concerns low regularity of the solution of Maxwell's equations. It is well-known that the solution of Maxwell's equations may have local singularities at corners and edges of the structure and material interfaces. Numerical methods based on uniform meshes are inefficient in resolving the local singularities. The adaptive finite element method (AFEM) based on the a posteriori error estimates has been successfully and widely applied in many other areas [3-6], which provides a systematic way to achieve the optimal computational complexity by refining the mesh according to the local a posteriori error estimator on the elements. Unfortunately, parallel implementation of the AFEM on distributed memory parallel computers is very difficult because of the complexities of the mesh management and load balance issues. Also, highly efficient numerical methods for solving the linear system resulting from finite element discretization are required. For facilitating implementing the AFEM, we have developed the toolbox PHG, *Parallel Hierarchical Grid* [7]. The motivation of this toolbox is to support the research on AFEM algorithms and development of AFEM codes. PHG deals with conforming tetrahedral meshes and uses bisection for adaptive mesh refinement and MPI for message passing. Using the idea of object oriented design, the details of complex mesh management and parallelism are hidden from users. PHG provides supports for adaptive finite element computations, such as finite element bases (including the Nédéléc edge elements for electromagnetic computations) [8], numerical quadrature, and basic operations with finite element functions. For building, assembling, and solving linear systems and eigenvalue problems resulting from finite element discretization, an unified linear algebra module for manipulating distributed sparse matrices stored in compressed sparse rows (CSR) and distributed vectors is provided,. Load balancing is achieved through mesh repartitioning and redistribution.

FEMAG is developed based on PHG. It solves both time-dependent and time-harmonic eddy current problems and can deal with nonlinear and anisotropic materials. For large-scale computing, the scalability of algebraic system solver plays the key role in parallel finite element codes. An efficient solver should possess two properties: (1) its convergence rate should keep quasi-uniform as the number of elements increases; (2) its convergence rate should not degenerate when using massive CPU cores. In FEMAG, the Maxwell solver for the discrete problem is a GMRES algorithm with the auxiliary space preconditioning [9]. And the auxiliary problems are solved by the conjugate gradient method with the Boomer algebraic multigrid preconditioning [10].

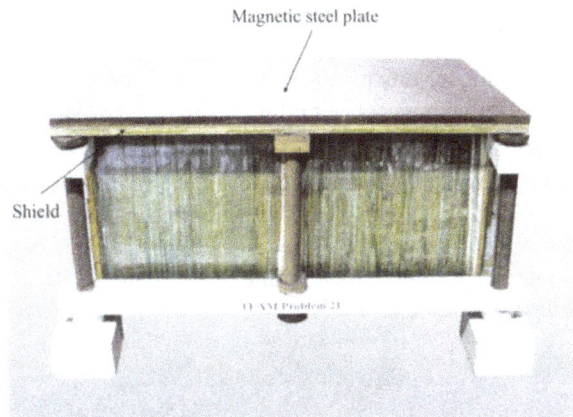

Figure 1. *A magnetic shield model (TEAM Problem 21^c-M1, The magnetic shield made of steel laminations).*

The most challenging task is to compute the iron loss and eddy current density in grain-oriented steel laminations. An iron core may consist of hundreds of even more than one thousand of laminations, each of which is only 0.18-0.35mm thick. But the thickness of the coating film over the laminations is only 2-5µm. Thus the lamination stack has multi-scale sizes and the ratio of the largest scale to the smallest scale can amount to 10^6 (see Figure 1). Full 3D finite element modeling is extremely difficult due to extensive unknowns from meshing both laminations and the coating film. There are very few papers on the computation of 3D eddy currents in the literature. In FEMAG, steel laminations are treated by two approaches:

1. Compute 3D eddy current density with an approximate eddy current model that omits the coating film and meshes each lamination respectively [11-12]. This approach yields accurate results but requires a fine mesh for the lamination stack.
2. Compute 3D eddy current density with effective or homogenized conductivity and permeability [13]. This approach is very fast but less accurate in computing iron loss in laminations than the previous one. But it is more favorable in the simulation of the whole of a power transformer.

In this paper, we present the numerical results for a power transformer by using homogenized conductivity and permeability.

In this paper, we present a numerical experiment on hp-adaptive finite element method for time-harmonic eddy current problem. The h-adaptive finite element method reduces the error by local mesh refinements. It has been well-studied in the a posteriori error analysis, mesh refinement algorithms, and optimal complexity. Using the idea of error equidistribution, the h-adaptive method based on a posteriori error estimates could yield a quasi-optimal approximation with algebraic convergence rate

$$\eta_h \approx C N_h^{-p/d},$$

where η_h is the a posteriori error estimate, d is the spatial dimension, p is the polynomial degree, and N_h is the number of degrees of freedom. However, due to the singularity of the solution, the quasi-optimality η_h will degenerate for higher-order finite elements since the constant C may blow up with increasing p.

The hp-adaptive finite element method reduces the error by both local mesh refinement and local improvement of polynomial degrees. It is more efficient than the pure h-adaptive or p-adaptive methods and could reduce the error exponentially. For example, the optimal convergence rate of the hp-adaptive method is

$$\eta_{hp} \approx C e^{-\delta N_{hp}^{1/3}},$$

for two dimensional elliptic problems [14], and is also conjectured to be

$$\eta_{hp} \approx C e^{-\delta N_{hp}^{1/5}},$$

for three dimensional elliptic problems [15], where C, δ are positive constants independent of h and p, η_{hp} is the a posteriori error estimate for the hp-adaptive method, and N_{hp} is the number of degrees of freedom. But for solutions with edge singularities, the meshes leading to the exponential convergence must be obtained by anisotropic refinements, that is, by using "needle elements" which are parallel to the edges (see [15] for more comments). The implementation of the hp-adaptive method is very challenging for higher dimensional problems. In this paper, we present an hp-type a posteriori error estimate for the time-harmonic eddy current model. Then based on the a posteriori error estimate, an hp-adaptive algorithm is proposed by the strategy of predicted error reduction. We implemented in FEMAG the parallel hp-adaptive method on unstructured tetrahedral meshes. The numerical experiment is performed on TEAM Problem 21^a-2. The hp-adaptive method shows exponential decay of the a posteriori error estimate and the numerical results agree well with experimental data.

The second numerical experiment is performed on TEAM Problem 21^b-MN which has a magnetic plate and a nonmagnetic plate. It shows that, on 12288 CPU cores, FEMAG has very good scalability for nonlinear eddy current problems. The last numerical experiment is performed on a homemade single-phase transformer. We use effective material parameters for iron core and iron yokes. The relative error of the iron loss is less than 10%.

2. The A-Formulation of Eddy Current Problems

We start by the $\mathbf{A}$-formulation of eddy current model for laminated conductors. Let $\mathbf{A}$ be the magnetic vector potential

satisfying $\nabla \times \mathbf{A} = \mathbf{B}$. Then the $\mathbf{A}$-formulation of eddy current model reads (cf. e.g. [11]):

$$\sigma \frac{\partial \mathbf{A}}{\partial t} + \nabla \times (\nu \nabla \times \mathbf{A}) = \mathbf{J}_s \quad \text{in} \quad \Omega ,$$

$$\mathbf{A} \times \mathbf{n} = 0 \quad \text{on} \quad \Gamma , \tag{3}$$

where Ω is the truncated domain with boundary Γ, $\mathbf{n}$ is the unit outer normal of Γ, and $\mathbf{J}_s$ stands for the excited current or source current satisfying $\nabla \cdot \mathbf{J}_s = 0$. In (1), σ is the conductivity, $\nu = \operatorname{diag}(H_1/B_1, H_2/B_2, H_3/B_3)$ is the nonlinear and anisotropic reluctivity, and $\mathbf{H} = (H_1, H_2, H_3)$ denotes the magnetic field and is determined by BH-curves and the magnetic flux $\mathbf{B} = (B_1, B_2, B_3)$.

Now we introduce some function spaces used in this paper:

$$\mathbf{L}^2(\Omega) = \{ \mathbf{u} : \int_\Omega |\mathbf{u}(x)|^2 \, dx < \infty \},$$

$$\mathbf{H}(\mathbf{curl}, \Omega) = \{ \mathbf{u} \in \mathbf{L}^2(\Omega) : \nabla \times \mathbf{u} \in \mathbf{L}^2(\Omega) \},$$

$$\mathbf{H}_0(\mathbf{curl}, \Omega) = \{ \mathbf{u} \in \mathbf{H}(\mathbf{curl}, \Omega) : \mathbf{u} \times \mathbf{n} = 0 \ \text{on} \ \partial\Omega \}.$$

The weak formulation of (1) reads: Find $\mathbf{A} \in \mathbf{H}_0(\mathbf{curl}, \Omega)$ such that, for all $\mathbf{v} \in \mathbf{H}_0(\mathbf{curl}, \Omega)$,

$$\int_\Omega \left[\sigma \frac{\partial \mathbf{A}}{\partial t} \cdot \mathbf{v} + \nu (\nabla \times \mathbf{A}) \cdot (\nabla \times \mathbf{v}) \right] = \int_\Omega \mathbf{J}_s \cdot \mathbf{v} \tag{4}$$

Let $0 = t_0 < t_1 < \cdots < t_N = T$ be a partition of the time interval $[0, T]$ and $\tau_n = t_n - t_{n-1}$ denote the n-th time step. Let $\Im_n$ be a tetrahedral triangulation of Ω at time t_n such that $\Im_n \big|_{\Omega_c}$ forms a tetrahedral mesh of Ω_c. We define the edge element space as follows

$$\mathbf{U}(\Im_n) = \{ \mathbf{v} \in \mathbf{H}_0(\mathbf{curl}, \Omega) : \mathbf{v} \big|_T \in P_k(T)^3 \ \text{for all} \ T \in \Im_n \}, \tag{5}$$

where $P_k(T)$ is the space of polynomials of order $k > 0$. The fully discrete approximation to (4) reads: Given $\mathbf{u}_0 = 0$, find $\mathbf{u}_n \in \mathbf{U}(\Im_n)$, $n \geq 1$ such that

$$a(\mathbf{u}_n, \mathbf{v}) = \int_\Omega \mathbf{J}_n \cdot \mathbf{v} \tag{6}$$

for all $\mathbf{v} \in \mathbf{U}(\Im_n)$, where

$$a(\mathbf{w}, \mathbf{v}) = \int_\Omega [\sigma \mathbf{w} \cdot \mathbf{v} + \tau_n \nu (\nabla \times \mathbf{w}) \cdot (\nabla \times \mathbf{v})], \qquad \mathbf{J}_n = \sigma \mathbf{u}_{n-1} + \int_{t_{n-1}}^{t_n} \mathbf{J}_s(t) \, dt.$$

3. Numerical Experiments for FEMAG

3.1. Scalability for Nonlinear Eddy Current Problem

In this subsection, we report the numerical experiment for the TEAM Problem 21^b[2]. The conducting region consists of a magnetic plate and a nonmagnetic plate (see Figure 2). The unit of dimensions in Figure 2 is millimeter. The thickness of the plate is 10mm. The conductivity is 6.484×10^6 Siemens/Metre. The height of each coil is 12mm and the radiuses of the inner arc and the outer arc at four corners are 10mm and 45mm respectively. The vertical distance between them is 24mm. The two coils carry the source currents of 3000 Ampere/Turn in opposite directions. The frequency of source currents is 50 Hertz.

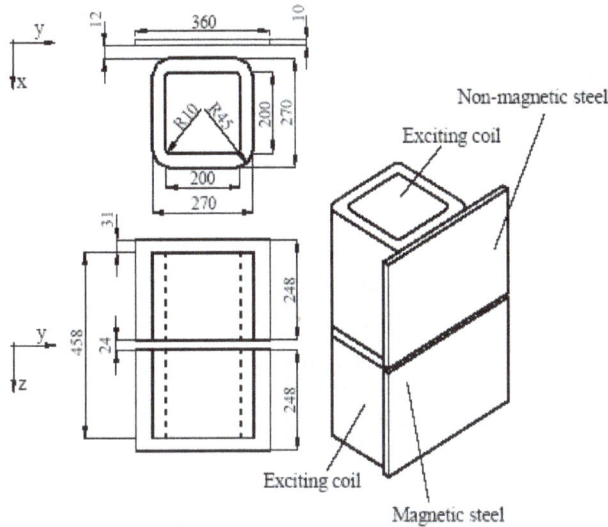

Figure 2. *Geometric illustration for TEAM Problem 21^b-MN.*

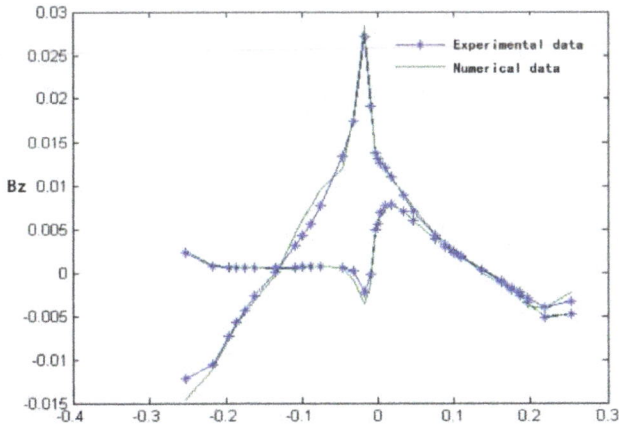

Figure 3. *Magnetic flux: numerical data versus experimental data.*

We carry out the computation on the super computer Tianhe-1A, Tianjin, China. We use the second-order edge element method of the second family [8] to solve the problem. The largest number of CPU cores used is 12288 and the finest mesh contains 0.44 billions of degrees of freedom. Table 1 shows that the weak scalability is larger than 70%. It also shows that, with 12288 CPU cores and 0.44 billions of unknowns, the solution time for the algebraic system is only 11 minutes. Table 2 shows the measured iron loss versus the calculated iron loss in the steel plates. They agree very well with each other. Figure 3 shows the experimental values and the numerical values of the magnetic flux density.

Table 1. *Weak scalability and computational time for solving the algebraic system.*

Number of CPU cores	Number of unknowns (million)	Computational time (S)	Weak scalability
768	29	501.0	100%
1,536	57	542.9	90%
3072	110	556.4	85%
6144	222	683.8	70%
12288	443	665.9	71%

Table 2. *Iron loss for TEAM Problem 21^b-MN.*

Calculated iron loss	Measured iron loss
7.005	7.03

3.2. Hp-Adaptive Finite Element Method for Time-Harmonic Eddy Current Problem

The second numerical experiment is to test the hp-adaptive finite element method for TEAM Problem 21^a-2 (see [2]). This problem consists of a non-magnetic steel plate with two slits and two racetrack shaped coils (see Figure 3). The steel plate has a conductivity of 1.3889×10^6 Siemens/Metre. The driving current for each coil is 3000 Ampere/Turn and has a frequency of 50 Hz. The driving currents for the two coils are in opposite directions. Since the material parameters are linear, we consider the time-harmonic eddy current problem

$$i\omega\sigma\mu\mathbf{A} + \nabla \times \nabla \times \mathbf{A} = \mu\mathbf{J}_s \quad \text{in} \quad \Omega \,,$$

$$\mathbf{A} \times \mathbf{n} = 0 \quad \text{on} \quad \Gamma \,,$$

where i is the imaginary unit, ω is the angular frequency, and μ is the magnetic permeability in the empty space.

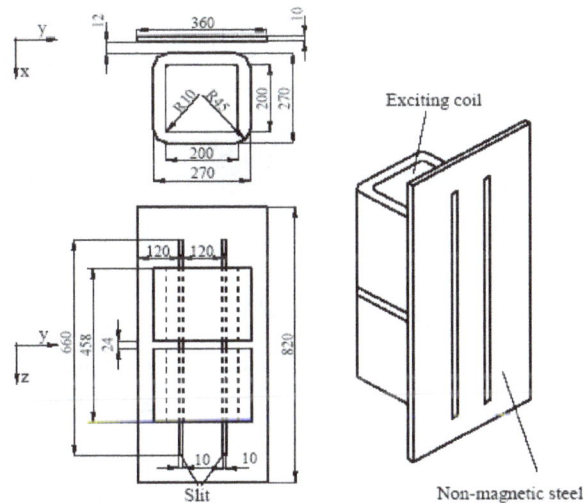

Figure 3. *Geometric illustration for TEAM Problem 21^a-2.*

Let $F(\Im_h)$ be the set of interior faces on the mesh $\Im_h$. For any face $F \in F(\Im_h)$ with $F = K_1 \cap K_2$, we denote the jump of a function v across F by $[v]_F = v|_{K_1} - v|_{K_2}$. For convenience in notation, we define the residual functions on each $K \in \mathfrak{S}_h$ and $F \in F(\mathfrak{S}_h)$ as follows

$$r_K = \mu \nabla \cdot (\mathbf{J}_s - i\omega\sigma\mathbf{A}), \qquad \mathbf{R}_K = \mu(\mathbf{J}_s - i\omega\sigma\mathbf{A}) - \nabla \times \nabla \times \mathbf{A}$$

$$j_F = \mu[(\mathbf{J}_s - i\omega\sigma\mathbf{A}) \cdot \mathbf{n}]_F, \quad \mathbf{J}_F = [(\nabla \times \mathbf{A}) \times \mathbf{n}]_F.$$

Then the error indicator is defined on each element as follows [16]

$$\eta_K = \frac{h_K}{p_K}\left(\|r_K\|_{L^2(K)} + \|\mathbf{R}_K\|_{\mathbf{L}^2(K)}\right) + \frac{1}{2}\sum_{F \in \partial K}\left(\frac{h_F}{p_F}\right)^{1/2}\left(\|j_F\|_{L^2(F)} + \|\mathbf{J}_F\|_{\mathbf{L}^2(F)}\right).$$

It is used to refine the mesh adaptively. The global and maximal a posteriori error estimates are defined by

$$\eta_{hp} = \left(\sum_{K \in \mathfrak{S}_h}\eta_K^2\right)^{1/2}, \qquad \eta_{max} = \max_{K \in \mathfrak{S}_h}\eta_K.$$

FEMAG uses the predicted error decrease strategy (PEDS) to refine the mesh. The algorithm assumes that the solution is locally smooth and the optimal convergence can be obtained by either local h-refinement or local p-refinement. Then on each element K marked for refinement, an error decrease factor λ_K is computed to judge which of the h-refinement and p-refinement should be performed. According to [16], λ_K can be defined heuristically as follows

$$\lambda_K = \left(\frac{p_T}{p_K}\right)^{p_T/2}\left(\frac{|K|}{|T|}\right)^{p_T/3},$$

where $T \supseteq K$ is the parent element of K. It may happen that

$T = K$ if the element is not refined at the previous adaptive iteration. The PDES is presented as follows:

Algorithm PDES. Given a tolerance $\varepsilon > 0$ and an initial mesh $\mathfrak{S}_0$ and an initial distribution of polynomial degrees P_0. Set $l = 0$ and $\theta_1 \in (0,1)$.

1. Solve for the finite element solution $\mathbf{u}_{hp} \in \mathbf{U}(\mathfrak{S}_l, P_l)$.

2. Compute the local error indicator η_K for all $K \in \mathfrak{S}_l$, the global error estimate η_l, and the maximal error estimate η_{max}.

3. While $\eta_l > \varepsilon$ do

   (1). Let $\hat{\mathfrak{S}}_l = \{K \in \mathfrak{S}_l : \eta_K > \theta_1\eta_{max}\}$ be the set of elements marked for refinement.

   (2). Refine $\hat{\mathfrak{S}}_l$ according to the predicted error decrease strategy, namely,

   (3). For any $K \in \hat{\mathfrak{S}}_l$, compute λ_K and η_T. If $\eta_K \leq \lambda_K\eta_T$, set $p_K \leftarrow p_K + 1$; otherwise, refine K using the bisection algorithm [17-18].

   (4). Solve the finite element solution $\mathbf{u}_{hp} \in \mathbf{U}(\mathfrak{S}_l, P_l)$ based on the new mesh and the new distribution of polynomial degrees.

   (5). Compute the local error indicator η_K for all $K \in \mathfrak{S}_l$, the global error estimate η_l, and the maximal error estimate η_{max}.

end while.

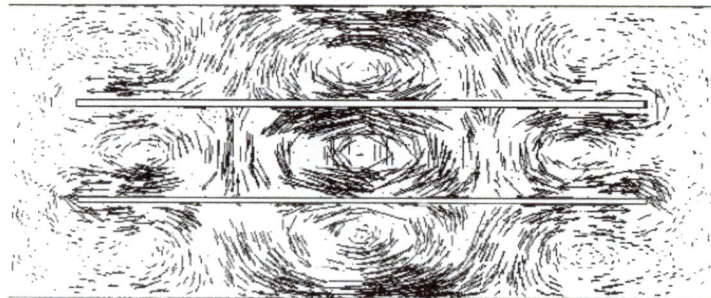

Figure 4. Eddy current distribution on a slice of the steel plate.

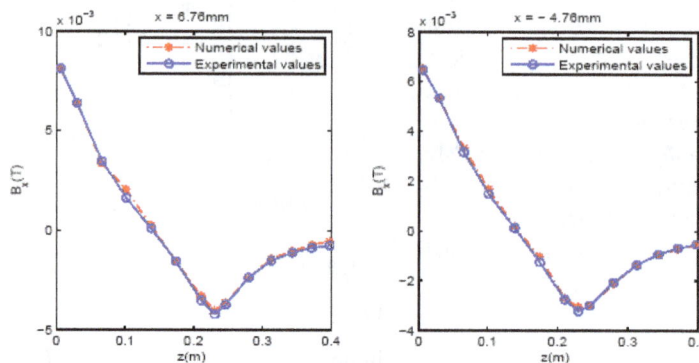

Figure 5. Values of the magnetic flux density on two lines: numerical data versus experimental data.

Figure 4 shows the eddy current distribution on a slice of the steel plate. The flow directions are clearly illustrated. Figure 5 shows the first component of the magnetic flux density B. The values are taken at some discrete points along two lines $\{(x,y,z): x = 0.00676, y = 0.0\}$ and $\{(x,y,z): x = -0.00476, y = 0.0\}$ which correspond to the two lines $\{(x,y,z): x = \pm 0.00576, y = 0.0\}$ in [2] respectively. The numerical data agree well with the experimental values. Figure 6 shows the reduction rates of the a posteriori error estimate by the h-adaptive method for p = 1, the PERS of the hp-adaptive method, and the Maximum Strategy of the hp-adaptive method used in [16]. We find that the hp-adaptive methods show great superiority over the h-adaptive method in reducing the error. It also shows that both of the algorithms yield exponential decay of the a posteriori error estimate:

$$\eta_{hp} \approx C e^{-\delta N_{hp}^{1/5}}.$$

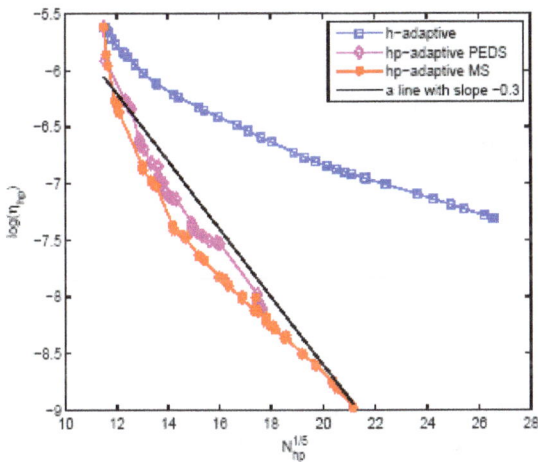

Figure 6. Decreasing rate of the a posteriori error estimate: hp-adaptive FEM versus h-adaptive FEM.

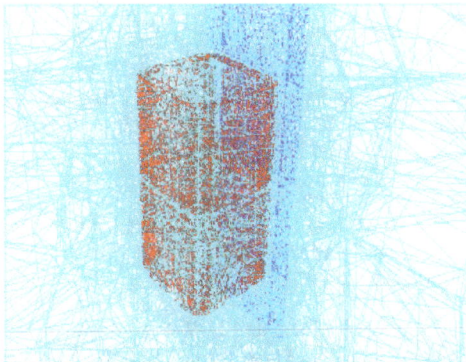

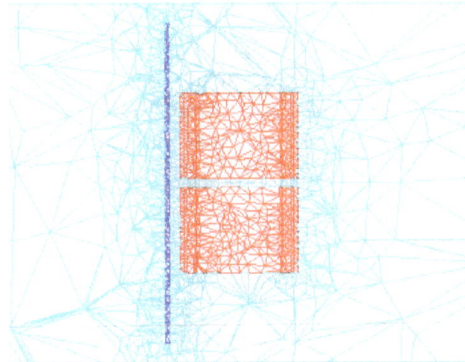

Figure 7 shows the hp-pair $(\Im_{47}, P_{47})$ by the PDES algorithm on the slice $\Sigma_1 = \{(x,y,z) \in \Omega : x = 0.0059999\}$, where the number of degrees of freedom is 1,729,148. From the figures, we find roughly that the hp-adaptive method use

1. lower order elements near the boundary where the error is very small,
2. lower order elements in the conducting domain where the solution varies rapidly, and
3. high order elements in the insulating region away from the boundary where the error is moderate.

Since the solution varies rapidly in the conducting domain, $\Im_{47}$ displays a fine mesh there in Figure 8.

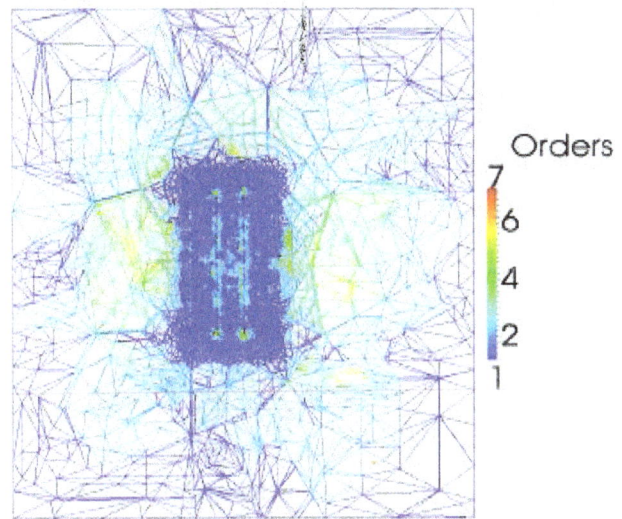

Figure 7. An adaptively refined mesh and the distribution of polynomial degrees on one slice after 47 refinements.

Figure 8. An adaptively refined mesh after 47 refinements.

3.3. Simulation for a Single-Phase Power Transformer

The last numerical experiment is carried out for a single-phase power transformer. The magnetic conductors are the wall of the oil tank, the iron core, and two iron yokes. Moreover, the model also contains nonmagnetic conductors and coils which carry exciting source currents (see Figure 9).

The computations are performed on the cluster LSEC-III of the State Key Laboratory on Scientific and Engineering Computing, Chinese Academy of Sciences.

We partition the computational domain into a tetrahedral mesh with 1,381,600 elements. The polynomial degree of the finite element space is set by $k = 2$. The total number of degrees of freedom is 13,147,086. We use 128 CPU cores in the computation. At each time step, the relative residual of the Newton method is reduced to be less than 0.0001. Table 3 shows the measured iron loss versus the calculated iron loss in the conductors. The relative error is about 10%. This demonstrates that FEMAG has the capability to simulate

complicated electromagnetic devices.

Table 3. Iron loss in the lamination and the magnetic palte (KW).

Calculated iron loss	Measured iron loss
4.10	4.55

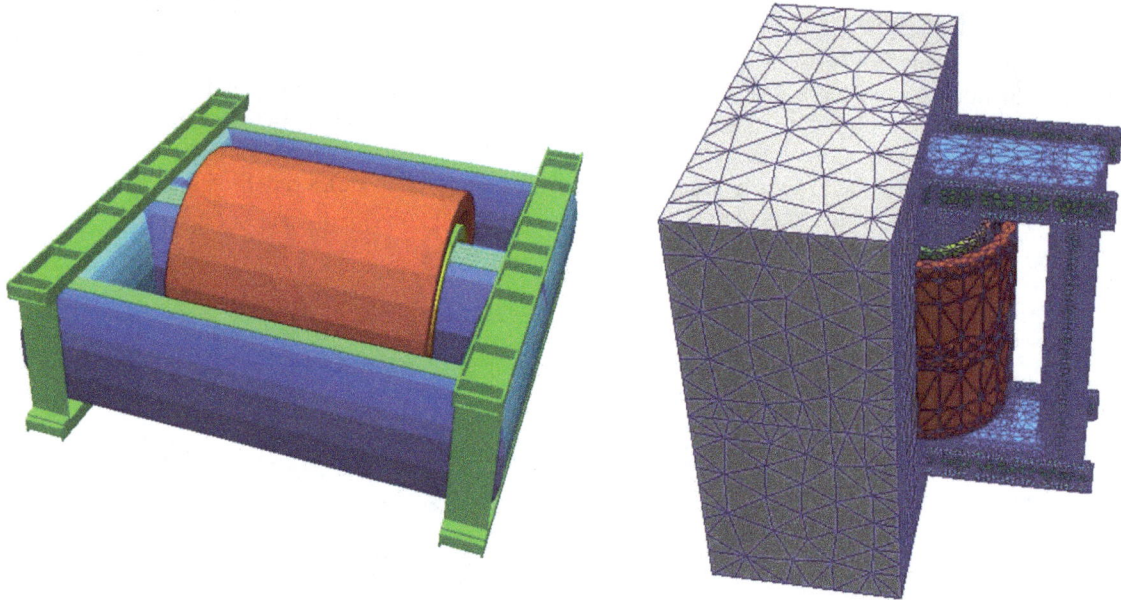

Figure 9. The geometric illustration of a single-phase transformer.

4. Concluding Remarks

In this paper, we report the parallel finite element toolbox, FEMAG (Fem for ElectroMagnetics on Adaptive Grids), for simulating large electromagnetic devices. Three numerical experiments are proposed to demonstrate the competitive performance of FEMAG, in terms of the scalability of parallel computation, the adaptivity for both mesh refinement and distribution adjustment of polynomial degrees, and the capability to deal with complex electromagnetic devices. All numerical results are verified by experimental data. The FEMAG is efficient for large-scale simulation of electromagnetic problems and has the potential in practical applications.

Acknowledgement

The authors would like to thank Professor Zhiguang Cheng of R & D Center, Baobian Electric Group, for his valuable discussions on the paper. The first author was supported in part by National 863 Project of China under the grant 2012AA01A309 and China NSF grant 11101417. The second author was supported in part by China NSF grant 11401040 and by the Fundamental Research Funds for the Central Universities 24820152015RC17. The third author was supported in part by China NSF grants 11031006 and 11171334, by the Funds for Creative Research Groups of China 11021101, and by the National Magnetic Confinement Fusion Science Program 2015GB110003.

References

[1]   H. Ammari, A. Buffa, and J. Nedelec, "A justification of eddy current model for the Maxwell equations", SIAM J. Appl. Math., 60 (2000), pp. 1805–1823.

[2]   Z. Cheng, N. Takahashi, and B. Forghani, "TEAM Problem 21 Family (V.2009)", approved by the International Compumag Society.   [Online]   Available: http://www.compumag.org/jsite/team.

[3]   I. Babuska and W. C. Rheinboldt, "Error estimates for adaptive finite element computations," *SIAM J. Numer. Anal.*, vol. 15, no. 4, pp. 736–754, Aug. 1978.

[4]   J. Chen, Z. Chen, T. Cui, and L.-B. Zhang, "An adaptive finite element method for the eddy current model with circuit/field couplings," *SIAM J. Sci. Comput.*, vol. 32, no. 2, pp. 1020–1042, Mar. 2010.

[5]   W. Zheng, Z. Chen, and L. Wang, "An adaptive finite element method for the H–ψ formulation of time-dependent eddy current problems", Numer. Math., 103 (2006), pp. 667–689.

[6]   Z. Chen, L. Wang, and W. Zheng, "An adaptive multilevel method for time–harmonic maxwell equations with singularities," *SIAM J. Sci. Comput.*, vol. 29, no. 1, pp. 118–138, Jan. 2007.

[7]   L.-B. Zhang, "A parallel algorithm for adaptive local refinement of tetrahedral meshes using bisection," *Numer. Math. Theory, Methods, Appl.*, vol. 2, no. 1, pp. 65–89, 2009.

[8]   J.N. Nédélec, "A new family of mixed finite elements in R^3", *Numer. Math.*, vol. 50, no. 1, pp. 57-81, 1986.

[9]   R. Hiptmair and J. Xu, Auxiliary Space Preconditioning for Edge Elements, *IEEE Trans. Magn.*, vol. 44, no.6, pp.938-941, 2008.

[10] V. E. Henson and U. M. Yang, "BoomerAMG: A parallel algebraic multigrid solver and preconditioner," *Appl. Numer. Math.*, vol. 41, no. 1, pp. 155–177, Apr. 2002.

[11] X. Jiang and W. Zheng, "An efficient eddy current model for nonlinear Maxwell equations with laminated conductors", SIAM J. Appl. Math., 72 (2012), pp. 1021–1040.

[12] W. Zheng and Z. Cheng, "An inner-constrained separation technique for 3D finite element modeling of GO silicon steel laminations", *IEEE Trans. Magn.*, 12 (2012), pp. 667–689.

[13] X. Jiang and W. Zheng, "Homogenization of quasi-static Maxwell's equations", Multiscale Modeling and Simulation: A SIAM Interdisciplinary Journal, 12(2014). pp. 152-180.

[14] W. Guo and I. Babuska, The h-p version of the finite element method. Part 2. General results and applications, Comput. Mech. 1 (1986) 203--226.

[15] I. Babuska, B. Andersson, B. Guo, J.M. Melenk, and H.S. Oh, Finite element method for solving problems with singular solutions, J. Comput. Appl. Math. 74 (1996) 51-70.

[16] X. Jiang, L. Zhang, and W. Zheng, Adaptive hp-finite element computations for time-harmonic Maxwell's equations, Comm. Comp. Phys., 13 (2013), pp. 559-582.

[17] W. Mitchell, *Optimal multilevel iterative methods for adaptive grids*, SIAM J. Sci. Stat. Comput, 13 (1992), pp. 146–167.

[18] I. Kossaczky´, A recursive approach to local mesh refinement in two and three dimensions, J. Comput. Appl. Math. 55 (1994) 275–288.

14

Loss Calculation and Optimization Design of High Frequency Transformer

Lin Li[*], **Keke Liu, Xiaoying Zhang, Ning Zhang**

Department of Electrical and Electronics Engineering, North China Electric Power University, Beijing, China

Email address:

lilin@ncepu.edu.cn (Lin Li)

Abstract: In this paper, the magnetization and loss properties are analyzed and compared of topical magnetic materials. The calculation methods are studied for the losses of the core and windings of the High Frequency Transformer (HFT). Based on the evaluation of the temperature increment, an optimization method is presented for the design of HFT. Finally, the losses of a test model of HFT is calculated and the results are compared with those tested.

Keywords: High Frequency Transformer, Magnetic Materials, Loss, Optimization

1. Introduction

In power electronic devices, such as DC/DC converter and solid state transformer, the operational frequencies of the transformers are chosen from several hundred Hz to several tens kHz in order to decrease their volumes and weights. This kind of transformers is called as High Frequency Transformer (HFT). The magnetization and loss mechanism of magnetic materials are related with the microscopic magnetization processes. In engineering application, the losses of soft magnetic material can usually be divided in three parts: eddy current loss, hysteresis loss and excess loss [1]. The Steinmetz formula is widely used to calculate the loss of magnetic core of transformer [2]. The losses of windings of high frequency transformer are related with the operation frequency, and the coefficient of alternative current resistance is used to reflect the skin effect and the proximity effect [3].

In this paper, the magnetization and loss properties are analyzed and compared of topical magnetic materials. The calculation methods of the losses of the core and windings of the HFT are analyzed. An optimization method is presented for the design of HFT. The losses of a test model of HFT is calculated and the results are compared with those tested.

2. Magnetization and Loss Properties of Topical Magnetic Materials

2.1. Silicon Steel Sheets

The superiorities of silicon steel sheets include: high saturation flux density, high stacking factor, high mechanical strength, good ductility, easy to cut, and low price, which let silicon steel sheets popularly used as the basic elements of lamination core of power transformer. But the eddy current and hysteresis losses of silicon steel sheets increase quickly with the increase of operation frequency, even if the sheets are ultrathin strips.

2.2. Amorphous Alloys

Amorphous alloys are also called as metallic glasses. When amorphous alloys are cooled from the liquid state and solidified as noncrystalline materials whose cooling rate is estimated to be in the range 10^5–10^6K/sec. The ribbons of amorphous alloys are usually a few millimeters wide, 25–35 nm thick, and meters to kilometers in length. The losses of amorphous alloys for unit volume are much lower than silicon steel sheets, especially in high frequencies. Therefore, amorphous alloys are usually chosen as the magnetic material for high frequency small capacity transformers and distribution power transformers. Maximum saturation magnetization is in the range 1.5–1.9 T, which is much larger than ferrites. Based on the design principle of transformer, rising the working flux density of the transformer core can decrease the turn number of the windings, and further the volume and weight of the high frequency transformer.

2.3. Nanocrystalline Alloys

A related class of nanocrystalline alloys is made by adding small amounts of Cu and Nb to an Fe–Si–B amorphous alloy. The most-studied composition is Fe74Si15B7Cu1Nb3. The Cu is believed to enhance nucleation of crystallites and the Nb to inhibit their growth. The saturation flux density of nanocrystalline alloys is high as amorphous alloys, and the high frequencies losses of nanocrystalline alloys for unit volume are lower than amorphous alloys. The price for unit mass of nanocrystalline alloys is so much higher than amorphous alloys that the material has not been applied in industry. In other hand, because the material is extremely brittle, it is difficult to manufacture magnetic core of nanocrystalline alloys with gaps.

2.4. Soft Ferrites

Ferrites have good properties of high frequencies losses. As a soft magnetic material, ferrites are widely used in small high frequency switching power supply, in which the operation frequency is from several kHz to several MHz. The saturation flux density of ferrites is lower as 0.2T so that the operation frequency chosen must be higher to make up the deficiency and lower the turn number of the windings of the transformer. Because ferrites are farinose, and unfavorable machining, the cores of ferrites are annular or closed UU shape without gaps.

2.5. Optimized Magnetic Material for High Frequency Transformer

The loss properties of silicon steel sheets, amorphous alloys, nanocrystalline alloys, and soft ferrites in 10 kHz and 50 kHz are shown in Fig.1. Loss properties and saturation flux densities of four magnetic core of silicon steel sheets, amorphous alloys, nanocrystalline alloys, and soft ferrites, are tested and listed in table 1. The saturation flux densities, losses for unit volume, machining properiesy and prices for unit mass are the main factors to choose magnetic material for high frequency transformers. From Fig.1 and table 1, it can be seen that the optimized magnetic materials are amorphous alloys or nanocrystalline alloys for high frequency transformer with several kHz operation frequency.

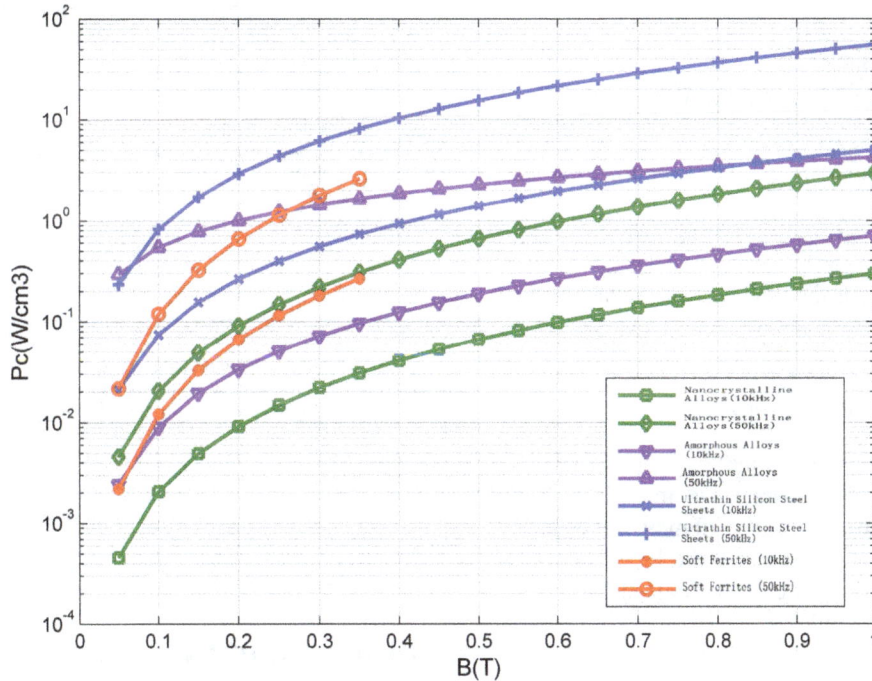

Figure 1. Loss properties of ultrathin silicon steel sheets, amorphous alloys, nanocrystalline alloys, and soft ferrites.

Table 1. Loss properties and saturation flux densities of four magnetic cores.

Magnetic material	Saturation flux densities (T)	Loss properties ([W/cm3])
Ferrites	0.35T	$P_c=0.1334B^{2.464}$[T]$f^{1.422}$[kHz]
amorphous alloys	1.2T	$P_c=0.0536B^{1.891}$[T]$f^{1.117}$[kHz]
nanocrystalline alloys	1.3T	$P_c=0.0111B^{2.161}$[T]$f^{1.428}$[kHz]
ultrathin silicon steel sheets	1.5T	$P_c=0.1593B^{1.827}$[T]$f^{1.496}$[kHz]

3. Losses of High Frequency Transformer

3.1. Loss of Magnetic Core

Usually, losses of magnetic core can be divided in three parts: eddy current loss, hysteresis loss and excess loss. The losses of magnetic core per unit mass can be calculated by [1]:

$$P_c = C_h B_m^\alpha f + C_e B_m^2 f^2 + C_{ex} B_m^{1.5} f^{1.5} \qquad (1)$$

where C_h, C_e, and C_{ex} are the coefficients of eddy current loss, hysteresis loss and excess loss, respectively. f is the operation frequency of high frequency transformer, and B_m is the maximum magnetic flux density of the core. α is an undetermined index, which is related with specified magnetic material. Another reduced core loss calculation formula is as follows [2]

$$P_c = C_w k B_m^{\alpha} f^{\beta} \qquad (2)$$

where C_w is the coefficients related with the waveform of the magnetic flux density, 1 for sinusoidal wave, $\pi/4$ for square wave, and 2/3 for triangle wave. k, α and β are undetermined indexes related with specified magnetic material. (2) is also called Steinmetz formula.

3.2. Losses of the Windings

The loss of winding of transformer can be calculated by:

$$P_w = k_r R_{DC} I_{rms}^2 \qquad (3)$$

where R_{DC} is the direct current resistance of the winding, and I_{rms} is the root mean square current. kr $(=R_{AC}/R_{DC})$ is the coefficient of alternative current resistance, which can be expressed as follows [3]:

$$k_r(X,m) = X[\frac{\sinh(2X)+\sin(2X)}{\cosh(2X)-\cos(2X)}$$
$$+\frac{2(m^2-1)}{3}\frac{\sinh(X)-\sin(X)}{\cosh(X)+\cos(X)}] \qquad (4)$$

where m is the number of layers of the winding, and $X = \dfrac{h_c}{0.071\sqrt{f}}$, in which h_c is the thickness of the conductor wires.

3.3. A Model of High Frequency Transformer

A model of high frequency transformer is made and shown in Fig.2, whose basic parameters are listed in table 2. The zero-load losses of the transformer for different frequencies are tested that can be used to represented the losses of core. The calculation formula of loss of core is obtained by means of the tested data, and it is as follows:

$$P_c = 0.00354 B_m^{1.75} f^{1.527} \qquad (5)$$

Table 2. Basic parameters of the model of high frequency transformer.

Magnetic material	Saturation flux densities (T)	Voltages of windings (V)	Frequenc y (Hz)	Capacity (kVA)
0.3mm silicon steel sheets	0.7	560/160	400	1.5

The calculated and test losses of core are listed in table 3. The calculated and test losses of windings are listed in table 4. It can be seen from table 3-4 that the errors are small between the calculated and the tested losses of the core and windings.

Figure 2. Model of high frequency transformer.

Table 3. Calculated and test losses of core.

f (Hz)	Calculated Pc (W)	Tested Pc (W)	Error(%)
50	0.740	0.7	5.71
60	0.935	0.9	3.89
100	1.900	2.0	-5.00
200	5.304	5.5	-3.56
300	10.470	10.35	1.15

Table 4. Calculated and test losses of windings.

f (Hz)	Calculated Pw (W)	Tested Pw (W)	Error(%)
50	32.5	31.05	4.67
60	32.5	31.05	4.67
100	32.8	31.25	4.96
200	33.3	32.68	1.90
300	35.3	35.10	0.57
400	38.2	38.13	0.18

4. Optimization Design of High Frequency Transformer

One of the optimization design objects of high frequency transformer is to determine the optimum magnetic flux density B_{opt} corresponding to the minimum total loss P_t of the transformer and the determined operation frequency. Let the derivative of the total loss P_t to the magnetic flux density B equal to zero, that is,

$$\frac{dP_t}{dB} = \frac{dP_c}{dB} + \frac{dP_w}{dB} = 0 \qquad (6)$$

We can obtained the optimum magnetic flux density B_{opt} as:

$$B_{opt}^{\beta+1} = [K_1 U_{in} \cdot 10^4 \cdot MLT \rho_1 I_1^2 + K_2 U_{out} \cdot 10^4 \cdot MLT \cdot \rho I_2^2] / 4.44 A_c f^{\alpha+1} k \beta V_c \quad (7)$$

where k, α and β are the indexes related with specified magnetic material and the same as that in (2). U_{in} and U_{out} are the voltages of the primary winding and the secondary winding, respectively. I_1 and I_2 are the currents of the primary winding and the secondary winding, respectively. V_c and A_c are the volume and across area of the magnetic core, respectively. K_1 and K_2 are the coefficient of alternative current resistance of the primary winding and the secondary winding, respectively. MLT is the average turn length of the windings and ρ is the resistivity of conductor of of the windings.

For one kind of amorphous alloys, the calculation formula of the loss of magnetic core is: P_c(W/cm^3)=$0.0306 B^{1.74} f(kHz)^{1.51}$. The optimization design results of a 10 kVA high frequency transformer are shown in table 5. For one kind of nanocrystalline alloys, the calculation formula of the loss of magnetic core is: P_c(W/cm^3)=$0.008 B^{1.982} f(kHz)^{1.621}$. The optimization design results of a 10 kVA high frequency transformer are shown in table 6.

5. Conclusions

Based on the study of the paper, we obtained the following conclusions:

(1) For the large capacity high frequency, the optimization magnetic materials are amorphous alloys and nanocrystalline alloys which have better magnetization and loss properties.

(2) The Steinmetz formula can be used to calculate the loss of core, and the method of coefficient of alternative current resistance to the losses of windings of HFT.

Table 5. Optimization design results of a 10 kVA amorphous alloys high frequency transformer.

f(kHz)	Bop(T)	Vc(cm3)	S/Vc	Pc(W)	Pcu(W)	Pt(W)
10	0.2939	259.9911	39.84205	30.5711	53.5436	84.1146
20	0.1574	241.6788	41.37723	27.3037	47.7399	75.0436
30	0.1106	229.5788	43.55803	25.8983	45.3952	71.2935
40	0.0951	203.5538	49.12706	26.3105	46.1546	72.4651
50	0.077	192.6796	51.89963	27.059	44.5896	71.6486
80	0.0472	232.6637	42.98049	26.1927	45.7791	71.9718
100	0.0375	250.1911	39.96945	26.4792	46.2359	73.705

Table 6. Optimization design results of a 10 kVA nanocrystalline alloys high frequency transformer.

f(kHz)	Bop(T)	Vc(cm3)	S/Vc	Pc(W)	Pcu(W)	Pt(W)
10	0.5633	129	77.51938	13.8365	27.6104	41.4469
20	0.3217	113.9947	87.7233	12.3767	24.6451	37.0218
30	0.2349	105.8522	94.4713	11.8885	23.5591	35.4476
50	0.1486	104.5664	95.6330	10.8533	21.8397	32.693
80	0.1011	104.5664	95.6330	10.8344	21.4794	32.3138
100	0.0799	112.2372	89.0970	10.4679	20.9212	31.3891

References

[1] Boglietti A, Cavagnino A, Lazzari M, et al. Predicting iron losses in soft magnetic materials with arbitrary voltage supply: an engineering approach [J]. IEEE Trans. Magn., 39(2): pp.981-989, March 2003.

[2] Bertotti G. General properties of power losses in soft ferromagnetic materials[J]. IEEE Trans. Magn., 24(1): pp621-630, January, 1988.

[3] P.L. Dowell. Effect of eddy currents in transformer windings[J]. Proc. Inst. Elect. Eng., 113, pp1387-1394, 1996.

[4] I. D. Mayergoyz, G. Friedman, and C. Salling, "Comparison of the classical and generalized Preisach hysteresis models with experiments," IEEE Trans. Magn., vol. 25, no. 5, pp. 3925–3927, Sep. 1989.

[5] E. Barbisio, F. Fiorillo, and C. Ragusa, "Predicting loss in magnetic steels under arbitrary induction waveform and with minor hysteresis loops," IEEE Trans. Magn., vol. 40, no. 4, pp. 1810–1819, Jul. 2004.

15

Research on Magnetostriction Property of Silicon Steel Sheets

Yanli Zhang, Yuandi Wang, Dianhai Zhang, Ziyan Ren, Dexin Xie

School of Electrical Engineering, Shenyang University of Technology, Shenyang, China

Email address:

zylhhjhyc_sy@163.com (Yanli Zhang), yuandiwang@sina.com (Yuandi Wang), zdh700@126.com (Dianhai Zhang), rzyhenan@163.com (Ziyan Ren), xiedx2010@163.com (Dexin Xie)

Abstract: When electrical devices such as electrical machines and transformers are in operation, the magnetostriction of the silicon steel sheets results in the deformation of the laminated iron core, which aggravates the vibration and noise of the iron core. Therefore, it is of significance to study the relationship between the magnetostrictive effect and external magnetic field. The aim of this paper is to report the researches finished, in progress, and to be done by the authors' group, so as to discuss the idea how to investigate the magnetostriction property of the silicon steel sheets, whether non-oriented or grain-oriented. Measurement method, property model formulation, and the application of the model to finite element analysis are described. The relative numerical results are given.

Keywords: Magnetostriction, Principal Strain, Anisotropic Property, Silicon Steel Sheet, Higher Harmonic Magnetic Field, DC Magnetic Bias, Finite Element Analysis

1. Introduction

Magnetostriction effect refers to the phenomenon that the size of ferromagnetic material at different directions elongates or contracts in supplied magnetic field, following the variation of the field. Silicon steel is a kind of typical ferromagnetic material. As a magnetic path silicon steel laminated iron core is used in electrical devices such as electrical machines, transformers, shunt reactors, etc. When these devices are in operation, the magnetostriction of the silicon steel sheets results in the deformation of the iron core, which aggravates its vibration and noise. The larger the product rating capacity is, the more obvious the effect is. In recent years, the capacity of power transformers and electric machines increases rapidly. As an example, the power transformers with their rating capacity and voltage up to 1000MVA and 1000kV have been put into operation. Although electrical devices have passed the delivery test standard in factory, the vibration and noise of some of the devices may excess the standard often in site. The reason is that the delivery test is carried out in standard test conditions, while certain complicated factors, involving DC magnetic bias, higher harmonics or overload do not be included in the test standard. Otherwise, these abnormal factors are difficult to be included in test conditions. Therefore, it is necessary to consider and analyze carefully the vibration and noise caused by magnetostriction of the silicon steel laminated core in the design process of the electrical devices with large scale, and to seize the mechanism and regularity of the magnetostriction property of the silicon steel sheet, thus to find the way of reducing the vibration and noise.

In fact, many papers have been published on decreasing the vibration and noise. However, it still needs to go deep into the research in this respect due to the complexity of the problem. The authors of this paper intend to report the researches finished, in progress, and will be done by the authors' group, to discuss the idea how to investigate the magnetostriction property of the silicon steel sheets, whether non-oriented or grain-oriented.

2. Measurement and Simulation of Magnetostriction Property of Silicon Steel Sheets Under Standard Magnetizing Forms

In general, research for magnetostriction property of the

silicon steel sheets includes property measurement, property model formulation, and coupling the property model with Finite Element (FE) analysis.

Tracing the researches of recent years, it is known that the magnetostriction property measurement of silicon steel sheets is developing from one dimensional (1D) to two dimensional (2D) measuring. The 1D measurement refers to the strain testing under alternating magnetization, i.e., the direction of magnetic field in one time period varies along a straight line, and only the strains at the magnetizing direction are measured, without regard to the intersection angle between the strain and magnetic flux density. For the measuring technique of anisotropic magnetostriction property, the research is just in initial stage. The 2D measurement means that the measuring is carried out under rotating magnetic field, which is supplied along the two directions perpendicular to each other, and the directions of the flux density can be controlled to be changed periodically. The 1D single sheet magnetostriction test system has been used extensively, and an international standard technical report, IEC/TR 62581, was issued at 2010 [1]. About the 2D measurement, reference [2] and [3] (Somkun S, Anthony J. Moses et al., 2010 and 2012) report the test technique and show related measuring results, while references [4-6] (M. Enokizono et al., 2011-1012) apply similar test method to measure the 2D magnetostriction property and research further the property under external stress. Constructing the 2D test system is a complicated task, so that the measurement and research are still in the stage of laboratory.

This section describes the authors' study on anisotropic magnetostriction property of silicon steel sheets with 1D magnetostriction test system firstly, then the numerical formulation of the model for the magnetostriction property and the relationship between the principal strain and the direction of magnetization using the triaxial measuring method [7-10], finally the primary 2D magnetostriction property measurement.

2.1. Study on Anisotropic Magnetostriction Property of Silicon Steel Sheets with 1D Measuring Equipment [8]

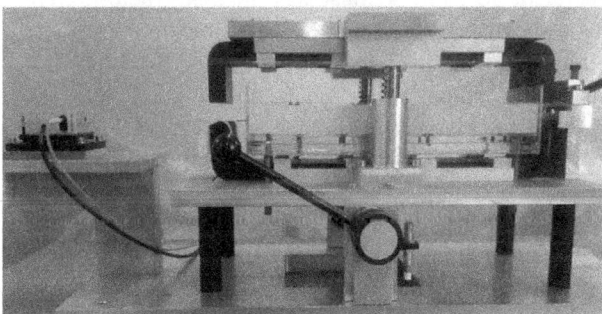

Figure 1. *Magnetostrictive measurement device.*

Figure 1 is a photo of traditional magnetostrictive measurement device MST500 for single silicon steel sheet specimen. The magnetostriction test unit consists of laser sender, receiver, and reflector. The laser sender emits light beam onto the reflector, and the reflector sends the beam back to the laser to test the infinitesimal displacement of the specimen. An external magnetic field is supplied along the longitudinal direction of the specimen. The advantages of the traditional device are as follows:

1) The laser sensor is with high resolution up to 10 nm/m.
2) It can provide the global magnetostrictive effect, while the traditional test by strain gauge only provides local deformation of the specimen.
3) The size of the specimen is a long strip of 500 mm×100 mm. Compared with that of 60 mm×60 mm in 2D magnetostriction test system, the influence of shear force in processing the specimen on the magnetostriction property can be reduced.

Considering that the test system can provide the magnetic field and test the strain only in the longitudinal direction, the specimens are cut in different angle, making their longitudinal direction at an angle with the rolling direction (RD) range from 0° to 90° with the interval of 15° to investigate the magnetostrictive property at different directions.

The alternating magnetic field of 50 Hz is created by the exciting winding, and the non-oriented silicon steel specimens are tested. The time-varying strain of the specimen is obtained with the laser sender and receiver device. Fig.2 (a) and (b) show the magnetostrictive strain curves along different magnetized directions in polar coordinate for the magnetic flux density of 0.6T and 1.4T respectively. It can be seen that the magnetostriction property of non-oriented silicon steel is also anisotropic, and the strain is related with not only the magnitude of magnetic flux density, but also its angle with RD.

To formulate the anisotropic magnetostrictive property, the conventional method uses the characteristic curves in RD and the transverse direction (TD), that is, expresses the linear magnetic pressure equation,

$$\lambda = k\mathbf{B} \tag{1}$$

As,

$$\begin{bmatrix} \lambda_x \\ \lambda_y \end{bmatrix} = \begin{bmatrix} k_x & 0 \\ 0 & k_y \end{bmatrix} \begin{bmatrix} B_x \\ B_y \end{bmatrix} \tag{2}$$

further. This formulation could be named orthogonal magnetostriction model. That means the magnetostriction property at arbitrary direction can be obtained by simplified computation using the property at RD and TD. At present, some commercial software is just based on this model to describe magnetostriction property of material.

However, the testing results shown in Fig.2 indicate that the orthogonal model is too simple to express the magnetostriction property at arbitrary direction. To improve the model, corresponding to the test method described in this section, the formulation of magnetostriction property of silicon steel sheet at arbitrary direction can be given as

$$\begin{cases} |\lambda| = f(\theta_B, B_{\max}) \\ \theta_\lambda = \theta_B \end{cases} \tag{3}$$

where the magnitude of the strain vector $|\lambda|$ is the peak-to-peak value of the magnetostriction λ_{pp}, and

$$\lambda_{pp} = \lambda_{max}^+ - \lambda_{max}^- \tag{4}$$

in which λ_{max}^+ and λ_{max}^- are the maximal elongation and contraction strain respectively; θ_λ and θ_B are the angles of the strain vector and magnetic flux density vector with the RD respectively.

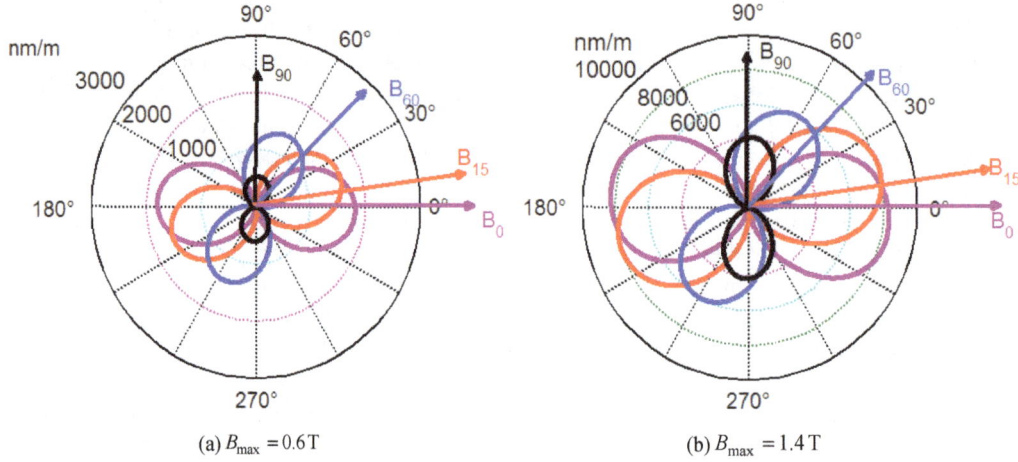

(a) $B_{max} = 0.6\,\text{T}$ (b) $B_{max} = 1.4\,\text{T}$

Figure 2. *Magnetostrictive strain curves along different magnetized directions in polar coordinate.*

2.2. Measurement of Principal Strain and Improvement of Magnetostriction Model [10]

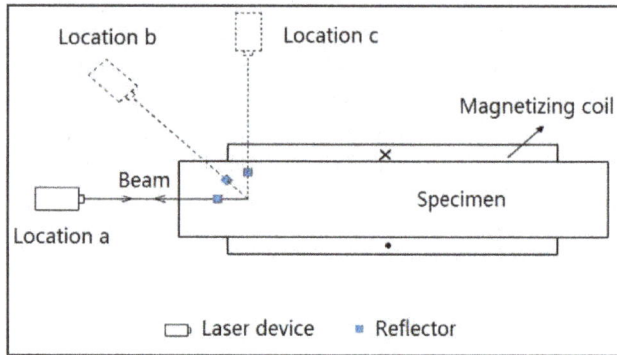

Figure 3. *Illustration of measurement device for triaxial laser method(Top view).*

It is known that from equation (3) the angles of strain and magnetic flux density, θ_λ and θ_B, are equal to each other. This is restricted by the test method described in Section 2.1, because the obtained magnetostriction is just the strain at the direction of magnetization, and not the principal strain. To acquire the magnetostriction principal strain of silicon steel sheet the "triaxial laser method" is used here, referring the triaxial strain gauge test technique in 2D magnetostriction measuring system. The method can be described as follows. At first, set the three test directions, making the angle between the directions and the longitudinal direction of the specimen as $0°$, $45°$, and $90°$, as shown with the location a, location b and location c in Fig. 3. Then put the laser and reflector on the three directions in turn, and measure the strains at each direction in one magnetizing period. After that, based on the knowledge of plane strain in material mechanics, substitute the measured strains λ_a, λ_b, and λ_c into (5) given as

$$\begin{bmatrix} \lambda_x \\ \lambda_y \\ \gamma_{xy} \end{bmatrix} = \begin{bmatrix} \cos^2\theta_a & \sin^2\theta_a & \sin\theta_a\cos\theta_a \\ \cos^2\theta_b & \sin^2\theta_b & \sin\theta_b\cos\theta_b \\ \cos^2\theta_c & \sin^2\theta_c & \sin\theta_c\cos\theta_c \end{bmatrix}^{-1} \cdot \begin{bmatrix} \lambda_a \\ \lambda_b \\ \lambda_c \end{bmatrix} \tag{5}$$

to calculate the orthogonal line strains λ_x, λ_y and a shear strain γ_{xy}. Finally, calculate the maximal principal strain λ and the direction θ_λ at which the λ occurs according to (6) and (7) given as

$$\lambda = \frac{\lambda_x + \lambda_y}{2} \pm \sqrt{(\frac{\lambda_x - \lambda_y}{2})^2 + (\frac{\gamma_{xy}}{2})^2} \tag{6}$$

$$\theta_\lambda = \frac{1}{2}\tan^{-1}(\frac{\gamma_{xy}}{\lambda_x - \lambda_y}) \tag{7}$$

The specimens are still cut at different angles making the longitudinal directions with the RD range from $0°$ to $90°$ with the interval of $15°$. Fig. 4 illustrates the 3 magnetostrictive loops of a specimen of $45°$ at the 3 test directions when the alternating magnetic field varying from 0.6T to 1.6T. The loops also called butterfly diagram. Contrasting the Fig. 4 (a), (b) and (c), it can be seen that the magnetostrictive pattern of the specimens at different directions under alternating magnetization is quite different.

The variation of the principal strain magnitude and direction along with the varying of the supplied magnetic field can be calculated using the test data of Fig. 4 combined with (6) and (7). Fig. 5 describes the magnetostriction principal strain magnitude and direction versus the magnetic flux density. Compared with Fig. 4, it can be seen that the magnetostriction is the largest one at the direction of principal strain, and the direction will change along with the variation of the magnetization. These results are consistent with theoretical analysis.

The magnitude and direction of the principal strain versus that of magnetic flux density are given in Fig. 6, from which we can see that the principal strain increases gradually along with the magnetic flux density going up, and when the flux density reaches to 1.6T the magnetostriction tends to saturation. It is also seen that from Fig. 6(b) for a certain direction of magnetization, the direction of principal strain θ_λ changes with the increase of magnetic flux density. The curve of θ_λ is relative gentle near θ_B of 0°and 90°, where principal strain direction is near the magnetostriction direction. That means when the silicon steel sheet is magnetized at the RD or TD, the maximal magnetostriction deformation occurs just at the magnetized direction, while the deformation deviates the magnetization direction when θ_B equals other angles. Therefore, the magnetostriction deformation at the magnetization direction given by using the conventional method is not the maximal deformation, which cannot truly reflect the magnetostriction effect of silicon steel sheets.

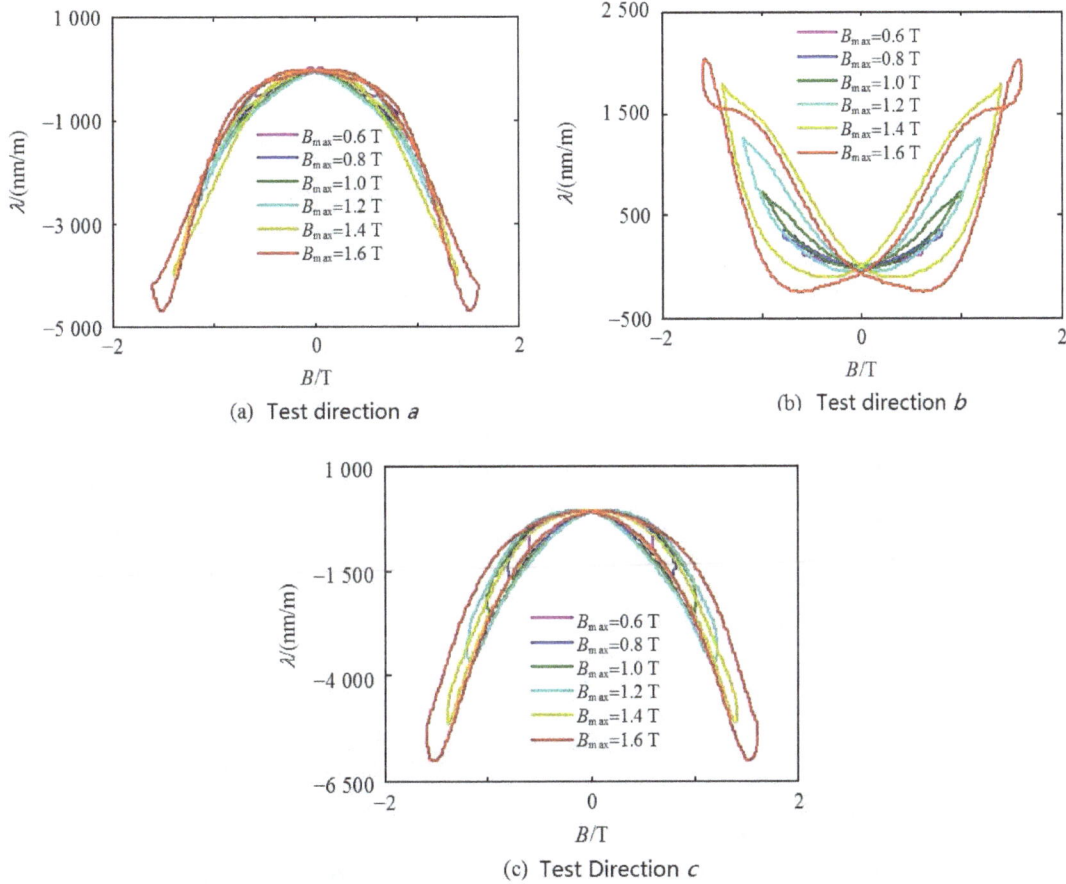

(a) Test direction a

(b) Test direction b

(c) Test Direction c

Figure 4. *Magnetostrictive loops along three detections.*

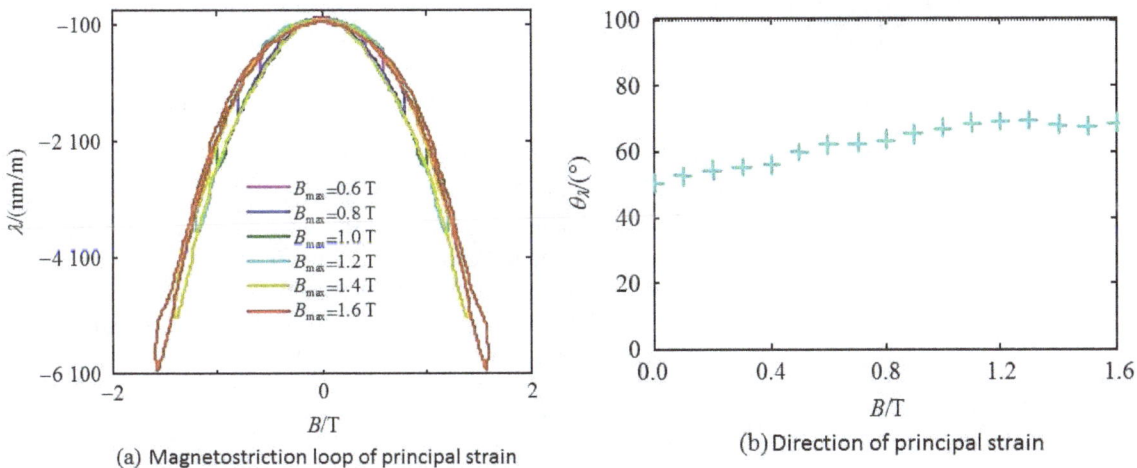

(a) Magnetostriction loop of principal strain

(b) Direction of principal strain

Figure 5. *Principal strain of magnetostriction.*

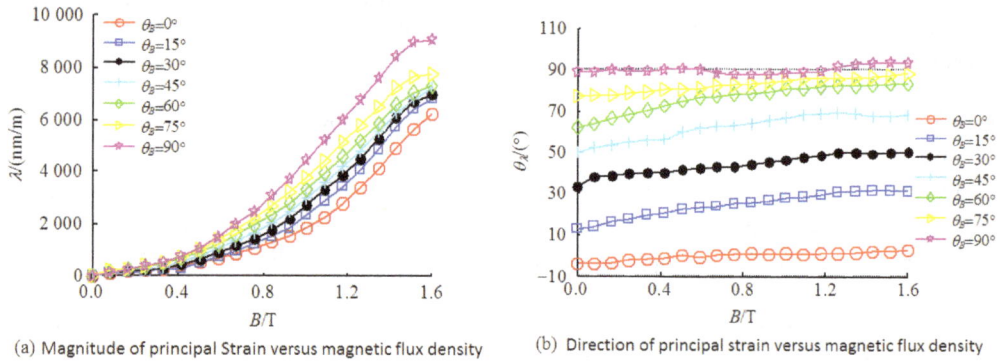

(a) Magnitude of principal Strain versus magnetic flux density   (b) Direction of principal strain versus magnetic flux density

Figure 6. Relationship curve between principal strain and magnetic flux density.

To sum up, the magnetostriction property of silicon steel sheets is vector-anisotropic, and the principal strain vector relates to not only the magnitude, but also the direction of magnetic field where the sheet locates in. Therefore, the vectorial magnetostriction property of silicon steel sheets expressed by principal strain has to be given as

$$\begin{cases} |\boldsymbol{\lambda}| = f(|\boldsymbol{B}|, \theta_B) \\ \theta_\lambda = f(|\boldsymbol{B}|, \theta_B) \end{cases} \tag{8}$$

Equation (8) is an improved formulation to (3), where $|\boldsymbol{\lambda}|$ is the magnitude of principal strain, which is the function of magnetic flux density and its direction angle θ_B; same as its magnitude, the direction angle of the principal strain θ_λ is also the function of magnetic flux density vector.

To incorporate the magnetostriction property model into FE analysis, the concrete computation steps are sketched now as follows:

- According to the relationship of principal strain vector and magnetic flux density vector described in (8), a three dimensional curved surface to depict the relation of the two vectors is obtained by using the B-spline function interpolation of measured data of Fig. 6, as illustrated in Fig. 7.

- Calculate the magnetic flux density, $|\boldsymbol{B}|$ and θ_B of each element by electromagnetic field FE analysis software, then find the magnetostriction principal strain of each element, $|\boldsymbol{\lambda}|$ and θ_λ based on the curve surface of Fig. 7 with interpolation. Finally, the distribution of the principal strain in whole computing region is obtained.

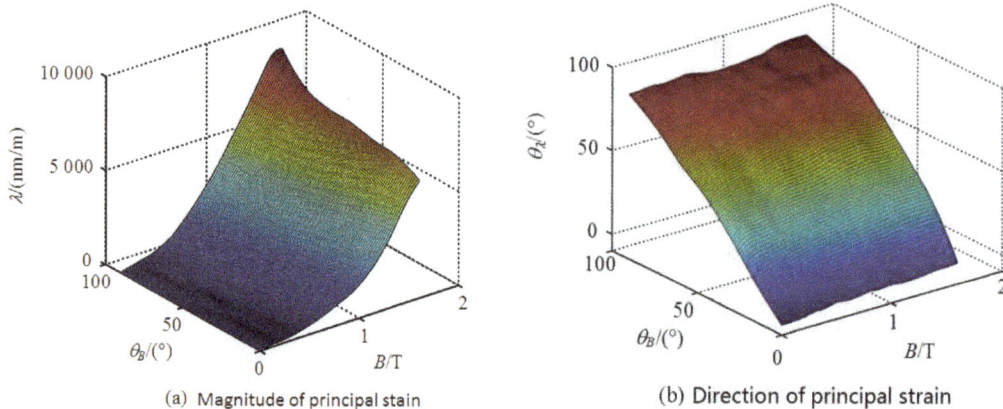

(a) Magnitude of principal stain   (b) Direction of principal strain

Figure 7. Curve surface of vector relationship between principal strain and magnetic flux density.

It is noted that the computational method above described is an indirect coupling one, and because that the deformation caused by the magnetostriction is too small to effect the whole structure of the analysed device, iterative calculation does not be needed.

2.3. Measurement of 2D Magnetostriction Property of Silicon Steel Sheets in the Condition of Rotational Magnetization

The measurement of magnetostriction described in Section 2.1 and 2.2 is carried out under alternating magnetization. However, in practical operation of electrical machines and transformers there exist local rotating magnetic fields, i.e., the situation that the local magnetic flux density vector changes its direction continually in a period. Therefore, the 2D magnetostriction property of silicon steel sheets has to be studied. This is the research we are doing now, but from the initial test results some enlightenment could be also gained.

Figure 8. 2D Magnetic field generator.

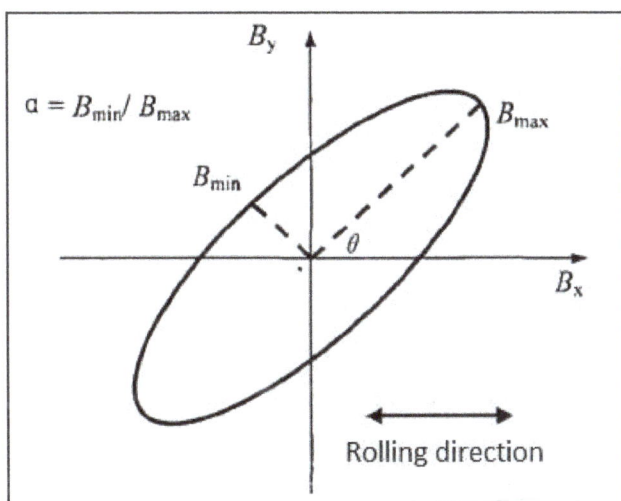

Figure 10. Structure sketch map of strain gauge.

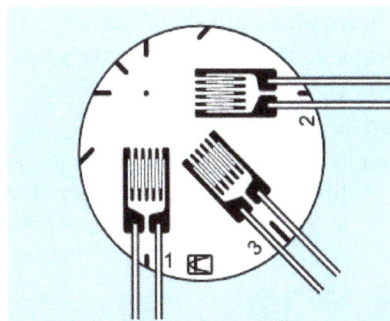

Figure 9. The ellipse created by the rotation of B vector in a time period.

The magnetostriction property of non-oriented silicon steel sheets with different B_{max}, θ, and α is measured by the author's group. Fig. 11 shows the loci of principal strain and magnetic flux density B in a time period for B_{max}=1.0 T, $\alpha = 0.25$ and θ equals $15°$, $45°$ and $75°$ respectively, where the black dotted line stands for the ellipse locus of B, red solid line for the locus of principal strain, blue arrow is for the long axis of the ellipse, and black arrow for the direction of the maximum principal strain. From the figure it can be seen that there are an intersection between the two arrows. A lot of tests indicate that the maximal principal strain vector is the function of the magnetic flux density magnitude, angle, and the ratio of ellipse long axis to minor axis. Therefore, more variables are introduced for the 2D magnetostrictive property in the condition of rotating magnetization. The measuring technique and model formulation are more complicated compared with the case of 1D magnetostrictive property. The further research is now under way.

3. Measurement and Simulation of Magnetostriction Property of Silicon Steel Sheets Under Abnormal Magnetizing Forms

3.1. Influence of Magnetic Field Harmonic Components on Magnetostriction and Noise [11]

There exist higher harmonic components in exciting voltage of electrical machines and transformers in practical operation due to the nonlinear loads in power system. The existence of the harmonics not only increase the additional losses, but also change the magnetostriction property of silicon steel sheets and intensify the vibration and noise of iron core. Therefore, it is significant to investigate the magnetostriction property of silicon steel sheets under the magnetic field with higher harmonics.

The authors' group measured the magnetostrictive waveforms of non-oriented grain electrical steel sheet by means of controlling the magnitude and phase angle of the third harmonic of magnetic field. Considering the third harmonic is the major higher harmonic in the resultant magnetic field, the magnetic flux density is expressed as,

Fig. 8 is the magnetic field generator for 2D magnetostriction property measurement. Its main magnetic circuit is the same as that of 2D magnetic property (B versus H) tester, which is used by some researchers in recent years. In the two exciting windings which are perpendicular to each other, the sinusoidal time-varying electric currents are controlled to produce a rotating magnetic field. The vertex locus of magnetic flux density vector in a period is an ellipse, see Fig. 9. Now express the components of the magnetic flux density, B_x and B_y as

$$\begin{cases} B_x = B_{max} \cos\theta \cos\tau - \alpha B_{max} \sin\theta \sin\tau \\ B_y = B_{max} \sin\theta \cos\tau + \alpha B_{max} \cos\theta \sin\tau \end{cases} \quad (9)$$

where B_{max} is the length of the ellipse long axis, θ is the angle between the long axis and RD of the specimen, α is the ratio of the long axis to the minor axis, $\tau = \omega t$. The ellipse locus is controlled by changing the three parameters B_{max}, θ, and α. The silicon steel sheet of 60mm×60mm locates in the center of the main magnetic path. The triaxial strain gauge test method described in Section 2.2 is still used for measuring the strain in the directions of $0°$, $45°$ and $90°$ (see Fig. 3). The principal strain is calculated according to (6) and (7). Fig. 10 illustrates the sketch map of the strain gauges.

$$B = B_1 \sin\omega t + B_3 \sin(3\omega t + \theta_3) \quad (10)$$

where B_1 and B_3 are the amplitude of fundamental and the third harmonic magnetic flux density, θ_3 is the phase angle of the third harmonic, $\omega = 2\pi f$, $f = 50Hz$. Different supplied magnetic field can be obtained by changing B_1, B_3 and θ_3. The measuring results indicate that in addition to the second harmonic (f = 100Hz), which account for the major

component, the fourth ($f = 200Hz$) and the sixth ($f = 300Hz$) harmonic also occupy a certain proportion of the magnetostriction, so that the influence of the harmonic magnetic field on the fourth and the sixth harmonic of magnetostriction is discussed.

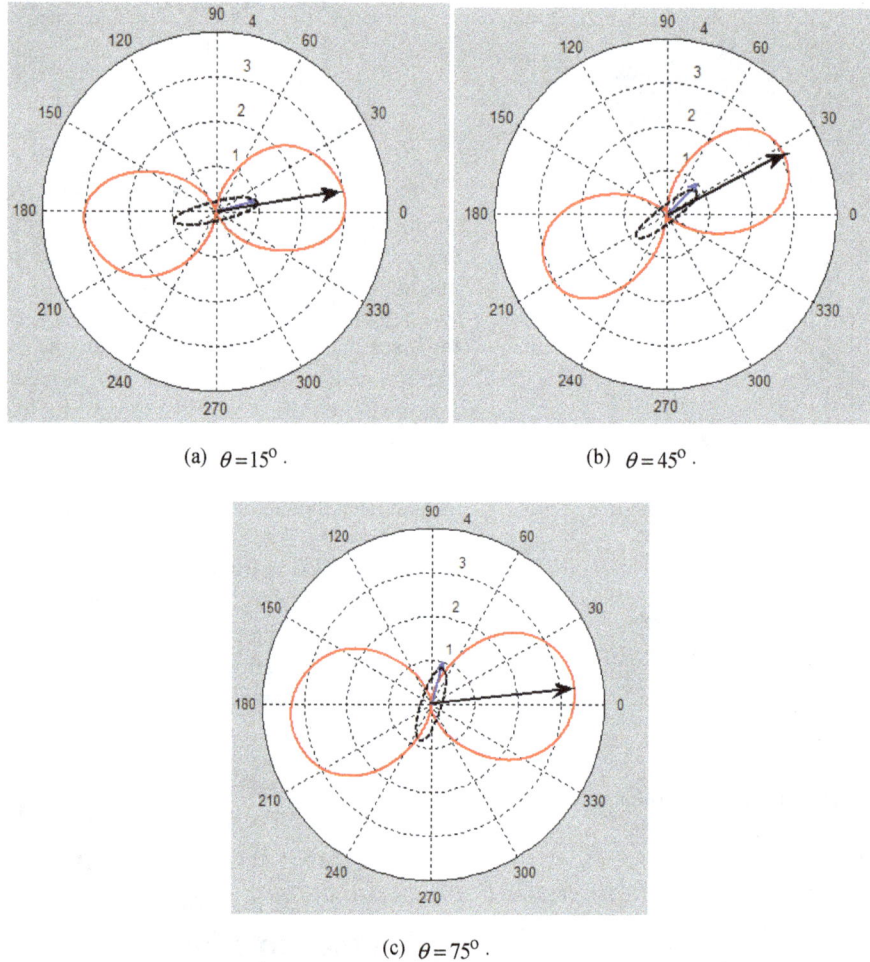

(a) $\theta = 15^{\circ}$.

(b) $\theta = 45^{\circ}$.

(c) $\theta = 75^{\circ}$.

Figure 11. *Loci of magnetostriction principal strain* $\lambda(\mu m/m)$ *and magnetic flux density.*

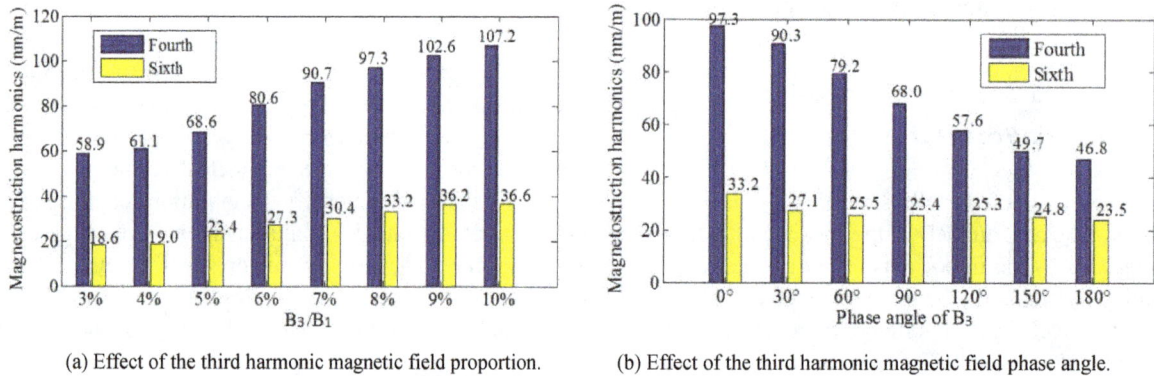

(a) Effect of the third harmonic magnetic field proportion.

(b) Effect of the third harmonic magnetic field phase angle.

Figure 12. *Variation of magnetostrictive harmonic components with the increase of resultant magnetic flux when B = 1.0T.*

The variation of magnetostrictive harmonic components with the increase of the resultant magnetic flux when B = 1.0T is investigated. The results are illustrated in Fig. 12. It is seen that along with the increase of the third harmonic component

of magnetic field the magnetostrictive harmonic components also increase obviously, while the increase of the third harmonic phase angle of the magnetic field makes the magnetostrictive harmonic components reduced, especially for the fourth harmonic of magnetostriction. The measured relation curve surface is shown in Fig. 13.

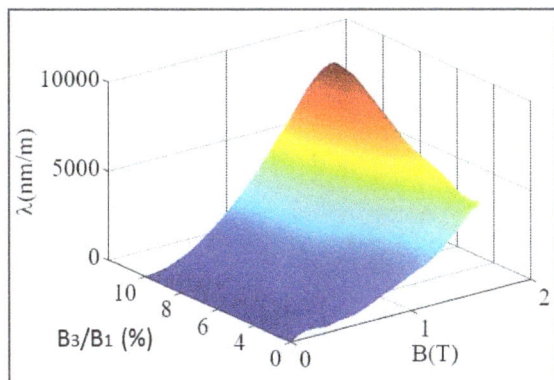

Figure 13. *Relationship between magnetostriction and magnetic flux density with the 3rd harmonic.*

To find the noise due to the magnetostraction in the magnetic field with the third harmonic included, firstly solve the electromagnetic field in electrical device with a FE analysis software and find the waveform of element magnetic flux density, then carry out the Fourier analysis and reserve the amplitude of the fundamental and the third harmonic of element magnetic flux densities. After that, get the magnetostriction principal strain distribution under the resultant magnetic field by interpolating the relation curve surface of Fig. 13. Finally the expression of the so-called A-weighted velocity level (AWV) [1]

$$L_{VA} = 20\log_{10} \frac{\rho c \sqrt{\sum_i \left[(2\pi f)_i \cdot \left(\lambda_i / \sqrt{2}\right) \cdot \alpha_i\right]^2}}{p_{e0}} \quad (10)$$

is used to calculate the decibel (DB) value L_{VA} of the noise, where ρ is the air density at room temperature, f is the fundamental frequency, λ_i is the i-th component of the principal strain, α_i is the corresponding AWV coefficient, p_{e0} is the minimal audible pressure. Figure 14 shows the noise caused by magnetostriction under the magnetic field with higher harmonic, it can be applied to get the distribution of noise in the electrical device by interpolation.

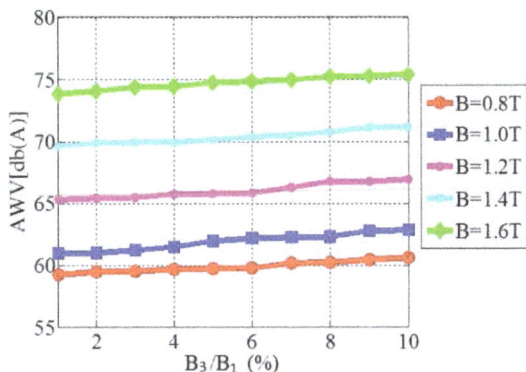

(a) Effect of the third harmonic magnetic field proportion.

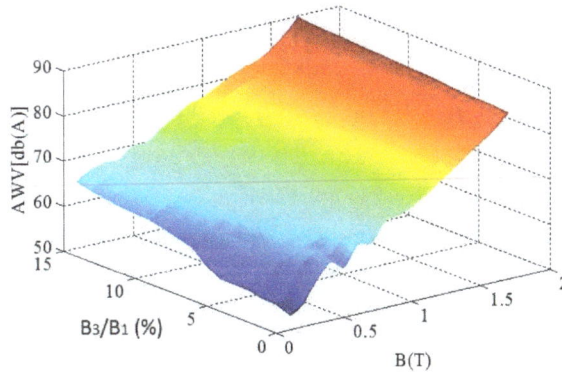

(b) Curve surface of noise versus harmonic magnetic field proportion.

Figure 14. *Noise caused by magnetostriction under higher harmonics of magnetic field.*

3.2. Influence of DC Bias Magnetic Field on Magnetostriction Property [12]

The external magnetic field may cover DC component when power transformers are in operation. It is necessary to identify how different magnetization patterns contribute to the magnetostriction noise of the iron cores. To measure the effect of the DC bias magnetic field on magnetostriction and noise, a DC component is superposed in the electric current of the exciting winding to create a DC bias magnetic field along the rolling or transverse direction of a grain-oriented silicon steel sheet. Using the method described in Section 2, the relationship of magnetostriction and magnetic field can be measured. Fig. 15 shows the waveforms of magnetic flux density and principal strain in a magnetization period along RD and TD when DC biased magnetic field intensity $H = 30$ A/m for B = 0.7, 0.9, 1.2 and 1.4T. It is worth noting that the waveform of magnetic flux density B is still sinusoidal time-varying without the DC component, shown as in Fig. 15(a) although the DC magnetic component exists, because

that the B-testing coil is winded in the test specimen and the DC magnetic field component cannot induce voltage in the secondary coil. From Fig. 15(b) it can be seen that the principal strain of RD and TD increases obviously along with the increase of the magnetization. To illustrate the effect of DC bias on principal strain further, the comparison of principal strain of a GO silicon steel sheet with alternating B of 1.4T under different DC biased magnetic field $H = 0, 10, 20$, and 30 A/m along RD. It is seen that compared the situation without the DC component, the waveform of principal strain λ with the DC bias is distorted and its waveform in the first half of period is different from that in the second half of period any more. That means the waveform of the λ includes higher harmonic components in addition to the basic component of f = 100 Hz. Furthermore, with the increase of biased field supplied, the contractive principal strain along the RD in Fig. 16 (a) increases slightly, while the elongation strain along the TD in Fig. 16 (b) is obviously bigger than that without DC biased field. More studies to the measured data shown in Fig. 16 indicate that the presence of a DC biased magnetic field not only increases the amplitude of the magnetostrictive strain, but also changes the property of the principal strain from

contractive characteristic to elongate one.

(a) B versus t.

(b) Principal strain along the RD.

(c) Principal strain along the TD.

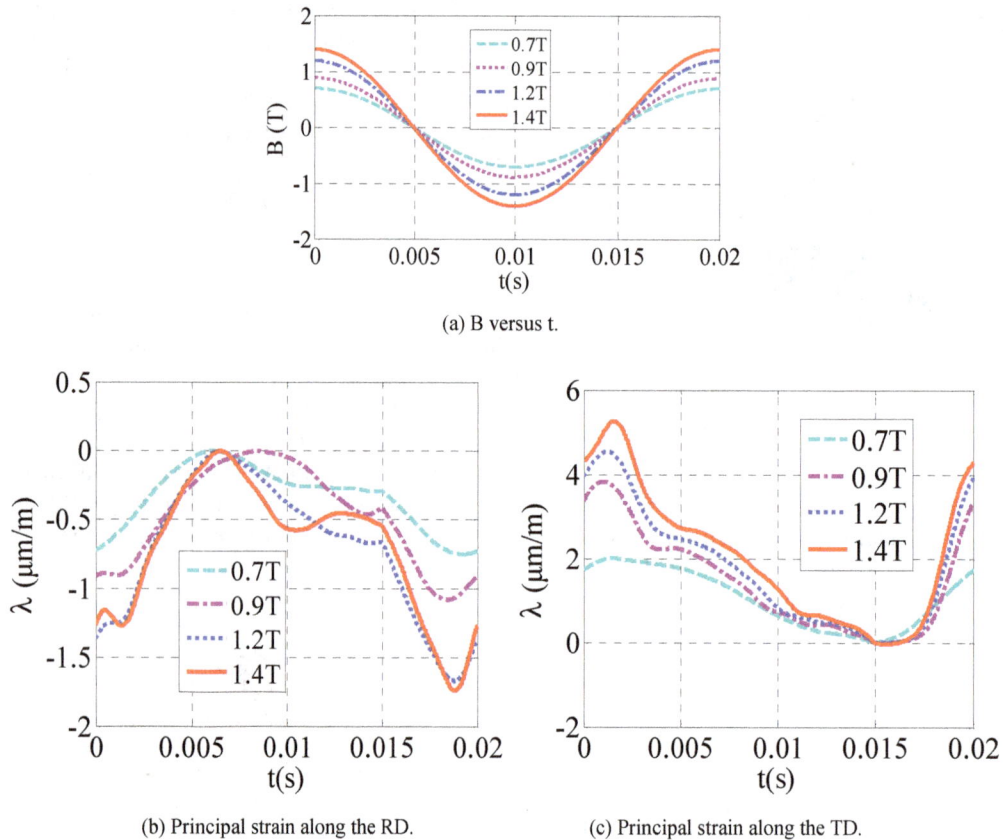

Figure 15. *Measured magnetic flux density and principal strain waveform in one magnetization period under 30A/m biased field applied along the RD in a GO silicon steel sheet.*

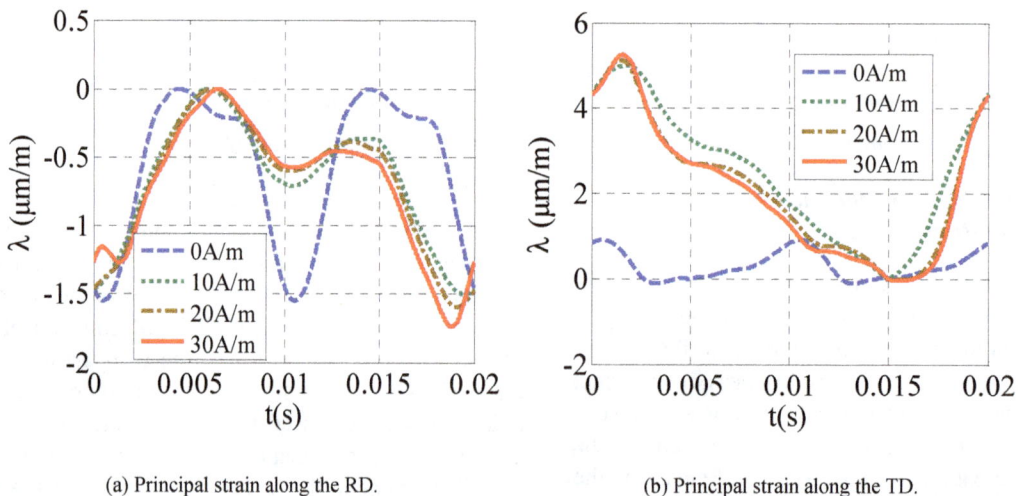

(a) Principal strain along the RD.

(b) Principal strain along the TD.

Figure 16. *Principal strain under different DC biased field in a GO silicon steel sheet with alternating B of 1.4 T.*

4. Conclusions

This paper introduces the research situation on magnetostriction property of silicon steel sheets in recent years initially, then mainly focuses the studies on this area by the author's group.

The research on anisotropic magnetostrictive property of

silicon steel sheets with 1D measuring equipment under alternating supplied magnetic field is described firstly, and the triaxial laser test method is used for measuring and calculating the principal strain. It can be concluded that the principal strain of silicon steel sheet, whether non-oriented or grain-oriented, have to be expressed by a vector with anisotropic characteristic, of which both the magnitude and direction angle are related with that of magnetic flux density vector. An improved magnetostrictive property model is proposed.

For the 2D magnetostrictive property of silicon steel sheets in the condition of rotational magnetization, some primary measured results are given to indicate the complexity of the property.

Otherwise, the simulation method combined the magnetostrictive property model and FE analysis is presented with its application to the research of the effect of the higher harmonics or DC bias magnetic field on magnetostriction and relative noise.

In fact, more researches need to be done on the magnetostrictive property of silicon steel sheets, such as better formulation of the property model, direct couple method of the model with FE analysis, and further research of the 2D magnetostrictive property under rotating external magnetic field.

References

[1] International Electrotechnical Commission. IEC/TR62581 Electrical Steel–Methods of measurement of the magnetostriction characteristics by means of single sheet and Epstein test specimens[S]. Switzerland: IEC Central Office, 2010.

[2] Somkun S, Moses A J, Anderson P I, et al. Magnetostriction anisotropy and rotational magnetostriction of a nonoriented electrical steel. IEEE Transactions on Magnetics, 2010, 46(2): 302-305.

[3] Somkun S, Moses A J, Anderson P I. Measurement and modeling of 2-D magnetostriction of nonoriented electrical steel [J]. IEEE Transactions on Magnetics, 2012, 48(2): 711-714.

[4] Kai Y, Todaka T, Enokizono M, et al. Measurement of the two-dimensional magnetostriction and the vector magnetic property for a non-oriented electrical steel Sheet under stress. Journal of Applied Physics, 2012, 111(7): 07E320-1-07E320-3.

[5] Daisuke W, Takashi T, Masato E. Three-dimensional magnetostriction and vector magnetic properties under alternating magnetic flux conditions in arbitrary direction. Electrical engineering in Japan, 2012, 179(4): 1-9.

[6] Daisuke W, Takashi T, Masato E. Measurement of three-dimensional magnetostriction on grain oriented electrical steel sheet [J]. Journal of Electrical Engineering, 2011, 62(3): 153-157.

[7] Yanli Zhang, Xiaoguang Sun, Dexin Xie, et al. Measurement and simulation of magnetostrictive properties for non-grain oriented electrical steel sheet. Transactions of China Electrotechnical Society, 2013, 28(11): 176-181. (in Chinese).

[8] Yanli Zhang, Xiaoguang Sun, Dexin Xie, et al. Modeling of anisotropic magnetostriction property of non-oriented silicon steel sheet. Proceedings of the CSEE, 2014, 34(27): 4731-4736. (in Chinese).

[9] Yanli Zhang, Jiayin Wang, Xiaoguang Sun, et al. Measurement and modeling of anisotropic magnetostriction characteristic of grain-oriented silicon steel sheet under DC bias. IEEE Transactions on Magnetics, 2014, 50(2): 7008804.

[10] Yanli Zhang, Yangyang Wang, Dianhai Zhang, Ziyan Ren, Dexin Xie, Baodong Bai, Jiakuan Xia. Vector Magnetostrictive Properties of Electrical Steel Sheet With Alternating Magnetization, to be published in Proceedings of the CSEE, 2015. (in Chinese).

[11] Yanli Zhang, Qiang Li, Yangyang Wang, Dianhai Zhang, Ziyan Ren, Baodong Bai, Dexin Xie. Analysis on Magnetostrictive Properties of Silicon Steel Sheet Under Harmonic Magnetic Field. Transactions of China Electrotechnical Society, 2015, 30(14): 544-549. (in Chinese).

[12] Yanli Zhang, Qiang Li, Dianhai Zhang, Baodong Bai, Dexin Xie, and Chang Seop Koh. Magnetostriction of silicon steel sheets under different magnetization conditions. IEEE Transactions on Magnetics, 2015, 51(11): 6101604.

16

Design and Built a Research AUV Solar Light Weight

Ali Razmjoo[1], Mohammad Ghadimi[2], Mehrzad Shams[3], Hoseyn Shirmohammadi[4]

[1]Department of Energy Systems Engineering, Faculty of Engineering, Islamic Azad University-South Tehran Branch, Iran
[2]Department of Mechanical Engineering, Islamic Azad University, roudehen branch Tehran, Iran
[3]Mechanical Engineering Faculty, Energy Conversion Group, K. N Toosi University of technology, Tehran, Iran
[4]Department of industrial Engineering, West Tehran Branch, Islamic Azad University, Tehran, Iran

Email address:

Razmjoo.eng@gmail.com (A. Razmjoo), m.ghadimi@riau.ac.ir (M. Ghadimi)

Abstract: Nowadays, renewable energy consumption especially solar energy and the number of vehicles using this kind of energy is increasing. One of the vehicles that can use solar panels to provide sufficient energy for movement is AUV. Autonomous underwater vehicle (AUV) is an unmanned underwater vehicle which is utilized to accomplish various missions autonomously. In this article constructing a solar submarine is studied. On the hull of the vehicle solar cells are installed, and then its velocity under the water is calculated. It's believed that the present research could result in an underwater vehicle which is able to move under the water and provide its own required electrical energy using solar cells.

Keywords: AUVs, Submarine, Diving and Climbing, Buoyancy Force, Solar Panel

1. Introduction

One of the important factors for economic growth and development is energy [1].The population of the world and its energy consumption is increasing, these ends up to use much more fossil fuel and inevitably it will be a world issue in the future; by the way, researchers decided to use sustainable energy resources. Nowadays, majority of countries have a plan for decreasing fossil fuel use and developing renewable energy. Note that many projects in different regions of the world are being done [2,-4]. Renewable energy is an essential alternative in order to reduce the CO2 emission which, in turn, leads to healthy environment. European Union Energy started some programs for decreasing greenhouse gases emission aiming to achieve up to 20% decrease by 2020 and 80%–95% decrease by 2050 by [5,6]. The fig 1. Shows Average Annual Growth Rates of Renewable Energy Capacity from 2007 to 2012.[7]

AUVs are one of the solutions for this purpose. They are vehicles which are able to help us in different fields such as missions in offshore oil and gas platforms [8,9]. AUVs are relatively small, self- propelled with the capability to be controlled from different places; furthermore, they can help us in various underwater missions [10]. Rapid progress in AUVs development is increasing steadily. Meanwhile, electrical AUVs are used in order to scope activity in the oceans [11-15].

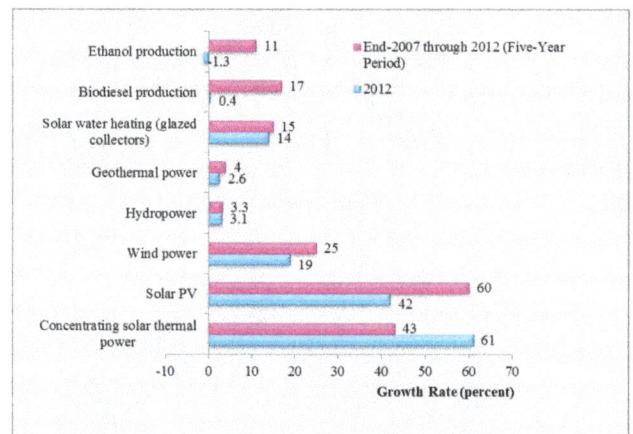

Fig. 1. Average Annual Growth Rates of Renewable Energy Capacity from 2007 to 2012[7].

1.1. Potential and Efficiency of Solar Energy

Today role of solar energy is very clear in our live and using of Photovoltaic systems is a main subject in order to obtain necessary energy such as electrical energy [2].

Solar energy is a known energy that has much more benefits. It will be an important renewable source for producing electricity in the future. In most countries a large number of studies has been done on this kind of energy and how to consume it, and it is increasing continuously. According to the statistics the potential of solar radiation in tropical areas is more than other places. As the published data shows the annual irradiation in Europe is about 1000 KWh/m^2, while in the middle east the value is approximately 1800 KWh/m^2 [16-23].

1.2. Efficiency of Solar Cells

Technology of Solar cells for producing electricity was first introduced in the late 1950s, and then gradually developed [24]. At present Concentrate of Solar Power technology implementation in all of the world and using of solar power in variety equipments is growing fast [25] . The following table shows different modules, technologies and efficiency of solar cells. In this research module BP 7190 with technology CZ, SI, S. P. J and efficiency 15.1 elected.

Table 1. *Efficiency and technology advances of solar cell [26].*

Module	Technology	Efficiency
Sun Power 315	Mono-Si (S. P. J)	19.3
Sanyo HIP-205 BAE	CZ-SI, HIT, S. P. J	17.4
BP 7190	CZ, SI, S. P. J	15.1
Kyocera KC 200GHT-2	MC, SI, Standard Junction	14.2
Solar worlds w 185	CZ-SI STD J	14.2
BPSX 3200	MC-SI STD J	14.2
Suntech STP 260S 24V/b	MC or CZ SI STD J	13.4
Solar WORLDS W225	MC-SI STD J	13.4
Ever Green Solar ES 195	String Ribbon-SI STD.J	13.1
Worth Solar WS 1100 7/80	CIGS	11
First Solar FS-275	CDLE	10.4
Sharp NA-901-WP	A-SI/NC-SI	8.5
GSE Solar GES 120-W	CIGS	8.1
Mitsubishi heavy MA100	A-SI-Single Junction	6.3
UNI-Solar PVL 136	A-SI-Triple Junction	6.3
Kaneka T-SC(EC)-120	A-SI-Single Junction	6.3
Schott Solar ASI-TM86	A-SI/A-SI Same band Gap	5.9
EPVEPV-42	A-SI/A-Si Same band Gap	5.3

There are several vital factors for solar cells which are dependent on many environmental features and weather parameters such as humidity, wind speed, sun intensity and so on. High temperature increases the conductivity of cells. Semiconductor properties define suitability of a material for being used in PV cells. One of these properties is called band gap, which is the energy gap an electron must cross to promote from the valence band to the conduction band. Low temperature reduces the band gap of the semiconductor. Fig. 2 illustrates the dependency of band gap on temperature and its efficiency with respect to the content elements of solar test. Recent studies called band gap method has proved that the efficiency of solar cells degrades as a result of increase in temperature. According to this method as temperature increases, band gap is reduced [27-29]. Fig. 2 shows band gap temperature and efficiency.

In order to provide required energy a 42.5 cm of solar panel was used. The solar panels charge two batteries. Efficiency of solar panels is 15 % (BP 7190 module with CZ-SI, SP. J technology). The proposed submarine is not strong enough to be used in deep water. Furthermore, it is a research

vehicle whose time of testing was brief, approximately 10 minutes. Therefore, the installed solar panel is suitable for short times and charging in limited times, to support a wider time span more panels are needed. According to the obtained figures, total efficiency is obtained as follows.

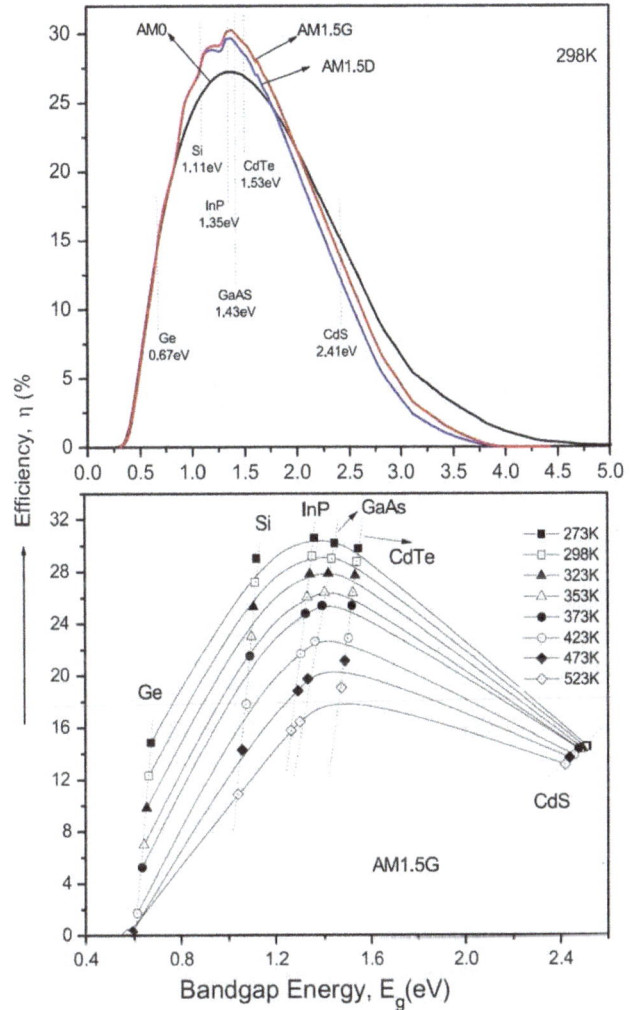

Fig. 2. *Dependency of the band gap temperature and efficiency [26].*

Considering the efficiency of 50 m A (Ampere) per solar panel, three panels (whose total efficiency would be equal to 0.1 m A) and the voltage required to charge the batteries (which is 6 volt), one may conclude that more solar panels must be used to meet our requirements; otherwise, the panels must be exposed to the sun for longer time period in order to properly provide the required power. It can be said that we require 120 solar panels to provide 6 volt (given the value of 50 m A), considering the economic costs, a new question arises; whether the project is economically feasible or not. To answer this question, we must say that a motorcycle works with a power of 6 or 12 volts; and in addition to air pollution, fuel costs must be considered to supply the amount of electricity that is obtained from the fuel. Although initial costs are higher in the projects that are done (or are in progress) on solar cells, these costs would be rational in long term view. Fig. 3 shows the solar panels installed on submarine.

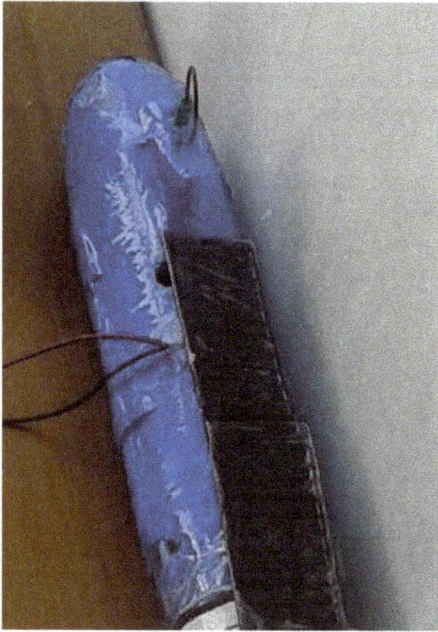

Fig. 3. Solar panels installed on submarine.

2. Mechanic of Designing the Vehicle

As mentioned before AUVs are small in size, typically capable and useful, all submarines have an outside and an inside hull. For this project the outside part is made of fiber glass, and has a high resistance against water pressure in the examined pool. Note that required power is supplied by batteries. Before the design is proved, the hull needs ring profile equation (Myring 1976), a known method to produce minimum drag force to a given fineness ratio (1/d). It could be said it is the ratio of its length to its maximum diameter. Figure 4 shows the sample design of the vehicle. [11, 31, 32]

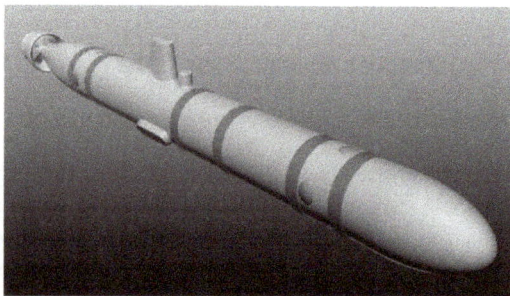

Fig. 4. Sample design of the outside hull of the vehicle [30].

2.1. Inside Layout of a Submarine

In addition to the outside hull equipment, a space is also needed in the internal equipment. The above mentioned design used 3 electrical motors which rotate. It also includes a shaft and a blade on the external body, to provide electrical power a cable is used to connect the motor to the battery, an empty space for the battery is then needed. In this project common vessels are used in the inside hull, in the main vessel the motor cable and the blade are located while battery is located under it.

2.2. Diving, Climbing and Move for Stability

As mentioned before, this system uses a common hull that has an inside part. It means when the hull becomes full of water vehicle can dive in water using its weight. Afterwards, it can climb via climb motor. As the figure below shows for inlet water a window is used together with a support in order to avoid extra water. Fig. 5 shows the water inlet window on the submarine.

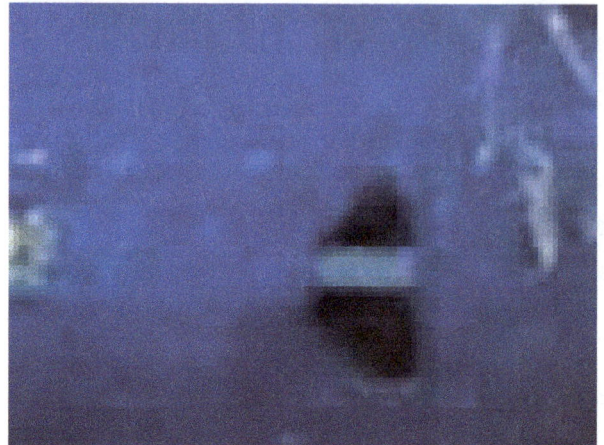

Fig. 5. Water inlet window.

In order to Control Surfaces as well as Depth Control Using Archimedes principle, weight of the vehicle was calculated in air and in water to achieve desired buoyancy of the vehicle in water.

2.3. Stability

Stability is an important issue regarding every sea vehicle and its associated rules need to be known. To be more detailed about this rule, according to Archimedes' principle, for any object immersed in a fluid, a force is exerted on the object by the fluid, which is equal to the weight of the displaced volume of the fluid, this is called buoyancy force. The fig. 6 shows a kind of instance for body and inertia coordinate systems which are an important law.

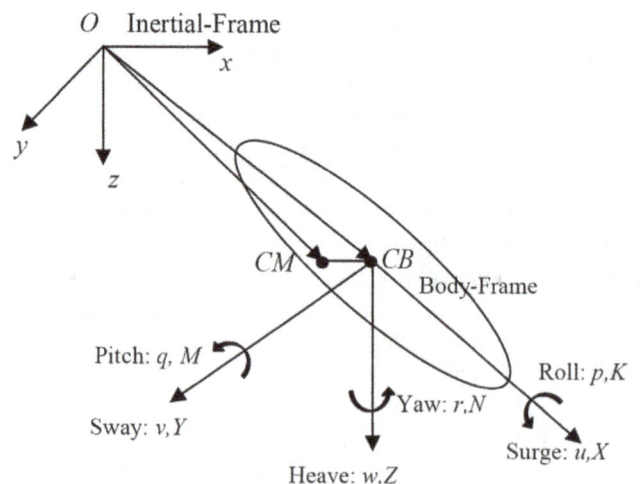

Fig. 6. Body and inertia coordinate systems [33].

Actually stability for everybody at first depend to integration all members. To be more detailed about this law, according to Archimedes' principle, for any object immersed in a fluid, a force is exerted on the object by the fluid, which is equal to the weight of the displaced volume of the fluid, this is called buoyancy force.

Positive buoyancy mode (W > B, weight smaller than buoyancy)

Neutral buoyancy mode (W = B, weight and buoyancy are equal)

Negative buoyancy mode (W < B, weight greater than buoyancy)

Buoyancy force depends on the size of the body's seal; the submarine balance can be controlled by changing the buoyancy force. In this case, the size BG should be large enough to avoid feeling the movement of the inner weight. Such bodies are inherently stable in the vertical direction. If the immersed neutrally buoyant body is raised or lowered to a different depth (disturbance), this body will remain in equilibrium at that location (Cengel & Cimbala, 2006). Also, from dynamic perspective size of BG has a great effect on the behavior of high-speed underwater submarine. Besides, BG builds a hydrostatic resistance of the body against the longitudinal momentum. BG is so important in the equilibrium of the submarine during diving and climbing [34]. Table 2 illustrates the characters of submarine include length, area, height, width, weight.

Submarine size:

Table 2. *Characters of submarine.*

character	amount
Length	75 cm
Area	180 cm
Height	7 cm
width	5 cm
Wight	1 kg

3. Velocity Calculation

Calculating the velocity of the submarine is an important step in order to estimate the power. Experiments are done in a pool with laminar flow in a pool located in Mosavi Street in Tehran, in 1 meter deep and in 3 states. These states include 1m, 5m and 10m distances. The whole experiments showed that this submarine could traverse a distance of 48 meter per minute, thus the velocity of an underwater submarine, according to maritime knots is equal to 1.5 knot. Table 3 shows important parameters for test of underwater vehicle in the pool.

Table 3. *Remarkable parameters for test of underwater vehicle in the pool.*

Parameter	Max	Min	Test time
T(water)	30	4	27
P	20	1	3
T (environment)	35	4	10
Deep	1.50	40	80

Also terminal is equal[35-39]:

$$vi=\frac{\sqrt{2mg}}{\rho Acd}$$

That in this article has:

$$vi = \text{velocity}$$

$$m = \text{mass}$$

$$g = 9.8 \text{ g /cm}$$

$$\rho = \text{density}$$

$$A = \text{Area}$$

$$cd = \text{coefficient drag}$$

3.1. Laminar and Turbulent Velocity Experiments

When an underwater vehicle moves under the water there are different situations, which affect the velocity, Pressure in different directions can even change the movements of the vehicle as well as reducing its speed. There is a table that shows calculated velocity for laminar and turbulent flow. Table 4 illustrates velocity terminal that calculated in laminar and turbulent flow in poor water, As table shows, velocity terminal in laminar and turbulent flows in pool water are calculated. Actually in this experiment evaluation is performed for two flows. Results demonstrate that there are an obvious difference in submarine movement. Moreover, the submarine in turbulent flow needs more energy.

Table 4. *Velocity terminal that calculated in laminar and turbulent flow in poor water.*

Time (s)	Laminar Flow m/s	Turbulent Flow m/s
6	4.869	2.841
12	9.738	5.682
18	14.607	8.523
24	19.476	11.364
30	24.345	14.205
36	29.214	17.046
42	34.083	19.887
48	38.952	22.728
54	43.821	25.569
60	48.690	28.410

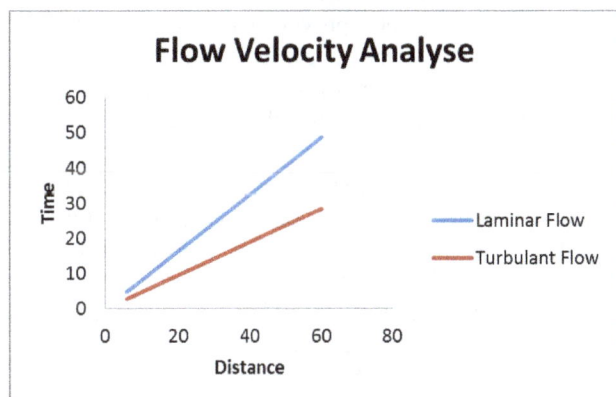

Fig. 7. *Velocity terminal that calculated in laminar and turbulent flow.*

Fig. 7 shows analysis of velocity in laminar and turbulent

flows. As can be seen there are a direct line for velocity in different points. In this figure, it is obvious that for laminar flow from first point to final point velocity is not reduced while in turbulent flow it decreases because the water has up and down flow and submarine cannot move appropriately.

A remarkable thing for each under water vehicle it is relation between Reynolds and movement in different levels of, turbulent fluids and resistance against water pressure fluid the fig 8 shows movement a vehicle in the water and different resistances.

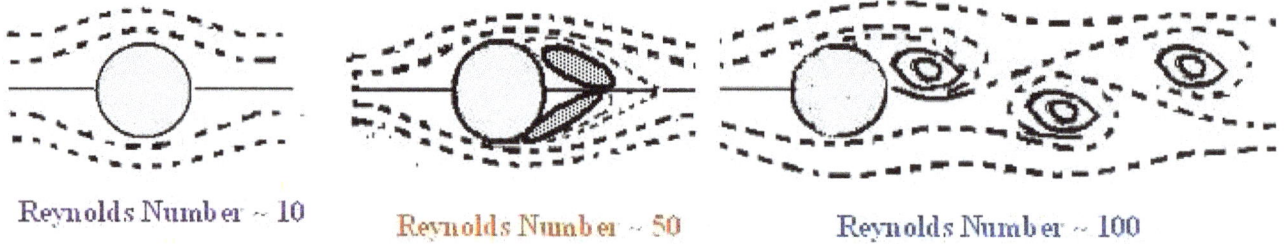

Reynolds Number ~ 10   **Reynolds Number ~ 50**   **Reynolds Number ~ 100**

Fig. 8. Resistance against water pressure for a underwater vehicle to Reynolds number [40].

As it obvious in Reynolds number 100 there are a pressure of all direction to vehicle from all directions.

After up points about move of vehicle it needs to explain about Member and weight vehicle, table 5 shows the member and weight them which consists servo motor, solar panel, rechargeable battery, hull, cover of battery, charger, of battery, blade, weight of beam, weight of keeper static and in the end total weight of submarine.

The following table shows the utilized equipment and the weight of each one.

Table 5. Members and weight them.

Serial	Description	Quantity	Weight (g)	
			Single	Total
1	Servo motor	3	35	105
2	Solar panel	3	10	30
3	Rechargeable battery	5	40	200
4	Hull	1	300	300
5	Cover of battery	1	75	75
6	Charger, of battery	1	90	90
7	Blade	6	16.6	100
8	Weight of beam	1	50	50
9	Weight of Keeper static	1	50	50
Total weight				1000

3.2. Submarine's Battery and Controller

Strong points of lithium batteries include longer lifespan, high energy density and providing electricity when discharging. A lithium battery can produce a voltage between 1.5 v to about 3.7 v. (iron disulfide [li – FeS2 (FR)] Propylene Carbonate, dioxolane, dimethoxymethane…v=1.4-1.8 types). As we know batteries are good power sources for many equipment and industrial applications. To provide electricity, solar panels are connected in series and convert the solar energy to electricity in order to charge batteries. To supply electricity of the proposed submarine, 1.5 V rechargeable batteries are used. four ultra-light lithium type batteries are exploited to obtain, a voltage source of 6V. The required power of 6 volts 50 m A is supplied through connecting batteries poles to the solar panels via power cables. There are some other vehicles that are powered by solid polymer electrolyte Fuel Cell (PEFC) (Hyakudome et al, 2001), aluminum/oxygen full Cell (vestgard et al, 2001) and

solar cells,[41-43]. The following figure shows control radio used for the submarine.

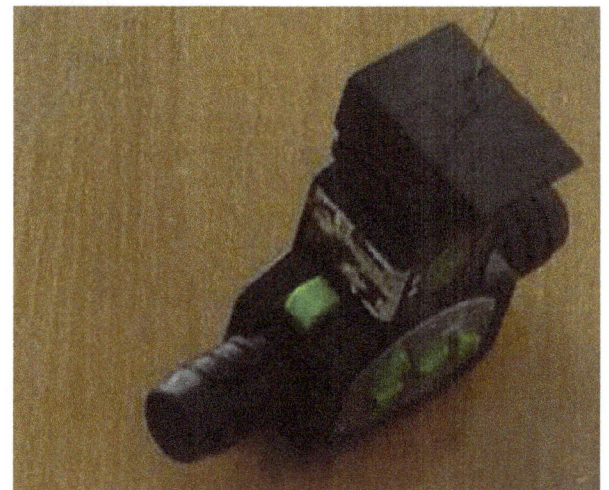

Fig. 9. Used radio controller type of solar submarine for near distance.

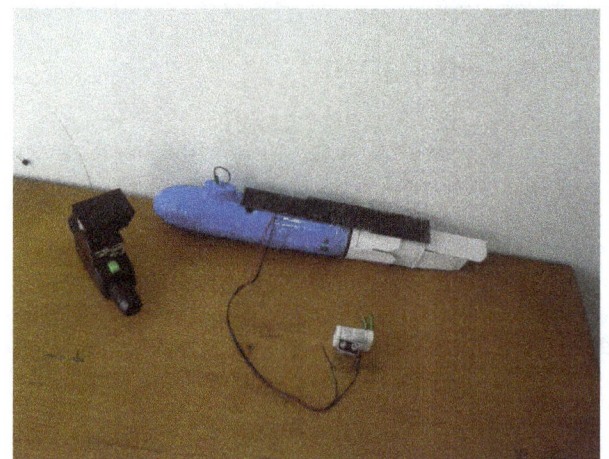

Fig. 10. Research solar underwater vehicle.

4. Conclusion

Nowadays much interest is shown in autonomous unmanned systems. To develop submarines solar cells are more beneficial. This research has three goals which are mentioned in the following.

First, making a light but strong against the water pressure submarine. Second calculating the velocity of submarine under the water and, third, calculating the amount of the battery charge for submarine movements. To construct the submarine's hull a kind of light but strong fiberglass is selected. Then, 3 panels of solar cells with 15% capability (BP 7190 module with technology CZ-SI, SP. J) are installed on the hull. Finally the amount of time needed for the charge to afford the movements and the velocity of submarine in the laminar flow in the pool are calculated.

References

[1] Orhan Ekren and Banu Yetkin Ekren Size Optimization of a Solar-wind Hybrid Energy System Using Two Simulation Based Optimization Techniques, Volume 85, Issue 11, pp 1086–1101, 2008.

[2] Jasim Abdulateef, Simulation of solar off- grid photovoltaic system for residential unit, International Journal of Sustainable and Green Energy; vol 4(3-1): pp, 29-33,2014.

[3] Farivar Fazelpour, Nima Soltani, Marc A. Rosen, Wind resource assessment and wind power potential for the city of Ardabil, Iran, International Journal of Energy and Environmental Engineering, Vol 3: pp 1-8, 2012.

[4] Mohammad-Ali Yazdanpanah Modeling and sizing optimization of hybrid photovoltaic/wind power generation system, journal of industrial engineering international, Vol 1 pp, 1-14, 2014.

[5] Warit Werapuna, Yutthana Tirawanichakul, Watsa Kongnakorn, Jompob Waewsak, An Assessment of Offshore Wind Energy Potential on Phangan Island by in Southern Thailand, Energy Procedia 52 pp 287 – 295,2014.

[6] Georg Felber and Gernot Stoeglehner, Onshore wind energy use in spatial planning a proposal for resolving conflicts with a dynamic safety distance approach, Energy Sustainability and Society, pp 4-22, 2014.

[7] Fahime Heidarzade1, Mohammad Hossein Morshed Varzandeh, Placement of Wind Farms Based on a Hybrid Multi Criteria Decision Making for Iran World Sustainability Forum the 4th sustainability forum, 2014.

[8] Chris Schwarz, Geb Thomas, Kory Nelson, Michael Mac Cray, Nicholas Schlarmann, toward autonomous vehicles, p 67, 2013.

[9] Zool H Ismail, Ahmad A. Faudzi and Matthew W. Dunnigan, Tracking control scheme for multiple autonomous underwater vehicles subject to union of boundaries, Procedia engineering pp1176-1182, 2012.

[10] Paul G. Fernandez, Pete Stevenson, Andrew S Brierley, Fredrick Armstrong And e John Simmonds, autonomous underwater vehicles: future platform for fisheries acoustics, ICEC journal of Marine Science, Vol 60 pp 684-691.2003.

[11] Tae- Hwan joung, karlsammut, Fangpo, He, And Seung-Keon Lee, Shape optimization of an autonomous underwater vehicle with a ducted propeller using computational dynamics analysis, International journal Navigation Archit ocean engineering Vol 4: pp 44-56, 2012.

[12] Hu Yuli, wang Jiajun, Study on power generation and energy storage system of solar powered Autonomous Underwater Vehicle (SAUV)2012 international Conference on Future Energy, Environmental, and material Energy procedia 16, pp. 2049-2053, 2013.

[13] Francisco Garcia-Cordova, Antonio Guerrero-Gonzales, Intelligent navigation for a solar powered unmanned underwater vehicle, international journal of advanced robotic systems, Int J Adv Robotic Sy, Vol. 10, 185:2013.

[14] Chunfeng Yue, Shuxiang Guo, Liwei Shi, hydrodynamic analysis of the Spherical underwater robot SUR-II, international journal of advanced robotic systems Vol 10,247,2013.

[15] Falmouth scientific, Inc, solar powered autonomous underwater vehicle (SAUV), September 2005.

[16] Alejandro Mendez, Teresa j Leo, And Miguel a Herreros, Current state of technology of full cell Autonomous Underwater Vehicle, 1st international e-conference on energy 14-31 march, 2014.

[17] M. Jamil Ahmad and G. N. Tiwari, Optimization of Tilt Angle for Solar Collector to Receive Maximum Radiation, the Open Renewable Energy Journal, pp 19-24 ,2009.

[18] T. Srinivas, B. V. Reddy, Hybrid solar–biomass power plant without energy storage Case Studies in Thermal Engineering pp75–81 , 2014.

[19] Andhy Muhammad Fathonia, b, N. Agya Utamab, Mandau A. Kristianto, A Technical and Economic Potential of Solar Energy Application with Feed-in Tariff Policy in Indonesia, Procedia Environmental Sciences pp 89 – 96,2014.

[20] Khai Mun Ng, Assessment of solar radiation on diversely oriented surfaces and optimum tilts for solar absorbers in Malaysian tropical latitude, International Journal of Energy and Environmental Engineering 5:5 doi:10.1186/2251-6832-5-5,2014.

[21] Pragya sharma, Tirumalachetty Harinariayna, Solar energy generation potential along national high ways, international of Energy and environmental Engineering, pp 4-16, 2013.

[22] Enda Flood, K. McDonnell, F. Murphy and G. Devlin, A Feasibility Analysis of Photovoltaic Solar Power for Small Communities in Ireland The Open Renewable Energy Journal, pp 78-92, 2011.

[23] Souvik Ganguli, Jasvir Singh, Estimating the Solar Photovoltaic generation potential and possible plant capacity in Patiala International a journal of applied engineering Research, Dindugul Vol 1, pp 253-260,2010.

[24] Xiaoming Wang, Jian zhongshang, Zirong Luo, Li Tang Xiangpo Zhang, Juan li. review of power system and environmental energy conversion unmanned for underwater vehicles, Renewable and sustainable energy review pp1958-1970, 2012.

[25] Ramadan Abdiwe, Markus Haider, Investigations on Heat Loss in Solar Tower Receivers with Wind Speed Variation,Vol 4(4): pp,159-165,2015.

[26] Farivar fazelpour, Majid Vafaiepour, Omid Rahbari, Considerable Parameters of Using PV cells for Solar-Powered aircraft, Renewable and Sustainable Energy Review, 22 pp 81-91, 2012.

[27] Singh P, Ravindra NM. Temperature dependence of solar cell performance an analysis. Solar Energy Materials & Solar Cells 2012; 101:36–45.

[28] Ali A. Sabzi Parvar, A simple formula for estimating global solar radiation in central arid deserts of Iran Renewable Energy pp1002–1010, 2008.

[29] Thomas huld, Ralph Gottschalg, and Hans Georg Beyer, Marko topic, mapping the performance of PV modules, effects of module type and data averaging solar energy 84, pp 324–338, 2010.

[30] S Gomariz, J. Prat, A. Arbos, O. Palares, C Vinolo, Autonomous vehicle development for vertical underwater observation, journal of maritime research vol 6 No 2, p 9,2009.

[31] Jenna Brown, Chris Tuggle, Jamie Mac Mahan, Ad Reniers, The use of autonomous vehicles for spatially measuring mean velocity profiles in rivers and estuaries. Intel Servo Robotics 4:pp 233–244,2011.

[32] Masoud Hekami Fard, Milad Bandegani, Mohammad Nasri, Mysam Yazdi, Design a solar submarine Montreal model, 13th conference nation nautical industrial,2011.

[33] M. H. Shafiei and T. Binazadeh, Movement control of a variable mass underwater vehicle based on multiple-modeling approach, Systems Science & Control Engineering: An Vol. 2, pp 335–341, 2014.

[34] Roy Burcher, Louis J Rydill, Concepts in Submarine Design Cambridge University Press. 170 publish 1995.

[35] Terminal Velocity, NASA Glenn Research Center. Retrieved March 4, 2009.

[36] Huang, Jian, Speed of a Skydiver (Terminal Velocity) , The Physics Factbook. Glenn Elert, Midwood High School, Brooklyn College, 1999.

[37] All About the Peregrine Falcon archived. U.S. Fish and Wildlife Service. December 20, 2007. Archived from the original on March 8, 2010.

[38] The Ballistician, Bullets in the Sky. W. Square Enterprises, 9826 Sagedale, Houston, Texas 77089, march 2001.

[39] Robert H. Perry; Cecil Chilton (eds.). Chemical Engineer's Handbook (fifth ed.). pp. 5–62. ISBN 978-0070494787.

[40] A. Yashodhara Rao, A. Sarada Rao, Appajosula S. Rao, Dynamics of Fluid Flow around Aerofoil, and Submarine: Effect of Winglets, The International Journal of Engineering And Science (IJES), Vol 2 pp 39-46, 2013.

[41] Uger Ecin, Josep Gilmartin, Jonathan Grilo, Minhaj Khan, Stephanie Limb, James Parsons. Multi-Disciplinary Design Project (MDDP) 2012/13 Underwater Vehicle for Submarine Cable Repair Final Report Monday, 14th January 2013.

[42] David A. Schoen Wald, AUVs, in space, Air, water, and on the Ground. IEEE control systems Magazine December vol 20 ,pp 15-18,No 6,2000.

[43] Khairul Alam, Tapabrata Ray, Sreenatha G. Anavatti. A Brief taxonomy of autonomous underwater vehicle design literature, University of new south Wales, UNSW Canberra, Act 2600 Australia, Ocean Engineering Vol 88, pp 627-630, 2014.

17

Release Activity and Potential Ecological Risk Assessment of Heavy Metals in Coal Gangue of Hancheng, China

Li Wan-peng, Sun Ya-qiao, Yao Meng, Dou Lin

College of Environmental Science and Engineering, Chang'an University, Xi'an, Shanxi, China

Email address:

1284993588@qq.com (Li Wan-peng)

Abstract: Focused on release activity of different weather degree gangue of Han Cheng, evaluate heavy metals potential release risks by Hakanson instrument. The results showed that the content of heavy metals were more than their background in unweathered gangue. The Sang shuping gangue were Cd，Cr and As, so did Cd, Pb, Cr, Cu, Ni and As in Xia yukou. Residue fraction were the main fraction of heavy metals in gangue. The release potential and activity of heavy metals decreased with the weather degree increasing. The release activity of unweathered and weathered one year of Sang shuping gangue from high to low were Cu, Pb, Ni, Cd, Cr, As and Cd, Pb, Cu, Ni, Cr, As respectively. The release activity of unweathered and weathered one year of Xia yukou gangue from high to low were Cu, Ni, Cr, Pb, Cd, As. The potential ecological risk of heavy metals were in slight degree, and the potential risk index of heavy metals decreased gradually with the increasing of weather degree. The potential risk index of heavy metals from high to low were Cd, Pb, Cu, Ni, As, Cr and Cd, Cu, Pb, Ni, Cr, As respectively.

Keywords: Coal Gangue, Heavy Metals, Release Activity, Potential Ecological Risk Assessment

1. Introduction

Coal gangue is a kind of solid waste generated in the washing and mining process [1-3]. At present coal gangue has been accumulated about five billion tons in China, at the same time it is increasing about 200 million tons year by year. Coal gangue is mainly used for power generation and manufacture of building materials in China [4]. A lot of open-air pile of coal gangue accumulatied, because of a low utilization rate about 30%. Soil and water environment even human health nearby were effected by heavy metals pollution element, which release more in long-term weathered and eluviated [5-11]. The coal gangue which deposited in the open air produced a large number of poisonous gases such as H2S, SO2, which harmed residents health and effected the growth of plants at Yangquan city, Shanxi Province [12]. The coal gangue which deposited outside for a long time resulted in plenty of salt to dissolve into water.And contaminated the groundwater at Yanzhou city, Shandong Province [13]. Finkelman pointed the release activity and toxicity of heavy metals have a lot of relations to its forms, which determing the ability of heavy metal elements release to the environment [14]. This article, which using Tessier five consecutive step extraction to confirm exchangeable form, carbonate bound, Fe-Mn oxides bound, organic bound and residual form heavy metal content and to research the release characteristic of different weathering degree of coal gangue and to evaluate the potential ecological risk.In order to provide a basis for the governance in mining pollution repairmen [15].

2. Materials and Methods

2.1. Sampling Area Profile

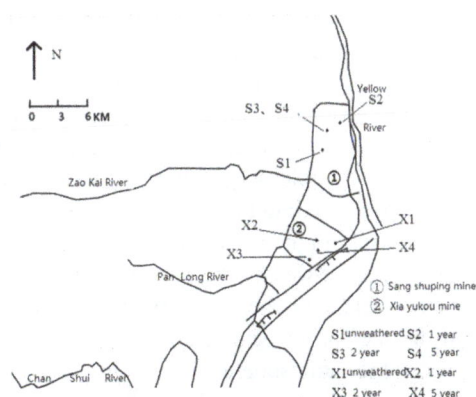

Fig. 1. *The schematic of sampling district.*

Hancheng city is located in the guanzhong basin and northern shaanxi loess plateau transition zone, YiChuan close to the north, HeYang joint to the south, the Yellow River across to the east the and Huanglong to the west. Mining area is located in the northeast of weihe coal field, according to the geological structure divided it into two parts.North part including Sang shuping, Liao yuan and Xia yukou three coal fields, south part including Xiang shan coal field. These mining fields stack more than 500 thound tons coal gangue and affect the surrounding ecological environment.

2.2. Experimental Method

Respectively collect Xia yukou and Sang shuping two different weathering degree of coal gangue heap of eight samples on April 27, 28, 2011 and May 14, 15, 2014. In the process of sampling use GPS positioning and the stainless steel shovelling 1 kg coal gangue in plastic bags.We take the unweathered and different weathered degree coal gangue back to the laboratory.

Then we place the samples at dry air to speed up the natural weathering.After smashed by mortar we get the final samples which could through the 100 mesh nylon mesh.

Using Tessier five consecutive step extraction to analyse heavy metals of exchangeable form, carbonate bound, Fe-Mn oxides bound, organic bound and residual form in the coal gangue. We use Inductively Coupled Plasma Optical Emission Spectrometer (ICP-OES) to analyse the heavy metals.

In order to reduce the random error and improve data reliability and accuracy of samples, this study conducted parallel experiment to confirm experimental data reliability.

3. Results

3.1. Heavy Metal Content in Gangue

From table 1 we see that the Cd, Cr and As contents in unweathered coal gangue of Sang shuping are much higher than China soil background value [16]. The times to the China soil background value are 31.75, 1.65 and 15.48 respectively. The Cd, Pb, Cr, Cu, Ni and As contents in unweathered coal gangue of Xia yukou are much higher than China soil background value.The times to the China soil background value are 568.75, 39.2, 2.83, 1.26, 1.76 and 45.56 respectively. The heavy metal contents in Xia yukou mine are obviously higher than that of Sang shuping mine.

Dates from figure 2 show that the heavy metal contents of weathering one year time in Sang shuping dropped sharply. By the time of weathering two years it rised slightly.This phenomenon show that the release potential of heavy metals of weathering one year is maximum.Then the release potential of heavy metals decline with the increasing of the weathering degree. The heavy metals of weathering five years (such as Pb, Cu, Ni) content are a little more than the unweathered samples. This is likely to be the earlier weathered samples releasing heavy metals into the atmosphere, where heavy metals particles sedimentating. Dates from figure 3 show that in addition to the Cr element the rest are obvious downward

trend in this five years in Xia yukou mine gangue. Continuous weathering in coal gangue make the heavy metal contents in surrounding environment accumulate. Affecting the environmental quality and harming human health [17-18].

Table 1. Concentration of heavy metals in Gangue.

		Cd	Pb	Cr	Cu	Ni	As
China soil background values		0.08	23.50	57.30	20.70	24.90	9.60
Sang Shu ping	unweathered	2.54	19.26	94.77	13.70	9.29	148.59
	1 year	0.04	2.28	35.74	6.20	6.03	45.56
	2 year	1.80	16.40	78.64	9.06	8.62	128.07
	5 year	0.53	21.40	87.50	18.24	27.02	67.57
Xia Yu kou	unweathered	45.50	921.13	162.38	26.13	43.75	437.38
	1 year	6.88	217.25	176.88	21.13	43.73	398.75
	2 year	19.63	276.13	170.88	23.63	41.63	442.13
	5 year	3.79	64.17	163.66	16.41	22.44	244.85

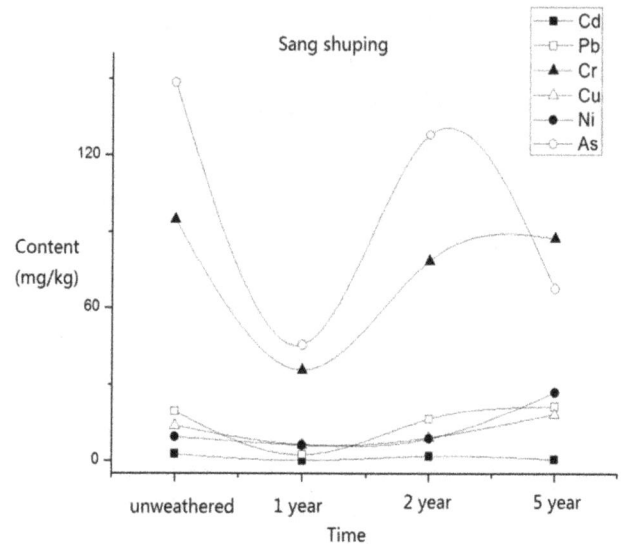

Fig. 2. Concentration of heavy metals in Gangue along with the weathering time variation of Sang shuping.

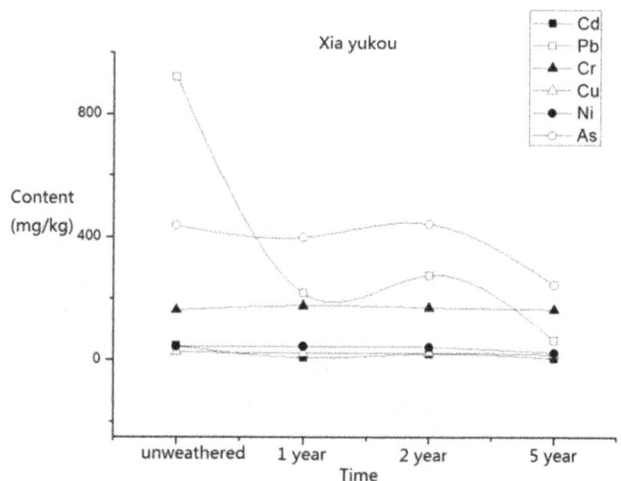

Fig. 3. Concentration of heavy metals in Gangue along with the weathering time variation of Xia yukou.

The release potential of heavy metals are the highest in the first year of Sang shuping and Xia yukou.Then the release potential decreases gradually with the increasing of weathering degree.The heavy metal contents of Xia yukou are much higer than the Sang shuping, whch have a obviously release like Pb and As during five years weathering.

3.2. Gangue Geochemical Forms of Heavy Metals

From table 2 we see that the mine heavy metal form is residual. Cd, Cr, Ni, As contents pass more than 90% and Pb, Cu also pass more than 85% in unweathered Sang shuping gangue. In weathering one year gangue, only Cr and As contents pass 90%. Cd and Pb are main in Fe-Mn oxides bound (44.12% and 42.05%). Cd in exchangeable form is 29.41%, Pb in organic bound is 23.30%, Cd in residual form is zero. Weathering two years and five years gangue are main in residual form.Only the exchangeable form of Cu are more than 10% (18.20%, 10.55% respectively). In weathering five years gangue, only Pb carbonate bound is more than 10% (11.67%). The other heavy metals in first four forms do not exceed 10%. Except the residue form the rest four forms of heavy metals all have biological effectiveness.This article analyzes the heavy metal release of the first four forms: exchangeable form, carbonate bound, Fe-Mn oxides bound and organic bound [19]. The heavy metals realse activity of unweathered and weathering two years from high to low are Cd, Pb, Ni, Cd, Cr and As.The weathering one year are Cd, Cu, Ni, Cr and As.The weathering five years are Pb, Cu, Cd, Ni, Cr and As.

Table 2. The percentage of various fraction of heavy metals in coal gangue of Sang shuping.

Time	Form	Cd	Pb	Cr	Cu	Ni	As
unweathered	exchangeable	0.34	0.14	0.01	0.01	0.83	0.06
	carbonate	0.15	2.42	0.03	1.24	0.98	0.00
	Fe-Mn	0.69	3.30	0.10	0.04	1.39	0.00
	organic	0.25	5.86	0.15	12.93	5.33	0.00
	residual	98.57	88.28	99.71	85.78	91.47	99.94
1 year	exchangeable	29.41	2.14	0.02	0.14	2.40	0.13
	carbonate	8.82	10.58	0.06	4.05	4.66	0.00
	Fe-Mn	44.12	42.05	0.12	1.09	9.64	0.00
	organic	17.65	23.30	0.48	18.12	4.54	0.00
	residual	0.00	21.93	99.32	76.60	78.76	99.87
2 year	exchangeable	0.56	0.30	0.01	0.15	0.55	0.06
	carbonate	0.69	4.21	0.07	6.91	3.19	0.00
	Fe-Mn	1.11	6.62	0.15	0.22	3.64	0.00
	organic	0.35	3.48	0.26	18.20	2.75	0.00
	residual	97.29	85.39	99.51	74.52	89.87	99.94
5 year	exchangeable	2.52	0.06	0.01	0.04	0.14	0.10
	carbonate	2.06	11.67	0.06	7.09	1.81	0.00
	Fe-Mn	2.52	6.62	0.21	0.46	2.47	0.00
	organic	1.15	3.11	0.21	10.55	1.72	0.00
	residual	91.75	78.54	99.51	81.86	93.86	99.90

From table 3 we see that the mine heavy metal form is residual form.Cr, Pb, Cd, As contents pass more than 90% in Xia yukou unweathered gangue.Cu is main in organic bound (76.97%) and carbonate bound is 12.92% however the residual form is only 4.31%. Ni is main in residual form (76.57%) and organic bound is 11.86%, the rest three forms are no more then 10%. The residual form are all more than 80% except Cu, which is main in organic bound (74.2%). Carbonate bound and Fe-Mn oxides bound are more than 10% in weathering one year. Except Cu and Ni the rest heavy metals in residual form are more than 90%. Cu is still main in organic bound (71.23%) and the residual form is (12.39%). Ni is main in residual form (77.98%) and the Fe-Mn oxides bound is 12.32% in weathering two years. The heavy metals in gangue of Xia yukou are over 90% in residual form.The release activity of heavy metals in unweathered, weathering one year and two years from hiagh to low are Cu, Ni, Cr, Pb, Cd and As. The weathering five years are Cu, Ni, Cd, Pb, Cr and As.

Table 3. The percentage of various fraction of heavy metals in coal gangue of Xia yukou.

Time	Form	Cd	Pb	Cr	Cu	Ni	As
unweathered	exchangeable	0.03	0.00	0.02	0.80	0.07	0.00
	carbonate	0.01	0.46	0.12	12.92	3.43	0.00
	Fe-Mn	0.08	0.88	3.18	5.00	8.07	0.00
	organic	0.20	0.93	2.31	76.97	11.86	0.00
	residual	99.68	97.73	94.37	4.31	76.57	100
1 year	exchangeable	0.51	0.00	0.01	1.63	0.00	0.00
	carbonate	0.51	1.08	0.10	11.02	2.73	0.00
	Fe-Mn	2.69	5.83	5.38	13.15	12.11	0.00
	organic	1.02	0.91	2.40	74.20	4.33	0.00
	residual	95.27	92.18	92.11	0	80.83	100
2 year	exchangeable	0.32	0.00	0.01	1.26	0.00	0.00
	carbonate	0.27	0.85	0.17	6.07	2.86	0.00
	Fe-Mn	0.57	2.76	3.60	9.05	12.32	0.00
	organic	0.69	1.19	4.40	71.23	6.84	0.00
	residual	98.15	95.20	91.82	12.39	77.98	100
5 year	exchangeable	0.40	0.37	0.01	3.33	3.97	0.04
	carbonate	0.30	0.06	0.03	1.75	1.25	0.00
	Fe-Mn	0.30	0.13	0.08	1.57	1.24	0.00
	organic	0.13	0.08	0.13	1.96	1.06	0.00
	residual	98.87	99.36	99.75	91.39	92.48	99.96

The release activity of Sang shuping and Xia yukou gangue are decreased with the increasing of the weathering degree. Both heavy metals release activity of two mines are very high during the first two years weathering.In the first year weathering, Cd and Pb are mine in Fe-Mn oxides bound, in which the oxidation stability condition is poor.As time goes on the potential of polluting to environment is increasing.Cd is main in exchangeable form, in which the heavy metal is extremely sensitive to environmental change.In the first two years weathering of Xia yukou, Cu is main in organic bound.In the condition of oxidation, this form can change into Fe-Mn oxides bound or carbonate bound, which can increase the release activity of heavy metals. Cu is more than 10% in carbonate bound, in which form the heavy metals particularly sensitive to pH change and easily to release under acidic water leaching.With the weathering degree increasing, heavy metals are main in residual form.In this form the gangue is stable.But with the weathering time increasing, the coal gangue lattice, which can result in the decrease of coal gangue residue form stability is easy broken.

3.3. The Potential Ecological Risk Assessment of Heavy Metals

This article uses Hakanson potential ecological harm index method to evaluate the potential risk of heavy metals in coal gangue. This method is widely used in the analysis of quantitative, types, levels of toxicity and risk assessment of heavy metal content [20-22]. According to the first four forms' dates, we evaluate the potential ecological risk assessment. The calculating formula is $RI = \sum_{i=1}^{n} E_r^i = \sum_{i=1}^{n} T_r^i C_r^i = \sum_{i=1}^{n} T_r^i C_a^i / C_n^i$

C_a^i means the actual measurment of heavy metal element content. C_n^i means the soil element background values(table 1). C_r^i means the single pollution index of heavy metal element. T_r^i means heavy metal toxicity response coefficient, which reflects the strength of toxicity and the sensitivity of heavy metal in water (Cd, Pb, Cr, Cu, Ni, As, respectively are 30, 5, 2, 5, 5, 10) [23]. E_r^i means the potential ecological harm coefficient of heavy metal. RI means the comprehensive index of potential ecological harm of heavy metal. (table 4)

Table 4. The classification standard of potential ecological risk assessment.

C_r^i	single element pollution level	E_r^i	potential ecological level	RI	Comprehensive potential ecological risk degree
<1	slight	<40	slight	<150	slight
[1, 3)	medium	[40, 80)	medium	[150, 300)	medium
[3, 6)	strong	[80, 160)	strong	[300, 600)	strong
>6	Very strong	[160, 320)	Very strong	[600, 1200)	Very strong
		>320	fortissimo	>1200	fortissimo

From table 5 we can see that the comprehensive potential risk index of heavy metals in Sang shuping and Xia yukou gangue are in slight degree.Figure 4 and figure 6 show that the potential ecological risk of heavy metals in Sang shuping are in slight degree.With the increasing of weathering degree, the potential ecological risk change is gradual small.The potential risk of Cd is the highest. Figure 5 and figure 7 show that in addition to Cd the other heavy metal elements potential risk index in Xia yukou gangue are in slight degree.With the weathering degree increasing, the degree of risk shows a downward trend. The potential risk of Cd is the highest, which has a strong dangerous degree of weathering in first two years. With the weathering degree increasing, the risk reduces to slight degree. The coefficient of potential ecological risk and the comprehensive potential ecological index of heavy metals in Xia yukou gangue are higher than that in Sang shuping gangue, which has a higher potential release risk of heavy metals.

The comprehensive potential ecological risk of heavy metals in Xia yukou gangue is in slight degree.With the weathering degree increasing, the potential ecological risk reduce to a slight degree, which still has a high potential release of heavy metals. The gangue which pile up long time can pollute the environment.So in the process of comprehensive utilization we need to pay attention to environmental protection.

Table 5. The results of potential ecological risk assessment.

time	E_r^i						RI	Risk level
	Cd	Pb	Cr	Cu	Ni	As		
Unweathered of Sang shuping	13.95	0.48	0.01	0.47	0.16	0.09	14.80	slight
1 year	15.94	0.38	0.01	0.35	0.26	0.06	17.00	slight
2 year	18.28	0.51	0.01	0.56	0.18	0.08	19.61	slight
5 year	16.88	1.05	0.01	0.80	0.34	0.07	19.15	slight
Unweathered of Xia yukou	54.38	4.44	0.01	6.04	2.06	0.00	66.92	slight
1 year	121.88	3.62	0.01	5.26	1.68	0.00	132.44	slight
2 year	135.94	2.82	0.01	5.00	1.84	0.00	145.61	slight
5 year	15.94	0.09	0.01	0.34	0.34	0.10	16.82	slight

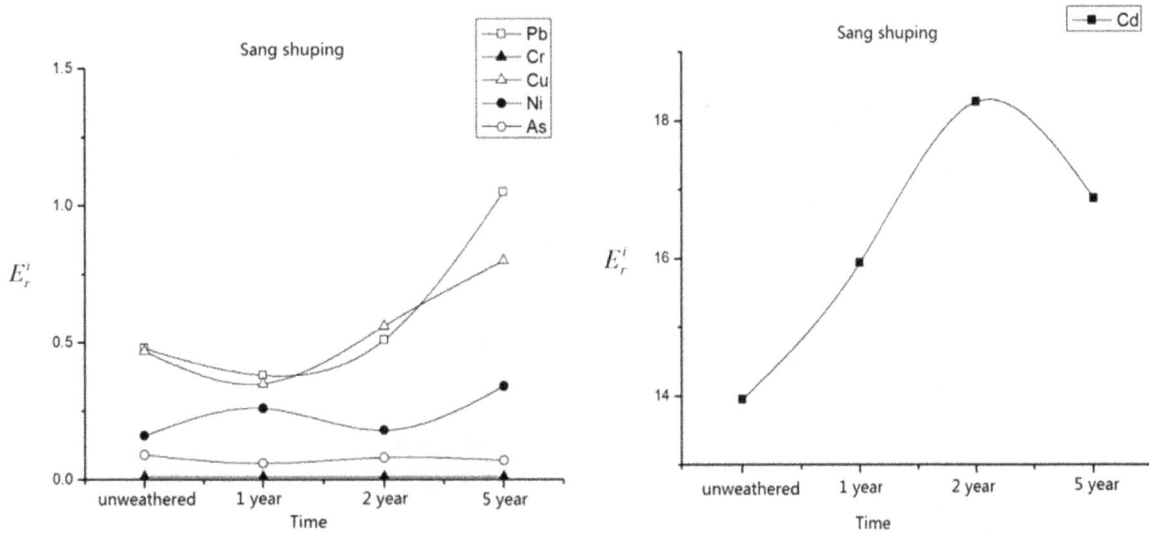

Fig. 4. *The potential ecological risk of heavy metals of Sang shuping coal gangue.*

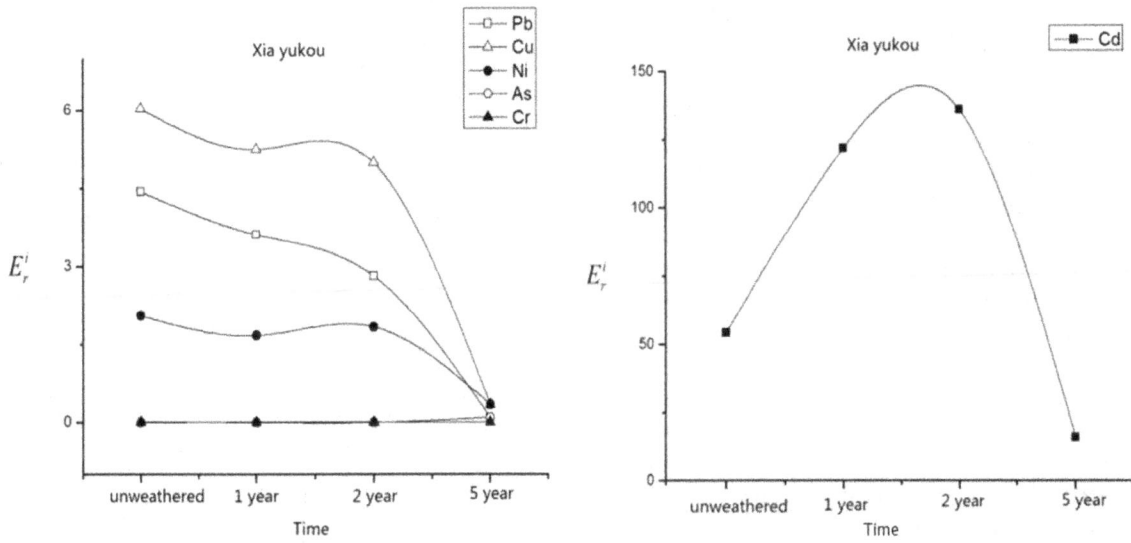

Fig. 5. *The potential ecological risk of heavy metals of Xia yukou coal gangue.*

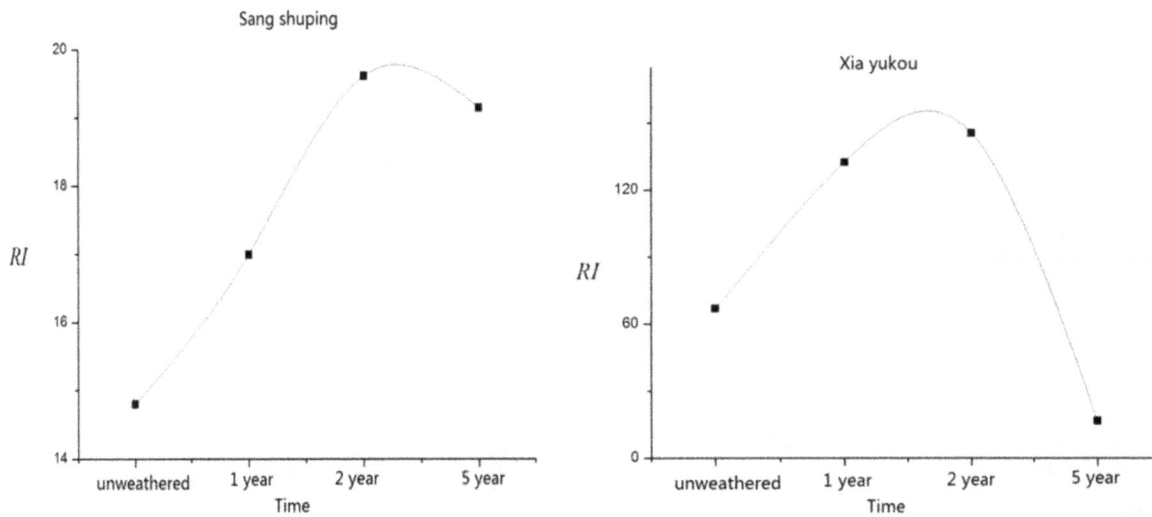

Fig. 6. *The synthetic potential risk index of Xia yukou coal gangue.*

4. Conclusion

(1) In Sang shuping and Xia yukou unweathered gangue, the heavy metals Cd, Cr, As and Cd, Pb, Cr, Cu, Ni, As are more than China soil background value respctively. The heavy metal release potential of Xia yukou mine is obviously higher than that of Sang shuping. In the first weathering year, the heavy metals have the biggest release potential. Then the content decrease with weathering degree increasing.

(2) With the weathering degreee increasing, the release activity of two mine gangue decrease.In first two years weathering, the release activity of heavy metals are high. In the first year weathering, Cd and Pb are mine in Fe-Mn oxides bound, in which the oxidation stability condition is poor. In the first two years weathering of Xia yukou, Cu is main in organic bound.In the condition of oxidation, this form can change into Fe-Mn oxides bound or carbonate bound, which can increase the release activity of heavy metals. Cu is more than 10% in carbonate bound, in which form the heavy metals particularly sensitive to pH change and easily to release under acidic water leaching. With the increasing of weathering degree, heavy metals are main in residual form, which is stable can not be easily replaced by the silicate mineral. But with the weathering time increasing, the coal gangue lattice, which can result in the decrease of coal gangue residue form stability is easy broken.

(3) The heavy metals potential ecological risk in Sang shuping from high to low is Cd, Pb, Cu, Ni, As, Cr.The heavy metals potential ecological risk in Xia yukou from high to low is Cd, Cu, Pb, Ni, Cr, As. The coefficient of potential ecological risk and the comprehensive potential ecological index of heavy metals in Xia yukou gangue are higher than that in Sang shuping gangue, which has a higher potential release risk of heavy metals.The potential ecological risk of heavy metals in two mine gangue are in slight degree.With the increasing of the weathering degree the potential ecological risk gradually reduce.

Acknowledgements

Fund program Natural Science Foundation of China (41002086, 41372258), Shaanxi Provincial Natural Science Foundation (2013JM5003), The central college scientific research fund (2013G1291065, 2013G1291067, 2013G1502036), The key laboratory of mine comprehensive utilization and resources exploration in shaanxi province, China sustenation funds (2014HB006).

References

[1] Zhou Chuang, Dan song. The dangers of coal gangue and resource utilization technology research [J]. Inner Mongolia Environmental Science, 2008, 20(4): 32-35.

[2] Liu di. The environmental hazards of coal gangue and comprehensive utilization research [J]. Journal of Meteorology and Environment, 2006, 22(3): 60-62.

[3] Zhang jianqiang. The environmental impact of coal gangue mountain and governance [J]. The Coke Coal Science and Technology of Shanxi, 2009(2): 44-46.

[4] Wang xinfeng, Gao mingzhong, Fang xiaomin. The new exploration of mining area comprehensive utilization of coal gangue [J]. Coal Technology, 2012(4): 50-53.

[5] Cui Longpeng, Bai Jianfeng, Shi Yonghong, etal. Study on soil heavy metals pollution from mining acitivity [J]. Acta Pedologica Sinica, 2004, 41(6): 898-904.

[6] Li Xuhua, Wang Xinyi, Yang Jian, etal. Review on heavy metal pollution to soil and corn near coal waste rock dump in Jiaozuo diggings [J]. Environmental Protection Science, 2009, 35(2): 66−69.

[7] Cai Feng, Liu Zegong, Lin Baiquan, et al. Study on trace elements in gangue in Huainan mining area [J]. Journal of China Coal Society, 2008, 33(8): 892-897.

[8] Wang Xinyi, Yang Jian, Guo Huixia. Study on heavy metals in soil contaminated by coal waste rock pile [J]. Journal of China Coal Society, 2006, 31(6): 808-812.

[9] Zhang Li, Han Guocai, Chen Hui, et al. Study on heavy metal contaminants in soil from coal mining spoil in the Loess plateau[J]. Journal of China Coal Society, 2008, 33(10): 1141-1146.

[10] Chai shiwei, Wen suomao, Wei xiange. The characteristics of heavy metal content in agricultural soil of suburbs and major cities in the pearl river delta [J]. Journal of China Coal Society, 2000, 17(2): 42-44.

[11] Hanson M L, Solomon K R. New technique for estimating thresholds of toxicity in ecological risk assessment [J]. Environmental Science Technology, 2002, 36: 3257-3764.

[12] Zheng guoqiang, Zhang chengliang, Zhang hongjiang, etal. Effect of temperature on water content of coal gangue and plant growth [J]. Chinese Journal of Soil Science, 2008, 6(3): 107-111.

[13] Wu xiaohua, Ye jin xia, Xia chunying, etal. Yanzhou coal mine geological environment field Situation and Countermeasures [J]. Coal Geology & Exploration, 2008, 36(1): 53-57.

[14] Liu guijian, Yang ping, Peng zicheng. Potentially harmful trace elements in coal gangue leaching precipitation research [J]. Geological Journal of China Universities, 2001.

[15] Tessier A, Campbell P G C, Bisson M. Sequen2 tail extraction procedure for the speciation of particulate trace metals. Anal. Chem, 1976(51): 844~850.

[16] Wei fusheng, Yang guozhi. The element background values basic statistics and characteristics of China soil [J]. Environmental Monitoring in China, 1991. 7(1)21-26.

[17] Wu daishe, Zheng baoshan, Kang wangdong. The leaching behavior and environmental influence research of coal gangue—use huainan pan xie mining area as an example [J]. Earth and Environment, 2004, 32(1): 55-59.

[18] Wei zhongyi, Lu liang, Wang qiubing. Research of open-pit mine and its surrounding soil heavy metal pollution of large coal gangue mountain in west Fushun [J]. Chinese Journal of Soil Science, 2008, 39(4)946-949.

[19] Xing ning, Wu pingxiao, Li yuanyuan etal. Dabaoshan tailings heavy metals speciation and potential migration of analysis [J]. Journal of Environmental Engineering, 2011, 5(6): 1370-1374

[20] Zhang mingliang, Yue xingling, Yang shuying. Release activity of heavy metals in coal gangue and potential ecological risk assessment of contaminated soil [J]. Journal of soil and water conservation, 2011, 25(4): 249-252.

[21] Chen jinghui, Lu xinwei, Zhai meng. The sources of soil heavy metal and potential risks in Xi 'an city roadside [J]. Chinese Journal of Applied Ecology, 2011, 7(22): 1810-1816.

[22] Ma xiaofeng. The soil heavy metal pollution and ecological risk assessment of Harbin city [D]. Harbin: Northeast Forestry University Harbin, 2009: 1-50.

[23] Lars Hakanson. An ecological risk index for aquatic pollution control. A sedimentological approach [J]. Water Research, 1980, 14(8): 975-1001.

18

Renewable Energy Policies and Practice in Tanzania: Their Contribution to Tanzania Economy and Poverty Alleviation

Halidini Sarakikya[1, *], Iddi Ibrahim[1], Jeremiah Kiplagat[2]

[1]Department of Electrical Engineering, Arusha Technical College, Arusha, Tanzania
[2]Department of Energy Engineering, Kenyatta University, Nairobi, Kenya

Email address:

Sarakikyazablon@yahoo.com (H. Sarakikya), ibrahiddi@gmail.com (I. Ibrahim), jeremykiplagat@gmail.com (J. Kiplagat)

Abstract: Tanzania is facing challenges in energy provision with a lot of people leaving in rural areas experiencing energy poverty exhibited by lack of access to electricity, therefore relying on traditional fuels for cooking and lighting. In Tanzania, the electricity access has risen from 18.4% in 2013 to 24% in 2015. Power generation remained generally stable in 2013 which contributed 7.3% to the growth of the National economy. In 2014 the estimates shows that the National economy grew by 7.2%, and is projected to reach 7.4% in 2015. It is reported that, the electricity demand in Tanzania is about 7% per year over the past 10 years. A large proportion of majority of rural population is located far away from the National grid and it is un economical to connect to the grid. The main objective of this paper is to examine renewable energy policies and practices in Tanzania and their contribution to Tanzania economy and poverty alleviation. The study focused on content analysis of projects reports and policy in 10 years. Tanzania has drafted renewable energy policies so as to shift dependence from hydropower which is many times affected by draught and weather patterns and petroleum that have been affected by price fluctuation to solar, wind, biogas and other biomass which are renewable. However, the adoption rate of these renewable energy technologies is low because of financial constraints, lack of awareness, lack of coordination between the Government, non Governmental organizations and private sectors. Existing renewable energy policies should be harmonized and the current practice should be evaluated so as to upgrade the adoption rate of renewable technologies.

Keywords: Tanzania, Energy Poverty, Renewable Energy, Policies, Practices

1. Introduction

Energy is one of the main components in the development of any country. Satisfying the energy demand through the use of renewable energy resources is one of the main issues now days because of the fossil fuel depletion and environmental impacts. Njiru et al [1], point out that, the world energy demand is expected to grow at the average annual rate of 1.8% between 2005 and 2030, where wealth generation and drivers for social economic development are the main reason for the increased global energy demand.

Africa has many energy resources such as hydropower, solar, wind and geothermal but only a small fraction is harnessed for domestic use. In the developing world, Africa has the lowest electrification rate, and the number of rural population without electricity in Sub Sahara Africa (SSA) is expected to increase [1]. Power utilities in Africa have failed to provide adequate levels of electricity services especially to poor societies living in rural areas [6].

In Sub Sahara Africa, 70 – 90% of primary energy supply and even up to 95% of total energy consumption is from tradition biomass energy [7], and only 8% of the rural population has access to modern energy services [2]. More than 650 million people in Sub Sahara Africa, rely on traditional biomass for cooking, heating and lighting, despite the effort done to promote electrification rate [5]. These areas of SSA have much renewable energy resources for decentralized renewable energy technologies which march the dispersed nature of settlements and which are also environmentally friend [1].

2. Energy Profile of Tanzania

2.1. Energy Consumption

The main source of energy in Tanzania is dominated by traditional biomass to satisfy the energy needs for rural households, where by other modern sectors such as transport, industries and commercial depend on imported petroleum. The country has abundant renewable energy resources in terms of hydro, sunlight, wind and biomass but only hydro is currently being exploited in a renewable manner. Other renewable energy sources such as solar, wind power, and biogas offer a small fraction source of energy [4]. About 24% of the total population has access to electricity services of which 7% is in rural area.

Demand for electricity is growing between 10% and 15% per annum and the country is aiming to increase connection levels by 30% by 2015. Power generation remained generally stable in 2013, contributing to the good performance of the manufacturing sector. The estimate shows that the economy grew by 7.2% in 2014, driven by good performance in manufacturing, services, mining and quarrying, and agriculture. This growth is projected to reach 7.4% in 2015 [10, 39].

It is estimated that, 90% of the population in Tanzania relies on traditional biomass such as wood fuel for cooking, because the majority of these people are rural, poor and cannot afford the cost of modern energy sources such as electricity [2]. Even for those rural minorities who could afford, it is not possible because electricity is not readily available, as connection to distant grids is too expensive to be cost effective for many rural areas and there is no priority to electrify those poor people living in rural areas [3]. Consequently, the country faces a lot of challenges caused by unpredicted level of use of firewood and charcoal as well as high and frequently unstable price of oil import [3, 32]. Janbert [8] highlight that, for those rural households willing to use modern energy services were discouraged by the high costs of connection in the past few years.

2.2. Electricity Access and Generation Capacity

In recent years, the demand for new electricity connections has increased due to the improved energy policy done by the government, by increasing electrification rate through Rural Energy Agency (REA). Before 2013, the connection charges were uniform for both rural and urban areas. The charges were Tsh 455,104.76 without pole, Tsh 1,351,883.52 with one pole and Tsh 2,001,421.60 with two poles. In 2013, the Government issued new electricity connection charges by lowering the connection charges in rural and urban areas as shown in Table 1.

The electrification rate has increased from 14% in 2010 to about 24% in 2015 [10]. The increase is due to identified additional power demands from existing new customers and a special electrification program which tallies with government policy statement of connecting 30% of population by 2015 [31]. The government's policy to advance electrification and the significant reduction of the connection fee in early 2013

are the main reasons behind the sharp increase [37]

Table 1. New Connection Charges [38].

	No pole- one way	One pole- one way	Two pole- one way
Urban	Tsh 320.000.00	Tsh 515,618.00	Tsh 696,670.00
Rural	Tsh 177,000.00	Tsh 337,740.00	Tsh 454,654.00

The update and improvement of energy policy has resulted to the increase of new customers from 1.2 million in the end of 2013 to about 1.5 million people in 2015, effectively raising electricity access from 17.7% in 2013 to about 24% of the total population in 2015 [9, 37, 38]. However, Tanzania energy sector and rural electrification level still lags behind that of middle income countries and is in line with low income countries and that of SSA averages [11].The application of renewable energy technologies have the potential to alleviate poverty that face rural population of Tanzania, and will be a viable option because they can easily be decentralized thus providing energy in areas far from the national grid.

The power generating capacity in Tanzania up to the end of 2014 was 1583 MW. In October 2015, another Gas fueled power plant, Kinyerezi I with 150MW capacity was commissioned, thereby increasing the total generating capacity to 1733MW. Out of these, 561MW is hydro, 677MW thermal and 495MW liquid fuel power plants. The main sources of electricity are hydropower and thermal generation, and the power system (interconnected grid) comprises of generation units owned by TANESCO and IPP's (permanent and rental) [10].

The new energy policy implemented by the government and Tanzania Development Vision- 2025 (TDV) is aiming at raising the generating capacity to 10,000MW by 2025. This will be done by the government itself through it power utility company (TANESCO), and Independent Power Producers-IPP. It is expected that, after the implementation, the electricity generating cost as well as electricity cost for commercial, domestic and industrial will be reduced [32]. The reduction cost of energy will stimulate the economic activities especially for people in rural areas when they access modern energy services.

Tanzania, along with the Sub-Saharan African countries has experienced a prolonged drought happened in 2003, 2006, 2009 and 2011. These dry spells have often depleted the entire hydropower reservoir system. The worst situation was in 2006 and in 2011 in such a way that the country was threatened by complete closure of Kidatu and Mtera hydropower plants, which accounted for an average of about 25 % to the entire power system installed capacity. This led to severe energy shortages which resulted in power rationing, an experience that made the government to start focusing on exploitation of other renewable energy sources, mainly solar, wind, biogas and other biomass which are not mainly weather dependent. This paper therefore, aims to evaluate how renewable energy policies have informed the energy practices in this country in order to overcome the challenges facing Tanzania.

2.3. Energy Sector in Tanzania

According to the structure of energy sector of Tanzania shown in Figure 1, the Government through the Ministry of Energy and Minerals (MEM) is responsible for policy formulation. The Energy and Water Utilities Regulatory Authority (EWURA) is an autonomous multi-sectoral regulatory authority under the MEM, established by the Energy and Water Utilities Regulatory Authority Act, Cap 414 of the laws of Tanzania. It is responsible for technical and economic regulation of the electricity, petroleum, natural gas and water sectors in Tanzania pursuant to Cap 414 and sector legislation.

Other functions of EWURA include among others, licensing, tariff review, monitoring performance and standards with regards to quality, safety, health and environment. EWURA is also responsible for promoting effective competition and economic efficiency, protecting the interests of consumers and promoting the availability of regulated services to all consumers including low income, rural and disadvantaged consumers in the regulated sectors.

Rural Energy Agency-REA is an autonomous body also under the Ministry of Energy and Minerals of the United Republic of Tanzania. Its main role is to promote and facilitate improved access to modern energy services in rural areas of mainland Tanzania. REA facilitates rural energy development by working in partnership and collaboration with private sector, Non Governmental Organizations, Community Based Organizations, and Government agencies. REA in Tanzania perform the following main functions:-

1. Promote, stimulate, facilitate and improve modern energy access for productive uses in rural areas in order to stimulate rural economic and social development.
2. Promote rational and efficient production and use of energy, and facilitate identification and development of improved energy projects and activities in rural areas.
3. Finance eligible rural energy projects through Rural Electrification Funds (REF).
4. Prepare and review application procedures, guidelines, selection criteria, standards and terms and conditions for grants allocation.
5. Build capacity and provide technical assistance to project developers and rural communities.
6. Facilitate preparation of bid documents for rural energy projects.

REA has already supported various off-grid projects in small hydro power projects, biomass cogeneration projects, and biomass gasification projects in various parts of the country such as Mafia and Mkonge energy project in Tanga Region. The supported projects are currently at various stages of implementation and the total expected capacity is 46 MW, with the expectation of about 7400 new connections. REA support fiscal incentives for rural energy projects and programs and count amongst the National aid initiatives attracting fiscal initiatives. On top of Government subsidy to REA, the agency is also allowed to take up to 5% surcharge on each unit of energy generated by commercial electricity

producer. REA subsidies also support solar PV Systems. However, the subsidy is limited to 100Wp for domestic use and up to 300Wp for Institutions.

At industry level, there is Tanzania Electric Supply Company Ltd (TANESCO), which dominates the sector in the Generation, Transmission and Distribution of Electric power. There are also Independent power producers- IPPs whose provide additional power capacity to the generating industry.

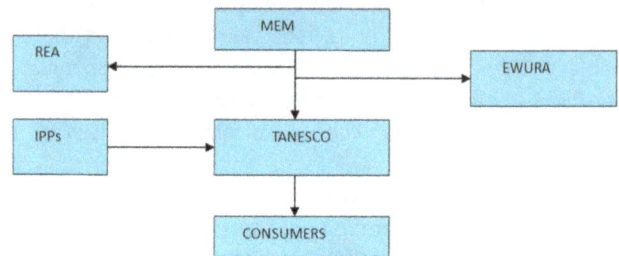

Figure 1. *The Structure of Energy Sector [15].*

All Independent Power Producers- IPPs generate and sell power to TANESCO, which connect it to the national grid. Electricity generation in Tanzania is from hydro, natural gas, coal, diesel oil, biomass and to a minor extent, solar photovoltaic [16].

2.4. Energy Policy

The first National Energy Policy (NEP) for the Country was formulated in 1992. Since then the energy sector has undergone a number of changes, necessitating adjustments to the initial policy. With various changes, the energy policy of 1992 was replaced in 2003.The objective of the 2003 NEP is to ensure availability of reliable and affordable energy supply and use in a rational and sustainable manner in order to support national development goals.

The National Energy Policy of 2003 aims to establish energy production, procurement, transmission, distribution and end-user systems in an efficient, environmentally sound, sustainable and gender-sensitized manner. Key objectives of the 2003 NEP regarding to Renewable Technologies (RT) and services include:

1. Encourage efficient use of alternative energy sources.
2. Facilitate Research and Development (R&D) and application of Renewable Energy for electricity generation.
3. Facilitate increased availability of energy service including off-grid electrification of rural areas.
4. Introduce and support appropriate fiscal, legal and financial incentives for Renewable Energy Technologies.
5. Ensure the inclusion of environmental consideration in energy planning and implementation.
6. Support Research and Development (R&D) in Renewable Energy Technologies
7. Establish norms, codes, of practice, standards and guidelines for cost-effective rural energy supplies and for facilitating the creation of an enabling environment

for the sustainable development of renewable energy sources.

8. Facilitate the creation of an enabling environment for sustainable development of Renewable Energy Sources.

9. Promote entrepreneurship and private initiatives for the production and marketing of products and services for rural and renewable energy.

10. Ensure priority on power generation capacity based on indigenous resources. The policy encourages public and private partnerships to invest in the provision of energy services. It also seeks to promote private initiatives at all levels and stresses the need to make local and foreign investors aware of the potential of the Tanzanian energy sector.

3. Renewable Energy Policies and Practices

The Government intends to develop these renewable energy sources so as to minimize production costs which will make electricity affordable to the majority of Tanzanians. Many times, Tanzania has improved its energy policy in order to encourage Independent Power Producers- IPP especially in remote locations because of the excessive cost of transporting electricity from large scale power plants to rural areas. The government also aims to contribute to at least 260MW of new renewable power generation being connected to the national grid by 2016 [31].

3.1. Biomass

Tanzania Government has issued ban for more than ten years ago on production and transportation of charcoal in order to stop illegal deforestation. This enforcement has not well succeeded because of delay and corruption in issuance of licenses for sustainable charcoal production and therefore allows illegal charcoal to dominate the market. Napendael [3] point out that, the average daily consumption of charcoal in Dar es Salaam is estimated to be 24,000 bags per day, and it has been revealed that only 10-20% of this amount passes through legal checkpoints and thus earning the government revenue. Biomass comes in a variety of forms, which can be utilized as an energy resource. It is possible to classify the material into two main groups; wood biomass and agro-forestry waste (that is, crop wastes, animal manure and forestry processing wastes).

These materials can be burnt directly or first converted into solid (charcoal), liquid (ethanol) and gaseous fuels (biogas, producer gas). The main source of fuel in both urban and rural areas of Tanzania is biomass in the forms of charcoal and fuel-wood. Charcoal is the energy source that is made from wood, while fuel-wood is collected and used directly from the field [4]. Biomass resources are mostly derived from forests, wood logging and agricultural residue, animal dung, solid industrial waste and landfill biogas.

The total forest area of Tanzania is about 39% which supply about 37% of the total biomass energy resources such as firewood. The other biomasses, agricultural residue, animal dung, solid industrial waste and landfill biogas cover the remaining part [35]. The main biomass energy use in most rural area of Tanzania includes charcoal making, direct firewood and dung. The utilization of this conventional biomass is still high due to poor conditions facing rural areas.

3.2. Biogas Practices

Biogas is a commonly used biofuel around the world and is generated through the process of anaerobic digestion or the fermentation of biodegradable materials such as biomass, manure, sewage, municipal waste, rubbish dumps, septic tanks, green waste and energy crops. This type of biogas comprises primarily methane and carbon dioxide, which is combustible and when burnt will produce heat. Domestic biogas in Tanzania was introduced by Small Industries Development Organization (SIDO) in 1975.

A number of other Non Governmental Organizations (NGOs), joined in the promotion of this technology all around the country whereby the involvement of CAMARTEC, later in cooperation with GTZ accelerated awareness and dissemination, particularly in the northern regions of the country. CAMARTEC and GTZ carried this work forward in the 1980s-1990s by developing, promoting and providing training in the biogas sector, where during those years, interested parties built around 6,000 biogas digesters [17].

It is estimated that, about than 7,133 domestic biogas plants have been built countrywide for domestic and commercial applications since 2009 as shown in Figure 2. However, as these new technologies get rolled out to more remote areas, biogas invariably encounters some isolated local cultures in few areas of the country. For example in predominantly Muslim households it is difficult to convince the community to use pig dung to generate energy. Studies have revealed that pig dung is more efficient fuel than cow dung [30].

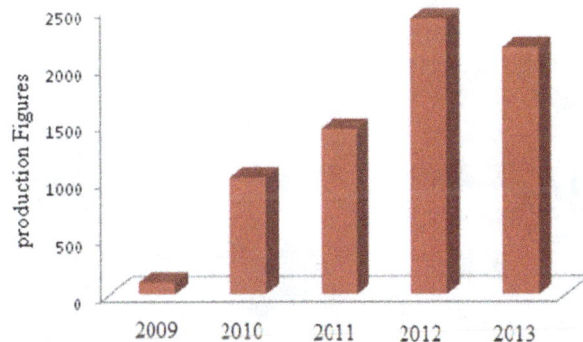

Figure 2. Annual production of biogas Plants in Tanzania since 2009 [17].

Based on the 2007 - 2008 feasibility study and the 2012 program implementation document, Tanzania Domestic Biogas Program (TDBP), estimates that the technical potential for domestic biogas in Tanzania is around 165,000 households [17].The potential is the number of household with basic requirements of enough availability of dung and water, the ambient temperature, the availability of construction materials, enough land (space) for plant installation, freedom from natural disasters like floods and earth quakes and availability of human resources for plant construction.

To date, different and affordable digester plants have been introduced and promoted in many parts of Tanzania. Recently, Dar es Salaam Institute of Technology (DIT) has developed a portable biogas plant shown in Figure 3, made from plastic containers which can be used by rural households. Biogas is used mostly for cooking and lighting by people who do not have access to grid connection in rural areas.

Figure 3. A biogas plant using plastic containers made by DIT [30].

Biogas technology has been disseminated and promoted by various NGOs, and by the support of Government through MEM purposely to increase the awareness to the general community in dissemination and promotion of this technology in order to reduce the dependence in fuel wood and charcoal as the only source of energy in rural areas of the country.

However, among the factors which seem to block the widespread initiation of biogas systems include the lack of enough education and information regarding their potential benefit, lack of construction skills, high cost of digester construction, poor construction design and maintenance and lack of financial support in the community [17].

3.3. Co – generation

In Tanzania, sugar industries use bagasse waste from sugar-cane processing to produce steam for running a turbine for electricity generation. Bagasse is a fibrous waste product produced after sugarcane has been crushed. Annual production is about 776,000 tones which is about 33% of the weight of the crushed sugarcane from all sugar Industries in the Country [13]. Table 2 shows the present installed capacity of sugar industries for generation from bagasse in Tanzania which is 40MW with estimated energy generation potential of 99.42GWh per year [14, 23].

Co-generation potential from bagasse, wood waste and from a sisal waste plant such as the one shown in Figure 4 in Tanzania is estimated at 395 MW. At present the installed capacity is 61.3 MW from sugar and wood-based industries shown in Table 3. Currently the generation capacity in most of the sugar and wood based biomass factories is designed to cover the requirements of the factory only, there is no export to the national grid and therefore, their contribution to the energy mix in the nation is low.

Table 2. Power generation and installed capacity from sugar industries [13, 14, 22].

Factory Name	Cane crushing capacity (tons/day)	Bagasse available (tons/day)	Installed capacity (MW)	Electrical energy generation (Gwh/yr)	Internal usage (Gwh/yr)	Export (Gwh/yr)
Kagera	60	3000	5	15.84	15.84	NIL
Mtibwa	350	2511	4	23.10	23.10	NIL
Kilombero	180	13,729	10.6	39.4	39.4	NIL
TPC	130	2674	20	21.09	21.09	NIL
TOTAL	720	21914	39.6	99.43	99.43	NIL

Figure 4. Sisal Energy (Biomass) Plant – Hale, Tanga [22].

Table 3. Existing Bio- mass fueled Power Plants in Tanzania [13, 22].

Station	Installed capacity(MW)	Energy source
Mtibwa Sugar Estate Ltd	4	Co-generation-bagasse
Tanganyika Planting Company Ltd-TPC	20	Co-generation-bagasse
Kilombero Sugar Company Ltd	10.6	Co-generation (bagasse)
Kagera Sugar Estate Ltd	5	Co-generation-bagasse

Station	Installed capacity(MW)	Energy source
Tanzania Sisal Board-Hale	0.3	Biomass-Sisal waste
Tanganyika Wattle Company Ltd	2.5	Biomass-Wood waste
Sao Hill Industries- Mufindi	15	Biomass-Wood waste
TANWAT-Njombe	2.5	Biomass- Wood-waste
Ngombeni Power Ltd-Mafia	1.4	Biomass- Wood-waste
Total	61.3	

Studies have revealed that, Tanzania has the potential of generating more than 395MW of electricity per annum from biomass sources. However, the contribution of biomass in the energy mix of the country is still small [30].

3.4. Liquid Bio-fuels

Liquid bio-fuels are liquid energy sources derived from plant materials specifically used to replace or supplement conventional petroleum-based fuels. Liquid biofuel can be used in existing vehicles with little or without any modification of engines and fueling systems. There is high potential for production of biodiesel and bio-ethanol in Tanzania [24]. Food and Agriculture Organization (FAO) and Government of Tanzania have identified a number of bio-fuel production scenarios using different feedstock crops and different types of downstream processing plants. In the analysis they focused on a subset options in order to capture the core difference in these crops.

The research shows that jatropha is a potential viable feedstock for biodiesel [24]. Jatropha curcas (JC) is a perennial small tree or large shrub, which can reach a height of up to 5 m, and is an ever green drought-resistant species that sheds its leaves during very dry periods. It is adapted to arid and semi-arid conditions, curently in Arusha and Moshi as shown in figure 5. The current distribution of Jatropha shows that introduction has been most successful in drier regions of the tropics with an average annual rainfall between 300 and 1000mm [39].

(a)   (b)

Figure 5. Jatropher plantation at Kikuletwa Moshi. (a) 1.2 m height, 8 months after planting, (b) 3.5 m height, 5 years old trees [39].

Only recently Tanzania has started production and marketing of straight vegetable jatropha oil for use in adapted car engines, but the output is still negligible. Nationally produced biodiesel is so far not available at competitive prices. Tanzania is highly dependence on oil imports, which places great strain on the country's balance of trade, especially in the time of soaring oil prices. The interest which has been shown by the Government for the production of oil liquid biofuels has great potential in terms of economic development for Tanzania by reducing the oil bill and support rural development [18, 23, 25].

3.5. Small Hydropower Practices

The estimation of small, mini and micro hydro potential in Tanzania is 4800MW at National level [20]. The detailed studies carried out on site surveys have identified more than 85 mini hydropower sites with a total potential of 187MW [26]. Many of these sites are in rural areas and are suitable for standalone systems for supplying power to small communities away from the grid. The existing small hydropower plants in Tanzania are given in Table 4 [26].

Table 4. Existing small, mini and micro hydro Plants [20, 21, 38].

Location	Year Installed	Turbine Type/manufacture	Installed Capacity(kW)	Owner
Sakare (Soni)	1948	Geisel Brecht	0.0063	Benedictine Fathers
Mbarali (Mbeya)	1972	Chinese	0.7	NAFCO/Govt
Ndolange(Bukoba)	1961	B. Maler	0.055	RC Mission
Ikonda (Njombe)	1975	CMTIP	0.04	RC Mission
Makumira (Arusha)	2011	Gross Flow/Ossberger	0.01	J Mungure
Ngarenanyuki (Arusha)	2011	Gross Flow/Ossberger	0.01	Ngarenanyuki Sec School
Tosamaganga (Iringa)	1951	Gilkes& Gordon/Francis	1.22	TANESCO
Kikuletwa (Moshi)	1937	Boving & Voith Reaction	1.160	TANESCO
Mbalizi (Mbeya)	1958	Gilkes& Gordon/Francis	0.34	TANESCO
Kitai (Songea)	1976	Gross Flow/Ossberger	0.045	Prison Dept/Govt
Nyagao (Lindi)	1974	N/A	0.0158	RC Mission
Isoko (Tukuyu)	1973	N/A	0.0155	Morovian Mission
Uwemba (Njombe)	1971	N/A	0.8	Benedictine Fathers
Bulongwa (Makete)	-	N/A	0.18	-
Kaegesa (S'wanga)	1967	N/A	0.044	RC Sumbawanga
Rungwe (Tukuyu)	1964	N/A	0.0212	Morovian Mission
Nyagao (Lindi)	1974	N/A	0.0388	RC Mission
Isoko (Tukuyu)	1973	N/A	0.0073	RC Mission
Ndanda (Lindi)		N/A	0.0144	RC Mission
Ngaresero (Arusha)	1982	Gilbsk	0.155	MH Leach
Mamba (Katavi)	1932	Gross Flow/Ossberger	0.01	Mamba Mission

Location	Year Installed	Turbine Type/manufacture	Installed Capacity(kW)	Owner
Mwenga Mini Hydro Ltd	2012	N/A	4.00	Community
Mapembasi hydro Power Ltd (Njombe)	-	N/A	12.00	Community
ACRA Tanzania	2013	N/A	0.3	Community
Andoya Hydroelectric Power Ltd	2013	N/A	1.0	Community
Kitonga Hydro (Ikololo)	-	N/A	10	Community
Ndola Hydro (Ruhuhu)	-	N/A	10	Community

Implementation of some of these projects started before independence but uptake has been slow because of the lack of encouragement and sufficient fund to support individuals and Non-Governmental Organizations in the uptake of this technology.

3.6. Solar Energy

Solar resources are good in the central portions of the country. This makes it naturally a suitable country for the application of solar energy as a viable alternative to conventional energy sources if efficiently harnessed and utilized. Both solar PV and solar thermal technologies are in development in the country. As grid electricity reaches about only 7 % of the rural population in Tanzania, the use of solar electricity seems to be an attractive option. The country average annual solar radiation levels are said to range between 4.2-5 kWh/m^2 per day. This solar energy is equivalent to 210 million tons of oil equivalent (Toe) [19].

The lowest annual average radiation value in the country is found to be $15 MJm^{-2}day^{-1}$ while the maximum value is $24 MJm^{-2}day^{-1}$. The lowest radiation value in many parts of the country is obtained in July (winter) which is sufficient to satisfy the needs of rural family demand. The important use of solar energy in Tanzania include solar thermal for heating, drying and photovoltaic (PV) for lighting, water pumps, refrigeration purposes and telecommunication. The solar energy market of Tanzania has grown and increased over the last few years.

The rapid increase of solar equipment market is due to partly, the need of providing electricity to houses and institutions located in remote areas, where there is no grid connection and also to supply power to equipments like water heaters which are used in domestic and commercial applications. The demand for SHS has been driven by the spread of broadcasting signals and the availability of TV sets and radios in rural areas, where they have become the biggest segment in Tanzania's solar market.

Currently, there are about 65,000 houses in Tanzania with solar PV panels ranging from 10-100kW per house. In 2008 the installed capacity was approximately $1MW_p$ and doubled to $2MW_p$ in 2009.This capacity was an estimate of 40,000 SHS in 2008 with annual sales of 4,000-8,000 SHS [28]. Individuals purchase solar equipments and accessories from whole sale dealers and retail shop owners who import them from abroad. Tanzania has free market, therefore solar equipments import is done by several companies, vendors and several installers.

Recent estimates on the installed capacity of PV systems in the country is about 1.7 MW, however market potential for solar PV countrywide is estimated to be 20.2MW. Tanzania is now experiencing significant growth in its PV market 350kW

in 2008 to about 500kW in 2012 [27].The Government is carrying out awareness and demonstration campaigns on the use of solar systems for domestic and industrial use, as well as supporting direct installation in institutions. VAT and import tax for main solar components (panels, batteries, inverters and regulators) have been removed to allow the end users to get PV systems at more affordable price.

Implementation of a solar PV for electrification of schools and other institutions which are far from the grid is currently done by the support Government and some NGOs as part of utilization of renewable source of energy to the overall energy supply [32]. It is anticipated that by 2025, about 800MW, will be fed to the national grid from solar power generation which will be located in the central part of the country. Although Tanzania has high levels of solar energy, ranging between 2800-3500 hours of sunshine per year, it seems the percentage contribution to the total energy mix is still small [32].

3.7. Wind Energy

Amjad et al [41] insist that, the beginning of twenty one century has been of an exciting time for wind energy. This is due to the various changes in technologies, policies, environmental concerns and challenges facing electricity industry infrastructure. Therefore, the coming years offer many opportunities for wind energy to emerge as a viable electricity source. Tanzania has large areas with average wind speeds of 5-7m/s. It is coupled with existing long coast line of about 800km with prevailing surface winds, moving from south east to north east.

There has been a trial by individual people in Tanzania to attempt to generate electricity from wind but there is no success. The wind energy development approach which has been used in Tanzania is perhaps wrong and that may have been the reasons for the failure to generate electricity from wind. Wind turbines and windmills have been installed without proper investigation on wind speed characteristics at the prospective sites [30, 34, 36].

Wind Energy has been used primarily in Tanzania for wind mills to pump water. However, their uses have declined due to lack of maintenance as well as the alternatively use of internal combustion engines, which are sometimes flexible and cheap compared to wind mills. It is estimated that about 101 wind mills have been installed in Tanzania for water pumping purposes [30]. Currently, interest of the community in wind power electricity generation is expanding due to factors such as the rising cost of oil, increased demand of power, and effect of long draught on hydropower.

Recent researches in Tanzania have shown that, wind farms for commercial plants appear promising at Makambako and Kititimo in Singida. Areas along rift valleys, the southern high lands and along Lake Victoria are reported to have some possibilities of potential wind sites [30]. At present, there are no grid connected wind turbines in Tanzania but the Government has shown much effort to ensure the utilization of this resource. There are about 7 potential wind sites located for electricity generation at Singida, Makambako and Mkumba. The constructions of the wind farms (300MW) in Singida have already started and are expected to generate in phases until its completion in 2018 [29, 33].

4. Conclusion

Tanzania has great potential of renewable energy which could supply about 50% of the electricity. Much focus has been on large hydropower and thermal power projects leaving sources such as biomass, solar, biogas and wind under exploited. There is a need to harmonize policies addressing issues on renewable energy exploitation in order to ensure timely implementation of the planned energy projects. The regular review of the existing policies to ensure encouragement of private investment in the energy sector and competitiveness should also be ensured, as renewable energy practices are in line with existing policies. However, adoption of renewable energy technologies has been slow mainly because, most of these require high initial costs. The current energy policy and the subsidy from the government though REA have encouraged private partnership in implementing solar and wind energy projects.

References

[1] Christine Njiru, Sammy Latema, Simon Maingi and Patrick Gichoi. Renewable energy policies and practice in Kenya and their contribution to appropriate Technology adoption. 6th International Conference on appropriate technology proceedings, Kenyatta University, (2014).

[2] Anders Larsen. Evaluating the development impacts of a solar PV projects in Tanzania. A Traineeship project submitted in Sustainable energy planning and management, Aalborg University, Sweden, (2007).

[3] Napendaeli S. Supply/Demand chain analysis of charcoal/firewood in Dar es Salaam and Coast regions and differentiation of target groups, (2004).

[4] David Banner, Melinder Sundell, Jacqueline Senyagwa and Jeremy Doyle. Sustainable energy markets in Tanzania. RENETECH, (2012).

[5] Lins, C. Global status report 2012 available at: http://www.thccngineei.co.UKjournals/2012/06/11/r/o/f/RenewableS-2012-GLOBAL-STATUS-REPORT, (2012).

[6] Karekezi. S and Kimani J. Status of power sector reform in Africa. Impact on the poor. Energy policy, 30, 923-924, (2002).

[7] Karekezi. S. Renewable energy in Africa- meeting the energy needs for the poor. Energy policy 30, 1059- 1069,(2002).

[8] Janbert Kiwia. River resource towards sustainable development of Tanzania. A contribution of Hydropower to the energy security in Tanzania. A case study, Rufiji River Basin.

Master's thesis submitted in the department of earth science, Upsala University, (2013).

[9] Christian Matyetele Msyani. Current status of energy sector in Tanzania. An executive exchange on developing an ancillary service market. USEA- WASHINGTON DC, 25th February-2nd March 2013, (2013).

[10] Ministry of Energy and Minerals (Government of Tanzania). electricity supply industry reform strategy and roadmap 2014-2025, (2014).

[11] Helene Ahlborg and Linus Hammer. Drivers and barriers to rural electrification in Tanzania and Mozambique – grid extension, off grid and renewable energy sources. World renewable energy congress, 8-13 May 2011, Linkoping, Sweden, (2011).

[12] President's office, Planning Commission (Government of Tanzania).Tanzania Development Vision. Unleashing Tanzanian's latent growth potentials, (2012).

[13] Gwang'ombe F. Renewable energy Technologies in Tanzania, Biomass Based Cogeneration. Second draft report, (2004).

[14] Ministry of Finance (Government of Tanzania). Handbook of Economic review, (2012).

[15] Ministry of Energy and Minerals (Government of Tanzania). Energy and Water Utilities Authority (EWURA). Annual report for the year ended 30th June, 2010. (2010).

[16] Casmiri, D. Energy Systems. Vulnerability- Adaptation-Resilience (VAR), - Regional Focus, Sub Sahara, (2009).

[17] Ngwandu et al. Tanzania Domestic Biogas Programme-(TDBP). Programme implementation document, Final Version, (2009).

[18] Ministry of Energy and Minerals. (Government of Tanzania). Guidelines for sustainable Liquid Biofuels development in Tanzania, (2010).

[19] United Nation Development Programme (UNDP). Opportunities and challenges in Tanzania, (2013).

[20] Hankins M. Tanzania's small hydro energy market. Target market Analysis, (2009).

[21] Justina Uiso. Rural energy and innovation in Delivery of modern energy services to rural areas, (2011).

[22] Annie D. Biomass generation at Saw Mill Plant. Building competitive advantage for IFC'S clients. A case study of Sao Hill Industry, Tanzania, (2009).

[23] Jasper V. Fueling Progress or Poverty? The EU and bio-fuels in Tanzania. Policy coherence for Development in Tanzania, (2013).

[24] Kiplagat et al. Renewable energy in Kenya, 'Resource potential and status of exploitation. Renewable and Sustainable Energy reviews- 15, page 2960 – 2973, (2011).

[25] Nepomuki et al, Economic viability of jatropha curcas L plantations innorthern Tanzania, (2009).

[26] Kato T and Florence G. Challenges in Small Hydropower Development in Tanzania. Rural Electrification perspective, International conference on small Hydropower – HydroSrilanka, held on 22th – 24th October, 2007, (2007).

[27] Tanzania Country Report. General operating environment. Energy and renewable energy and Environmental Governance. Information available at www./aurea.fi/en/connect/../Tanzania, (2011).

[28] Janosch O. The Sun rises in the East Africa. A comparison of development and status of the solar energy markets in Kenya and Tanzania. Working paper FNU-197, (2011).

[29] Mwihava et al. Draft Country Study Research report. Ministry of Energy and minerals, (2011).

[30] Mashauri A. A review on the renewable Energy Resources for rural application in Tanzania. Renewable energy – Trends and Applications, (2012).

[31] Ministry of Energy and Minerals (Government of Tanzania). Power System master plan, 2012 update, (2013).

[32] Ministry of Energy and Minerals (Government of Tanzania. Inverstment plan for Tanzania. Scaling up Renewable energy program. Investment plan for Tanzania. A Report submitted to World Bank, (2013).

[33] Nzali, H and Mushi, S. Wind Energy Utilization in Tanzania. PREA workshop, (2006).

[34] Kainkwa R. Wind Energy as an alternative source to alleviate the shortage of electricity that prevails during dry season. A case study of Tanzania. Renewable energy – 18, page 167-174, (1999).

[35] Ministry of Food and Agriculture (Government of Tanzania). Global Forest Resources assessment, FRA – 2010- country report, Tanzania, (2010).

[36] Kainkwa R. Wind speed and the available wind power at basotu in Tanzania. Renewable energy – 21, page 289 – 295, (2000).

[37] Ministry of Energy and Minerals(Government of Tanzania). National Electrification Program Prospectus-ENNEXES, (2014).

[38] Ministry of Energy and Minerals (United Republic Of Tanzania)-Tanzania Electric Supply Company Ltd-TANESCO, 2014.

[39] Mathis K. Jatropha production in semi arid areas of Tanzania. RLDC feasibility study, June, 2007, (2007).

[40] Prosper Charle and Rogers Dhliwayo. African economic outlook. TANZANIA, 2015, information available at: www.africaeconomicoutlook.org, (2015).

[41] Amjad Ali, Furqan Habib and Sheraz Alam Malik. Wind energy development policies in Developing Countries and their effects: Turkey, Egypt and prospects for Pakistan. American Journal of Energy and Power Engineering. Vol.2, No.5, 2015, pp. 56-61, (2015).

19

Performance of Solar Still with Different Phase Change Materials

Naga Sarada Somanchi, Anjaneya Prasad B, Ravi Gugulothu, Ravi Kumar Nagula, Sai Phanindra Dinesh K

Department of Mechanical Engineering, JNTUH College of Engineering, Kukatpally, Hyderabad, Telangana State, India

Email address:

nagasaradaso@gmail.com (N. S. Somanchi), ravi.gugulothu@gmail.com (R. Gugulothu), ravikumarnagula145@gmail.com (R. K. Nagula)

Abstract: Water is basic necessity of man. Fresh water sources are considered to be rivers, lakes and underground water reservoirs. Although, more than two-third of the earth is covered with water and remaining of the earth is land. However, the use of water from such sources is always not good, because of the polluted environment. All over the world, accessing of portable water to the people is narrowing and decreased day by day. Most of the human diseases are due to polluted or un-purified water. Nowadays, each and every country facing a problem of huge water scarcity because of pollution created by manmade activities. Under these circumstances, search for other sources becomes a must. A system is needed which supplies pure water without effecting the ecosystem and environment friendly. Adequate quality and reliability of drinking water supply is a fundamental need of all people on the earth. Fresh water, which was obtained from rivers, lakes and ponds, is becoming scarce because of industrialization and population explosion. The sun is regarded as the source of energy for its constant duration and hygienic state and its remarkable efficiency of not polluting the environment, as other kinds of energy like coal, oil that cause the pollution of atmosphere and environment. Water purification using solar energy has become more popular because it is eco-friendly and cost effective. A solar still is commonly used device for water purification and it doesn't require any electricity for distillation of water. A variety of solar distillation devices have been developed with different materials and in different shapes in different location to improve the efficiency of solar distillation. This article communicate about the distillation of solar still by using different methods in different areas to improve the efficiency of the solar still.

Keywords: Renewable Energy, Water, Solar Energy, Phase Change Materials, Energy Storage Materials

1. Introduction

Energy is an essential factor for the social and economic development of the societies. Renewable energy is accepted as a key source for the future on this earth. The combined effects of the deflection of fossil fuels and the gradually emerging consciousness about environmental degradation have given the first priority to the use of renewable alternative energy resources in the 21st century. All of renewable, solar thermal energy is considered to be practically unlimited in the long term and is a very abundant resource in the world. Many conventional and non-unconventional techniques have been developed for purification of saline water. Among these water purification systems, solar distillation proves to be economical and eco-friendly technique.

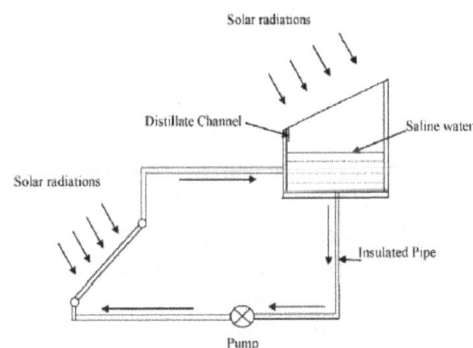

Fig. 1. *Active type solar still.*

There are two different types of solar systems; those are active type and passive type. The parameters which are affecting the solar still are; water depth in the basin, material of the basin, wind velocity, solar radiation, inclination angle of glass cover and ambient temperature. The yield of water from the solar still is depend on the temperature difference between the water in the solar still or basin and glass cover inner side temperature. The yield from solar still is directly proportional to the temperature difference of water in solar still and in side of the glass cover.

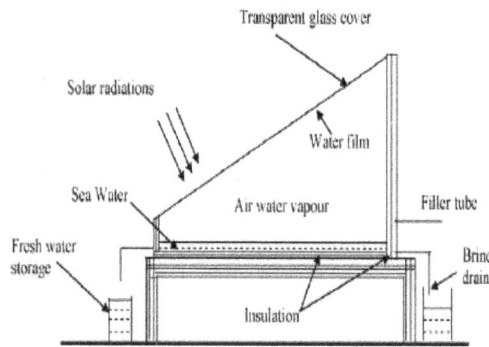

Fig. 2. Passive type solar still.

In a passive solar still, the solar radiation is received directly by the basin or solar still water and is only source of energy for raising the water temperature, so the evaporation leading to a lower productivity of pure water. In active type solar system extra thermal energy is supplied to the basin through an external mode to increase the evaporation rate and productivity of pure water. Passive solar systems give lower yield when comparing with active solar systems.

Among the non conventional methods to disinfect the polluted water, the most prominent method is solar distillation. The solar distillation method is more attractive than other methods. This method require simple technology as no skilled workers needed, low maintenance and it can be used anywhere without problems. The work done by previous researchers in obtaining distilled water using solar energy is listed below:

1.1. Literature Review

Al Hamadani A.A.F and Shukla S.K (2011) conducted experimental investigations on a solar still with lauric acid as phase change material (PCM). They found that the higher mass of PCM with lower mass of water in solar still basin increases the daily productivity and the efficiency. The distillate productivity at night and on day for solar still without PCM 30% to 35% and with PCM increased by 127%.

El Sebaii A.A (2009) et al, studied the still performance with and without the stearic acid as PCM by computer simulation on summer and winter days. He concluded as after sunset, the stearic acid (PCM) as a heat source for the basin water until sun rise in the early morning hours of the next day. The PCM becomes more effective at lower masses of basin water during the winter. On a summer day, the daily productivity of the still is higher with PCM.

Mona M Naim and Mervat A Abd El Kawi (2002) had

constructed a single stage solar still that made use of phase change energy storage mixer. They found that the use of an energy storage material led to a larger productivity of distilled water and that the larger the concentration of the saline water, lower the productivity. Also higher flow rate and high inlet saline water temperature improved the still efficiency.

Nijmeh S, Odeh S and Akash B (2005), experimentally studied a single basin solar still using various absorbing materials like violet dye, charcoal, potassium permanganate ($KMnO_4$) and potassium dichromate ($K_2Cr_2O_7$). The best result obtained by violet dye i.e. 29%.

Swetha K and Venugopal J (2011), experimentally studied on a single slope single basin solar still by adding a heat reservoir under the liner of the basin using Lauric Acid as a phase change material. They observed that 13% increment when the still is used with sand as heat reservoir and 36% increment when the still is used with Lauric Acid as PCM.

From the literature review it is observed that, experimental investigations were conducted on solar distillation by previous researchers using energy absorbing materials like gravels, sponge, charcoal etc. But the investigations using phase change materials are less. The advantage of using phase change materials is, they are better in energy absorbing as well as release of energy. The energy absorbed by PCM during day time is reduced after sunset, to maintain constant temperature of water in solar still. This helps in increasing the productivity of solar still. Hence, present experimental investigations are conducted using PCMs sodium sulphate (Na_2SO_4), sodium acetate ($C_2H_3NaO_2$) and potassium permanganate (KMO_4) used.

1.2. Experimental Work

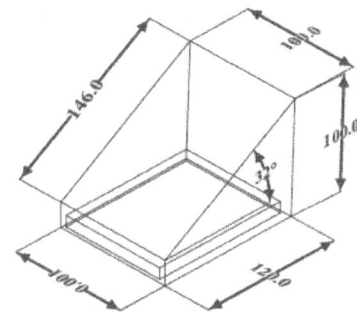

Fig. 3. Single slope solar still.

Figure 3 presents a schematic diagram of the solar still used in the present experimental study. It consists of a stainless steel basin which has an effective area of $1m^2$. This solar still is made of stainless steel with all dimensions in cm as shown in figure 3. The stainless steel sheet has a thickness of 0.8mm. It consists of a top cover of transparent glass with a tilt of 32^0 and is coated with black paint to absorb the maximum possible solar energy. This solar still faces south direction. The entire assembly is made air tight with the help of rubber gasket and clamps. Water enters the basin through an inlet valve.

To maintain constant water level of 8 cm, a floater is arranged inside the solar still. The distilled water is condensed on the inner surface of glass cover and runs along its lower edge. The distillate was collected in a bottle and measured by a graduated cylinder. Thermocouples were located at different places of the solar still to measure temperatures such as outside glass cover, inside glass cover, basin water temperature, vapor temperature and ambient temperature. In this experiment, potassium dichromate ($K_2Cr_2O_7$), sodium acetate (CH_3COONa) and Potassium Dichromate (KMO_4) are used as phase change materials. To enhance the performance of solar still, all the experimental works conducted in the month of February in Hyderabad, India.

1.3. Principle of Solar Desalination

A basin of solar still has a thin layer of water, a transparent glass cover that covers the basin and channel for collecting the distillate water from solar still. The glass transmits the sun rays through it and saline water in the basin or solar still is heated by solar radiation which passes through the glass cover and absorbed by the bottom of the solar still. In a solar still, the temperature difference between the water and glass cover is the driving force of the pure water yield. It influences the rate of evaporation from the surface of the water within the basin flowing towards condensing cover. Vapour flows upwards from the hot water and condense. This condensate water is collected through a channel.

1.4. Measuring Instruments

Pyranometer, Multimeter, Glass beaker, Pt-100 type thermocouples and infrared thermometer. Pyranometer is used to measure the direct solar radiation and diffused radiation. Glass beaker is used to measure the distillate water from the solar still. Pt-100 type thermocouples are used to measure the temperature of water which is in the basin or solar still, inclined glass cover inside and outside temperatures. Infrared thermometer is used for measuring the atmospheric temperature.

1.5. Results and Discussion

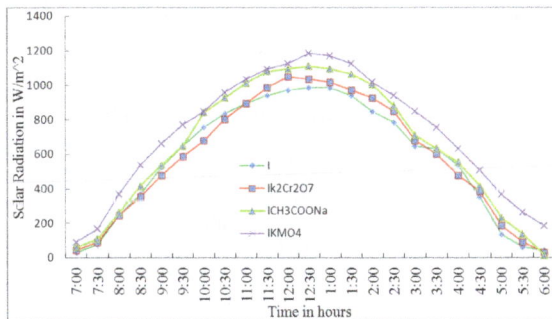

Fig. 5. Variation of Basin Temperature with Time.

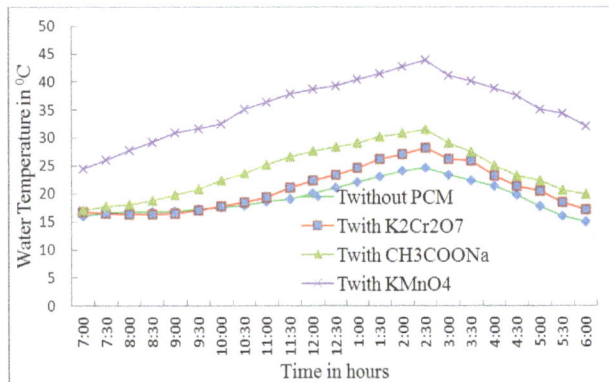

Fig. 6. Variation Water Temperature with Time.

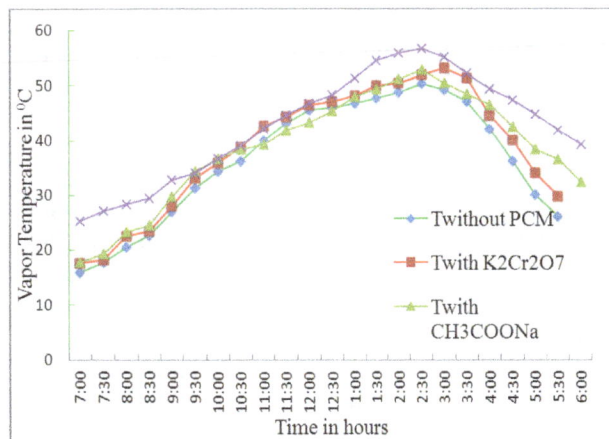

Fig. 7. Variation of Vapour Temperature with Time.

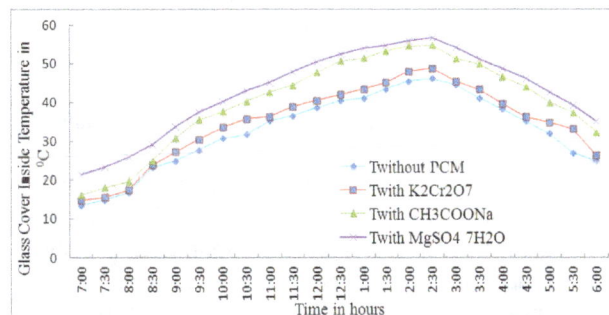

Fig. 4. Variation Solar Radiation with Time.

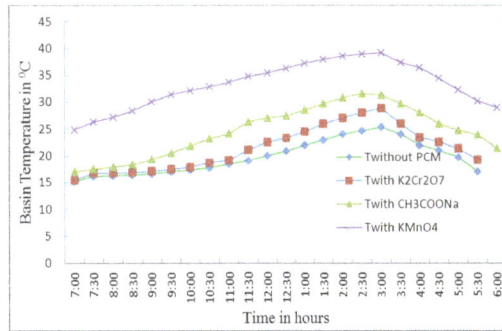

Fig. 8. Variation of Glass Cover inside Temperature with Time.

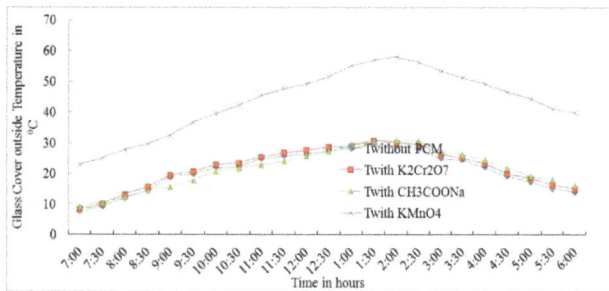

Fig. 9. *Variation of Glass Cover outside Temperature with Time.*

2. Conclusions

Energy and water are the basic necessity for all of us to lead a normal life on this beautiful earth. Solar energy technologies and its usage are very important and useful for the developing and under developed countries to sustain their energy needs. The use of solar energy in distillation process is one of the best applications of renewable energy. The solar stills are user friendly to the human being in the nature.

The present study focuses on the design and fabrication of efficient solar distillation system. It is more economical, therefore, to store water rather than store energy. It is beneficial in the cases of unavailability of electrical energy / fuel energy. Producing fresh water by a solar still with its simplicity would be one of the best solutions to supply fresh water with no technical facilities.

In this present experimental study, presence of Potassium Permanganate (KMO_4) in water could provide better yield when compared to that of Sodium acetate ($C_2H_3NaO_2$) and Potassium Dichromate ($K_2Cr_2O_7$) used as Phase Change Material. This may be due to the melting point temperature of Potassium Permanganate (KMO_4) is 240^0C, Sodium Accetate (324^0C) is higher than that of sodium acetate (324^0C). This experimental work conducted with Potassium Dichromate ($K_2Cr_2O_7$), Sodium Acetate ($C_2H_3NaO_2$) conducted in winter season and with Potassium Permanganate (KMO_4) was conducted in summer season. So, in summer temperature is more and more comparing with winter.

It has been demonstrated that the productivity of a solar still can be greatly enhanced by the use of phase chance materials. Future work should be directed towards preheating the saline water, preheating the saline water at different temperatures, solar still with flat plate collectors and solar still with flat plate collector using different phase change materials.

From the literature review, it is observed that, previous researchers conducted experimental investigations on obtaining distilled water with the help of solar energy in the presence of flat plate collectors, Phase Change Materials (PCM), mirrors, dyes, reflectors, cooling systems etc. But the work using combination of sun tracking system (single axis or double axis) coupled with PCM, dyes, sponges and nano-materials are limited. Hence, there is a scope to conduct experimental investigations on this topic.

References

[1]　Al-Hamadani.A.A.F and Shukla.S.K, "Water Distillation Using Solar Energy System with Lauric Acid as Storage Medium", International Journal of Energy Engineering 1(1): 1-8, 2011.

[2]　El-Sebaii.A.A, Al-Ghamdi.A.A, Al-Hazmi.F.S and Adel S Faidah, "Thermal performance of a single basin solar still with PCM as a storage medium", Applied Energy, 86, 1187-1195, 2009.

[3]　Hima Bindu Banoth, Bhramara Panithapu, Ravi Gugulothu, Naga Sarada Somanchi, Devender G, Devender V and Banothu Kishan, "A Review on Performance of Solar Still with Different Techniques", Proceedings of International Conference on Renewable Energy and Sustainable Development (ICRESD 2014), pp: 393-398, 2014.

[4]　Mona M.Naim, Mervat A.Abd El Kawi, "Non conventional solar stills part 2. Non conventional solar stills with energy storage element", Desalination 153, 71-80, 2002.

[5]　Nagasarada Somanchi, Hima Bindu Banoth, Ravi Gugulothu and Mohan Bukya, "A Review on Performance of Solar Still Coupled with Thermal Systems", Proceedings of 2014 1st International Conference on Non Conventional Energy (ICONCE 2014), pp: 114-118, 2014.

[6]　Naga Sarada Somanchi, Hima Bindu Banoth, Ravi Gugulothu and Mohan Bukya, "A Review on Parametric Performance of Solar Still", International Conference on Industrial Engineering Science and Applications-2014 (IESA-2014), Organized by Department of Electrical Engineering, NIT Durgapur, held on 2nd to 4th April, 2014. pp: 167-172, ISBN: 978-93-80813-27-1.

[7]　Naga Sarada S, Banoth Hima Bindu, Sri Rama Devi R and Ravi Gugulothu, "Solar Water Distillation Using Two Different Phase Change Materials", International Mechanical Engineering Congress 2014 (IMEC-2014), held at Department of Mechanical Engineering, NIT Tiruchurappalli, held on 13th – 15th June, 2014.

[8]　Naga Sarada S, Banoth Hima Bindu, Sri Rama Devi R and Ravi Gugulothu, "Solar Water Distillation Using Two Different Phase Change MAterials", Applied Mechanics and Materials, Volume: 592-594, pp: 2409-2415, 2014.

[9]　Naga Sarada Somanchi, Sri Rama Devi R, Hima Bindu Banoth and Ravi Gugulothu, "A Review of Solar Water Distillation Techniques", National Conference on Renewable and Sustainable Energy (NCRSE-2014) Organised by IST and JNTUH College of Engineering Hyderabad, held on 27th -28th June, 2014.

[10]　Naga Sarada Somanchi, Ravi Gugulothu, Sri Lalitha Swathi Sagi, Thotakura Ashish Kumar and Vijaya Koneru (2014), "Experimental Study of Solar Water Distillation Using Epsom Salt as Phase Change Material", Fourth International Conference on Hydrology and Watershed Management, Organized by Centre for Water Resources, Institute of Science & Technology, Jawaharlal Nehru Technological University Hyderabad, held on 29th October to 1st November, 2014, pp: 470-474.

[11]　Nijmeh.S, Odeh.S and Akash.B, "Experimental and theoritical study of a single basin solar still in Jordan", International Communications in Heat and Mass Transfer, 32, 565-572, 2005.

[12] Rajendra Prasad.P, Padma Pujitha.B, Venkata Rajeev.G and Vikky.K, "Energy efficient Solar Water Still", International Journal of ChemTech Research (IJCRGG), Vol.3, No.4, pp:1781-1787, Oct-Dec 2011.

[13] Ravi Gugulothu, Naga Sarada Somanchi, Sri Rama Devi and Devender Vilasagarapu, "Experimental Study of Solar Still with Energy Storage Material", International Conference on Industrial, Mechanical and Production Engineering: Advancements and Current Trends. Organized by Department of Mechanical Engineering, Maulana Azad National Institute of Technology (MANIT), Bhopal, held on 27th – 29th November, 2014, pp: 489-494.

[14] Ravi Gugulothu, Naga Sarada Somanchi, R.Sri Rama Devi and Devender Vilasagarapu (2014), "Experimental Study of Solar Still with Energy Storage Material", Journal of Sustainable Manufacturing and Renewable Energy, Volume 3, Number 1-2, ISSN: 2153-6821, Nova Science Publishers, Inc.

[15] Ravi Gugulothu, Naga Sarada Somanchi, Sri Rama Devi R and Kishan Banothu (2014), "Solar Water Distillation Using Three Different Phase Change Materials", Proceedings of "International Conference on Advances in Design & Manufacturing", at NIT Tiruchirappalli, Tamil Nadu, India, held on 5th –7th December, 2014, pp: 961-965.

[16] Ravi Gugulothu, Naga Sarada Somanchi, R.Sri Rama Devi and Hima Bindu Banoth (2015), "Experimental Investigations on Performance Evaluation of a Single Basin Solar Still Using Different Energy Absorbing Materials", Proceedings of International Conference on Water Resources, Coastal and Ocean Engineering (ICWRCOE'2015), Organized by Department of Applied Mechanics and Hydraulics, National Institute of Technology, Karnataka, Surathkal on 12th-14th March, 2015.

[17] Ravi Gugulothu, Naga Sarada Somanchi, R.Sri Rama Devi and Hima Bindu Banoth (2015), "Experimental Investigations on Performance Evaluation of a Single Basin Solar Still Using Different Energy Absorbing Materials", Proceedings of International Conference on Water Resources, Coastal and Ocean Engineering (ICWRCOE'2015), Aquatic Procedia 4 (2015), pp: 1483-1491.

[18] Ravi Gugulothu, Naga Sarada Somanchi, Devender Vilasagarapu and Hima Bindu Banoth (2015), "Solar Water Distillation using Three Different Phase Change Materials", 4th International Conference on Materials Processing & Characterization, Organised by Department of Mechanical Engineering, Gokaraju Rangaraju Institute of Engineering and Technology, during 14th- 16th March, 2015.

[19] Ravi Gugulothu, Naga Sarada Somanchi, Devender Vilasagarapu and Hima Bindu Banoth (2015), "Solar Water Distillation using Three Different Phase Change Materials", Materials Today, Processsdings of 4th International Conference on Materials Processing & Characterization.

[20] Swetha K and Venugopal, "Experimental Investigation of a Single sloped still using PCM", International Journal of Research in Environmental Science and Technology, 1(4), 30-33, 2011.

[21] Vijaya Kumar Reddy K, Naga Sarada Somanchi, Hima Bindu Banoth and Ravi Gugulothu (2014), "Experimental Study of Solar Still with Energy Storage Materials", Proceedings of the ASME-2014, 12th Biennial Conference on Engineering Systems Design and Analysis (ESDA14), Organized by ASME at Copenhagen, Denmark, held on 25th- 27th June, 2014, ISBN: 978-0-7918-4584-4.

[22] Vijaya Kumar Reddy K, Naga Sarada Somanchi, Bellam Sudheer Premkumar, Hima Bindu Banoth, Ravi Gugulothu and Bellam Samuel Naveen, "Experimental Investigation on Performance Evaluation of A Single Basin Solar Still Using Different Energy Absorbing Materials", Proceedings of the ASME 2014, 8th International Conference on Energy Sustainability, ES2014, held on Boston, Massachusetts, USA on 30th June – 2nd July, 2014.

20

A Simulation Study on the Energy Efficiency of Gas-Burned Boilers in Heating Systems

Zaiyi Liao[1, *], Wei Xuan[2]

[1]Dept of Architectural Science, Ryerson University, Toronto, Canada
[2]Dept of Architecture, Hefei University of Technology, Hefei, China

Email address:
zliao@ryerson.ca (Zaiyi Liao), xuanwei417@163.com (Wei Xuan)

Abstract: The energy efficiency of gas-burned boilers in space heating systems is sensitive to how the boiler is controlled. This study is aimed to investigate how the overall energy performance of a heating system can be optimized using best boiler control scheme. This is to be achieved through experimental studies and simulation studies. This paper presents the latter. A simplified boiler is proposed and integrated in a heating system modeling platform for the simulation study. The results show that the boiler in a heating system should be controlled according to the heating load in order to achieve the highest long-term energy efficiency while maintaining desired comfort.

Keywords: Energy Efficiency, Boiler, Simulation Study, Heating Systems

1. Introduction

In common with many other types of building services equipment, boilers in heating systems are often considerably oversized in order to provide a substantial margin of capacity [1] [2] [3] [4]. As a result, most boilers can generate sufficient heating capacity but often do so inefficiently especially when they are operated under part-load [5] and controlled by conventional boiler controllers, such as thermostats and weather compensators [6]. There are a broad range of boiler controllers used in current practice to maintain a satisfactory performance of heating systems. A conventional weather compensator changes the set-point of water temperature according to the external temperature such that the system can be operated at lower water temperature when the heating load is low [6]. Liao and Parand developed a boiler controller that can measure the heating load and accordingly determine the optimal water temperature at which the boiler efficiency can be maximized whilst sufficient heat can be delivered to the building [5]. Liao and Dexter developed a novel boiler controller, referred to as Inferential Control Scheme (ICS), that varies the water temperature according to an estimate of the average air temperature in the building [7] [8] [9] [10] [11]. One of energy saving strategies employed by these boiler controllers is to maximize the energy efficiency of the boilers through varying water temperature according to load or optimizing the mixture of oxygen and fuel, in additional to minimizing the heat loss throughout the heat distribution system and avoiding the overheating in the controlled spaces. The scientific credibility of these control techniques relies on a good understanding on how the energy efficiency of boilers is influenced and can be optimized in both short-term and long-term.

2. Boiler Energy Efficiency

It is well understood that there are at least three definitions of boiler efficiency:
- Combustion efficiency: how efficiently the combustion takes place in the burner. Higher combustion efficiency means that more heating capacity can be generated by consuming the same amount of fuel.
- Steady-state efficiency: how efficiently the heat is transferred from the combustion gases to the water when the boiler is running under full load.
- Seasonal efficiency: how efficiently the fuel is used by the boiler over the entire season.

These definitions are related with each other and equally important. However the seasonal efficiency is most important because it determines how much fuel is consumed over the entire heating season.

The seasonal efficiency depends on the steady-state efficiency, the combustion efficiency and the downtime losses that occur when the boiler is not operating. The downtime losses are affected by the boiler structure, type of application and design of the system.

The steady-state efficiency declines when the water temperature increases. This is because the temperature difference between the combustion chamber and water is higher when the water temperature is lower. In order to maximize the efficiency, it is always desirable to operate the boiler at as low a water temperature as possible, e.g. when the system is operating under part load.

A boiler continues to lose heat when it is turned off as follows:

- Radiation through the boiler shell or jacket.
- Convection between the boiler and the air that is drawn by the chimney draft and continues to flow through the boiler.

The more often the boiler cycles, the greater the downtime losses and the lower the seasonal efficiency are.

Katrakis and Zawachi studied the relationship between the seasonal efficiency and the load of a steam boiler through a field experiment [12]. They concluded that the seasonal efficiency of the boiler was highly sensitive to the control of boiler and the characteristic of the heating load. Higher seasonal efficiency can be achieved if the system is designed such that the off-cycling of the boiler is minimised.

Anglesio gives a relationship between the seasonal efficiency (η) and the load factor [13]. The load factor (L_f) is defined as the ratio of produced power (Q) to the maximum power (Q_{max}) of the boiler.

$$\eta = 0.9/(1 + 0.02/L_f) \qquad (1)$$

where: $L_f = Q/Q_{max}$

Based on Equation 1, Cardinale and Stefanizzi investigated the seasonal efficiency of boilers when different control schemes were used to determine the water temperature [14]. They concluded that the annual distribution of the load factor

was sensitive to how the water temperature was determined.

3. Simulation Study on Boiler Energy Efficiency

An experimental study on boiler energy efficiency has been reported in [15]. This paper presents a simulation study.

A boiler model has been developed and validated using the experimental data obtained from the boiler test rig. Figure 1 shows an electronic analogue of a boiler model. The boiler model consists of five major components: the combustion and flame passage, the inner shell, the water passage, the outer shell, and the insulation. The boiler model is based on the following governing equations:

$$C_{is}\frac{dT_{is}}{d\tau} = K_{flame}(T_{flame} - T_{is}) - K_2(T_{is} - T_w) \qquad (2)$$

$$C_w\frac{dT_w}{d\tau} = K_2(T_{is} - T_w) - \dot{m}_w\rho_w(T_{w\_out} - T_{w\_in}) - K_4(T_w - T_{os}) \quad (3)$$

$$C_{os}\frac{dT_{os}}{d\tau} = K_4(T_4 - T_{os}) - K_5(T_{os} - T_i) \qquad (4)$$

$$C_i\frac{dT_i}{d\tau} = K_5(T_{os} - T_i) - K_6(T_i - T_o) \qquad (5)$$

The relevant parameters are: (1) The total thermal capacity of the inner shell (Cis), the outer shell (Cos), the water content (CW), and the insulation (Ci).

(2) The thermal conductance between the flame and the inner shell (K1), inner shell and the water (K2), the water and the outer shell (K4), outer shell and the insulation (K5), and insulation and ambient (K6).

Cis, Cos, CW and Ci are calculated from the technical data of the boiler [7]. The value of K1, K2, K3, K4 and K5 are determined using experimental data through a commissioning procedure [7]. Then a different set of experimental data is used to validate the model. Figure 2 shows the validation results. The results show that the boiler model can accurately simulate the dynamics of the boiler.

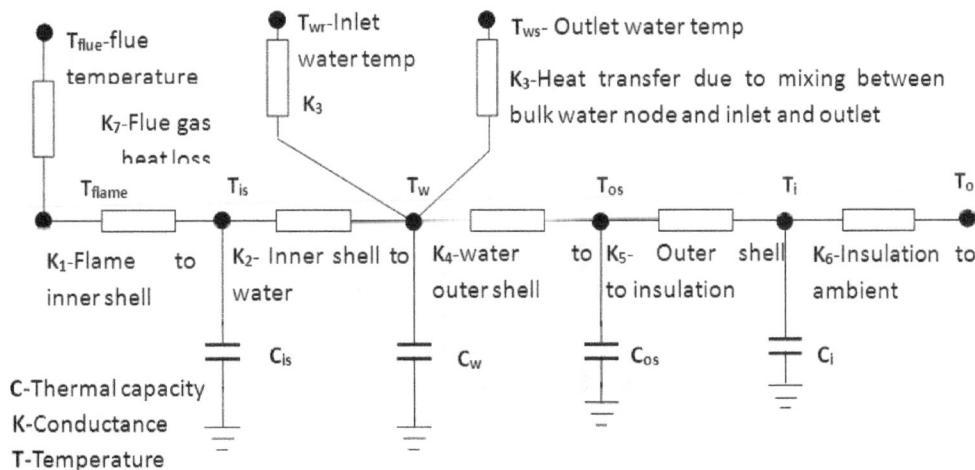

Figure 1. Electronic analogue of the boiler model.

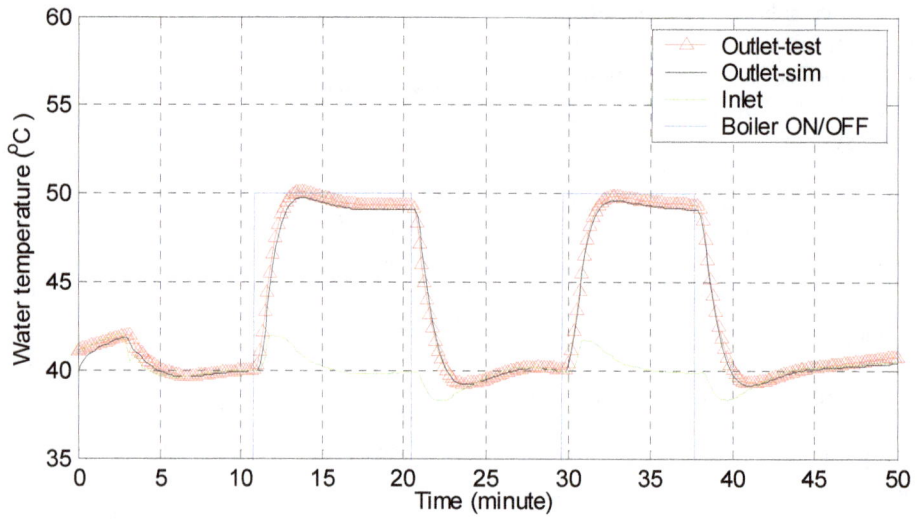

Figure 2 (a). Validation of the boiler model (constant inlet temperature).

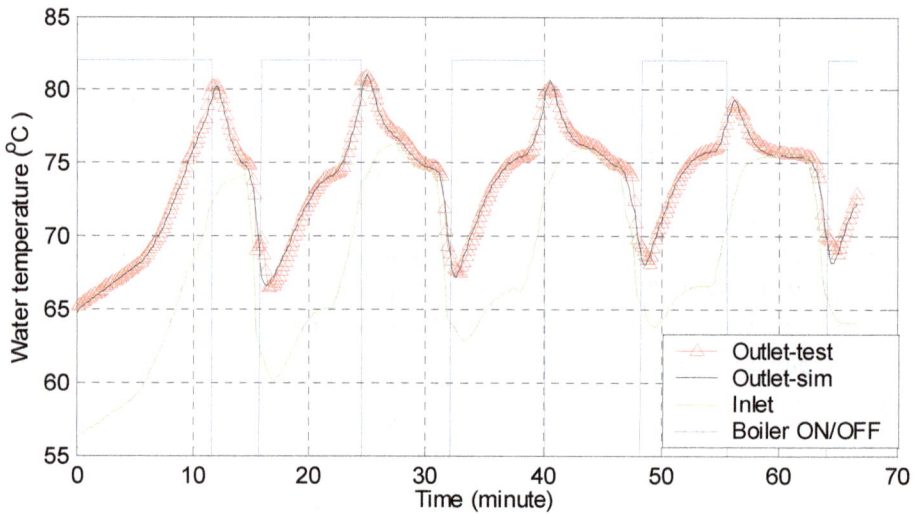

Figure 2 (b). Validation of the boiler model (variable inlet temperature).

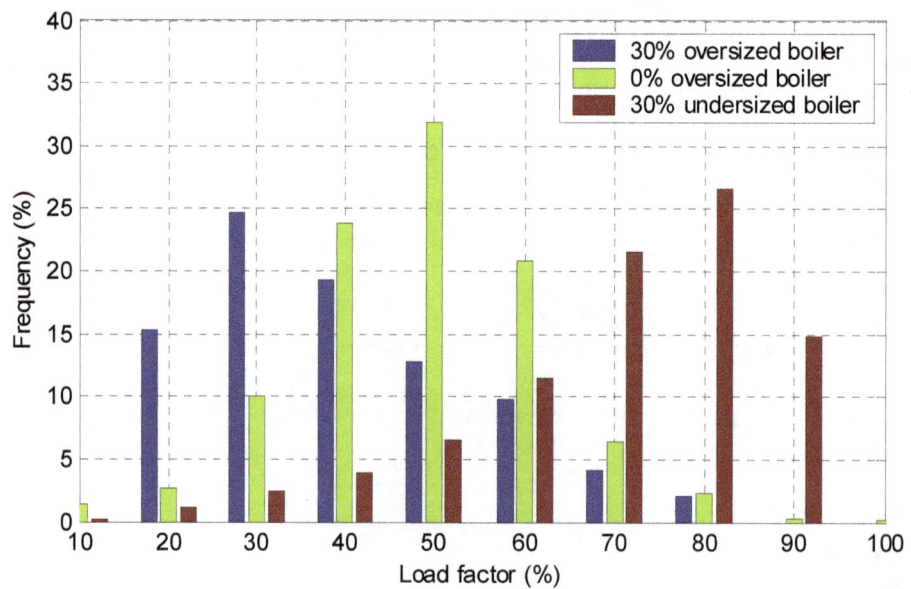

Figure 3. Distribution of the annual load factor for differently sized boilers.

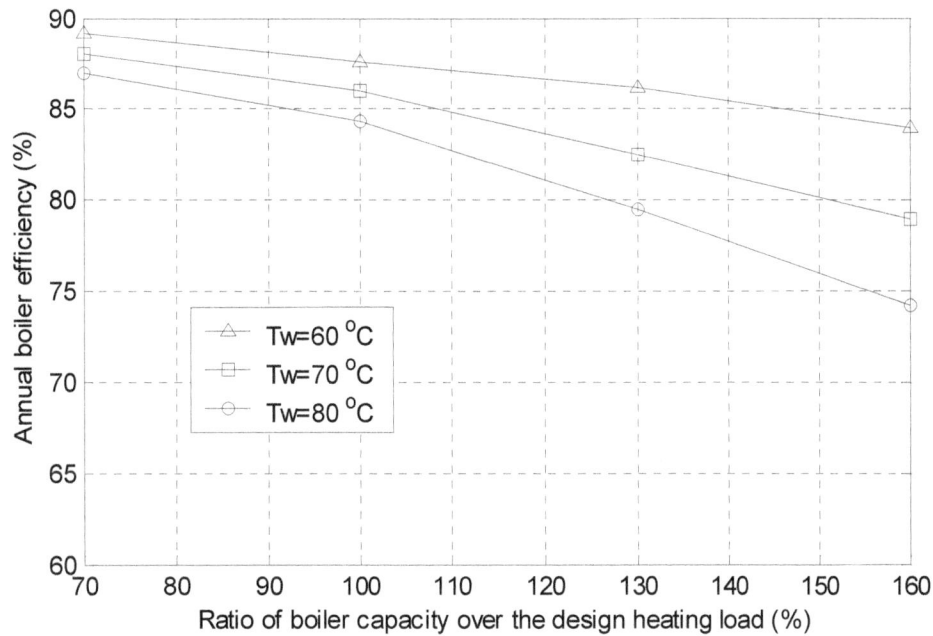

Figure 4. Annual boiler efficiency for differently sized boilers at constant water temperatures.

The boiler model is integrated with a heating system simulator to investigate the long-term, or annual, boiler efficiency under different operating conditions.

Figure 3 shows the annual distribution of the load factor in systems with differently sized boilers. When the boiler is 30% oversized, the load factor more frequently falls into the low range, which, according to the results presented in Section 2, means that there is a bigger potential for improving the boiler efficiency by reducing the water temperature. This potential is much less when the boiler is 30% undersized because the load factor more frequently falls into the high range.

Figure 4 shows that the annual boiler efficiency is more sensitive to the water temperature when the ratio of the boiler capacity over the design heating load is higher. This means that more energy can be saved through appropriately controlling the water temperature in heating systems with oversized boilers. In heating systems with undersized boilers, there is a very limited opportunity for saving energy through changing the water temperature.

4. Strategies to Optimize Energy Efficiency of Boilers in Heating Systems

Both simulation and experimental results show that the boilers in heating systems can be operated more efficiently if the water temperature is reduced. However, in order to produce and deliver sufficient heating capacity to the building, the supply water temperature must be maintained above a certain minimum level, which varies with the changing weather and the use of the building. Therefore the best strategy to maintain high energy efficiency for long-term is based on the ability to detect the minimum heating requirement. We have investigated the performance of some

boiler control techniques in actual heating systems over the last ten years. These control techniques are designed to optimize the operation of the boilers in space heating systems for the best long-term energy efficiency. The objective of this field study is to find out if this has been actually achieved. The result is summarized as follows:

- Weather compensators [6] [14] [16] [17]

The importance of determining the set-point of the supply water temperature according to the varying climatic condition was first recognised in early 1980s. Linear compensations are commonly used in medium and large buildings to obviate installing many individual valves and sensors throughout the building. A weather compensator varies the set-point of the water temperature according to the outside temperature linearly.

It was believed that this compensator was able to stop excess heat loss due to open windows. Being a time-independent controller, this compensator strongly relies on good design of the heating system and accurate hydraulic balancing because it attempts to match the heat input into the system to the steady-state heating load. The control performance of this compensator is very sensitive to the commissioning, which often require extensive monitoring. Dexter investigated this problem and developed a self-adaptive weather compensator [18], which relies on on-line measurement of both the external and indoor temperature. Yet the commissioning has been proven to be a very difficult, if not impossible, procedure in practice. As a result, very few of this kind of compensator are commissioned properly [5].

- Heating load compensators [5] [19] [20]

A weather compensator does not consider the impact of the solar radiation, infiltration, internal heat gain, and variation of the desired room temperature on the heating load. All these elements are taken into account by a heating load compensator, which can estimate the heating load and accordingly determine the optimal water temperature at which the efficiency of the boiler is maximized and sufficient heating capacity is produced.

- Inferential Control Scheme (ICS) [7] [9] [11]

An ICS controller can estimate the average room temperature of the building based on the information available to normal boiler controllers, including the solar radiation, the external temperature, and the boiler control signals. The estimated value of the average room temperature is then compared with the desired room temperature by a PI controller, which determines the best value for the water temperature set-point. A conventional ON/OFF control logic is then used to decide how the boilers should be operated.

The details of ICS have been reported in detail in other papers [7] [9] [11].

- Model-based predictive ICS [21] [22]

In practice, the overall performance of the ICS controller is sensitive to the parameters of the PI algorithm used to determine the water temperature set-point according to the desired and estimated average room temperature. The Model-based predictive ICS (MPICS) was developed to resolve this problem. It is a more robust version of ICS, meaning that it is much easier to commission and less sensitive to the varying operating conditions.

5. Conclusion

The following conclusions can be drawn:

- The energy efficiency of boilers in heating systems is influenced by a number of factors, including the boiler control strategy, the water temperature and the load factor.
- The energy efficiency of boilers declines if the water temperature increases.
- It is possible to maintain high energy efficiency when the load is low by changing the inlet water temperature appropriately. However a high water temperature is needed when the heating load is high.
- The lower the load factor, the higher the potential for improving the boiler efficiency by reducing the water temperature.
- When the boiler is oversized, there is a bigger potential for improving the boiler efficiency by reducing the water temperature. This potential is much less when the boiler is undersized.
- There are a number of boiler control schemes used in current practice to maintain high long-term energy efficiency by changing the water temperature set-point according to an estimation of either the heating load or the average room temperature in the building.

Currently the following tasks are being carried out to further investigate the problem:

- Investigating the energy efficiency of boilers in a broader range of heating systems, such as the systems with different terminal devices or distribution systems.
- Improving the accuracy of the boiler model.
- Developing a method and a system to test the

performance of boiler controllers.

Acknowledgements

The work presented in this paper was partially funded by FBE, UK (Foundation for the Built Environment). And the preparation of this paper is supported by Ryerson University, Canada.

References

[1] ASHRAE (American Society of Heating, Ventilation and Air-conditioning Engineers), ASHRAE Handbook: Fundamentals (SI Edition), Atlanta: ASHRAE, 2013.

[2] CIBSE (Chartered Institution of Building Services Engineers), CIBSE Guide A: Environmental Design New 2015, London: CIBSE, 2015.

[3] CIBSE, CIBSE Guide Vol. A, The Chartered Institution of Building Services Engineers, London: CIBSE, 1986.

[4] P. Gardner, Energy management systems in buildings, London: Energy Publications, 1984.

[5] Zaiyi Liao and Forutan Parand, "Controller efficiency improvement for commercial and industrial gas and oil fired boilers: A CRAFT (Cooperative Research Action for Technology) project, contract JOE-CT98-7010. 1999-2001," Building Research Establishment (BRE), Watford, 2001.

[6] Geoff Levermore, Building energy management systems: applications to low-energy HVAC and natural ventilation control, New York: E&FN Spon, 2000.

[7] Zaiyi Liao and Arthur Dexter, "An experimental study on an inferential control scheme for optimising the control of boilers in multi-zone heating systems," *Energy and Buildings,* vol. 37, no. 1, pp. 55-63, 2005.

[8] Zaiyi Liao, Arthur Dexter and Micheal Swainson, "On the control of heating systems in the UK," *Buildings and Environment,* vol. 40, no. 3, pp. 343-351, 2004.

[9] Zaiyi Liao and Arthur Dexter, "A simplified physical model for estimating the average air temperature in multi-zone heating systems," *Buildings and Environment,* vol. 39, no. 9, pp. 1009-1018, 2004.

[10] Zaiyi Liao and Arthur Dexter, "The potential for energy saving in heating systems through improving boiler controls," *Energy and Buildings,* vol. 36, no. 3, pp. 261-271, 2004.

[11] Zaiyi Liao and Arthur Dexter, "An inferential control scheme for optimising the operation of boilers in multi-zone heating systems," *Building Services Engineering Research and Technology,* vol. 24, no. 4, p. 245~256, 2003.

[12] J. T. Katrakis and T. S. Zawacki, "Field-measured seasonal efficiency of intermediate-sized low-pressure steam boilers," *ASHRAE Transaction,* vol. 99, no. 2, pp. 429-439, 1993.

[13] P. Anglesio, "Rendimento in funzione del carico di sistemi caldaiabruciatore per resicaldamento," *La Termotecnica,* vol. 10, pp. 79-90, 1982.

[14] N. Cardinale and P. Stefanizzi, "Heating-energy consumption in different plant operating conditions," *Energy and Buildings,* vol. 24, pp. 231-235, 1996.

[15] Zaiyi Liao and Wei Xuan, "An Experimental Study on the Energy efficiency of Ga-burned Boilers in Heating Systems," *International Journal of Scientific Research,* vol. 4, no. 11, pp. 267-269, 2015.

[16] Chris Underwood, HVAC control systems: modeling, analysis, and design, London and New York: E&FN Spon, 1999.

[17] John. L Levenhagen, HVAC control system, design diagrams, New York, London: McGraw-Hill, 1998.

[18] Arthur Dexter, "Self-adaptive control of hot-water space-heating systems," in *Proceedings of IASTED*, Athens, Greece, 1983.

[19] Surinder Jassar, Zaiyi Liao and Lian Zhao, "A Recurrent Neuro-Fuzzy System and its Application in Inferential Sensing," *Applied Soft Computing,* vol. 11, no. 3, pp. 2935-2945, 2011.

[20] Surinder Jassar and Zaiyi Liao, "Improve the control of Residential Heating Systems," *Applied Mechanics and Materials,* Vols. 52-54, pp. 1571-1576, 2011.

[21] Zaiyi Liao, "An inferential control scheme for optimizing the control of boilers in multiple-zone heating systems, PhD thesis," the University of Oxford, Oxford, 2004.

[22] Zaiyi Liao and Arthur Dexter, "Model-based Predictive Control of Boilers in Hot-water heating Systems," *IEEE Transactions on Control Systems Technology,* vol. 18, pp. 1092-1102, 2010.

21

Effect of Controllable Distorted Flux on Magnetic Loss inside Laminated Core

Yang Liu[1,2], Guang Ma[1], Yana Fan[2], Tao Liu[2], Lanrong Liu[2], Chongyou Jing[2]

[1]China State Grid Smart Grid Research Institute, Beijing, China
[2]Institute of Power Transmission and Transformation Technology, Baobian Electric Co., Ltd, Baoding, China

Email address:

liuyang_white@hotmail.com (Yang Liu), maguang@sgri.sgcc.com.cn (Guang Ma)

Abstract: This paper investigates the magnetic properties of grain oriented electrical steel under the controllable distorted flux conditions based on a product-level core model, and examines the effects of the distorted flux on magnetic loss inside the laminated cores with different excitation conditions, involving different harmonic phase difference, harmonic order and harmonic contents.

Keywords: Modeling distorted flux, Magnetic loss, Magnetic property, Transformer core

1. Introduction

In a great deal of applications, some transformers are operated under non-ideal conditions. The input voltage of transformer is not always sinusoidal, which results in a distorted flux in the transformer core. Usually the specific total loss of electrical steel is measured under the condition of sinusoidal waveform for the flux density [1-3]. However, the specific total loss values obtained by the standard method are not applicable when estimating the iron loss in a transformer under the distorted flux conditions [4-6], potentially causing a large increase in the iron loss.

Therefore, in order to estimate the iron loss of a transformer, it is very important to accurately measure magnetic properties under the distorted flux condition. Due to the difficulty of establishing the corresponding measurement system, until now, it is hard to find a detailed report of such magnetic properties measurements of a real laminated core.

In this paper, a measurement system for the magnetic properties of grain oriented (GO) silicon steel lamination under distorted flux conditions is presented. The B-H properties and specific total loss of a laminated core model (LCM) with different distorted fluxes are measured, and the effects of harmonic phase difference, harmonic content and harmonic order on magnetic loss are investigated in detail.

2. Experimental Setup

The magnetic properties measurements were carried out based on a LCM, with 45° mitred step-lap joints, made of GO silicon steel B27R095, BAOSTEEL, as shown in Fig.1. The detailed structural dimensions and some individual design parameters of the LCM are shown in Fig.1 and Table 1.

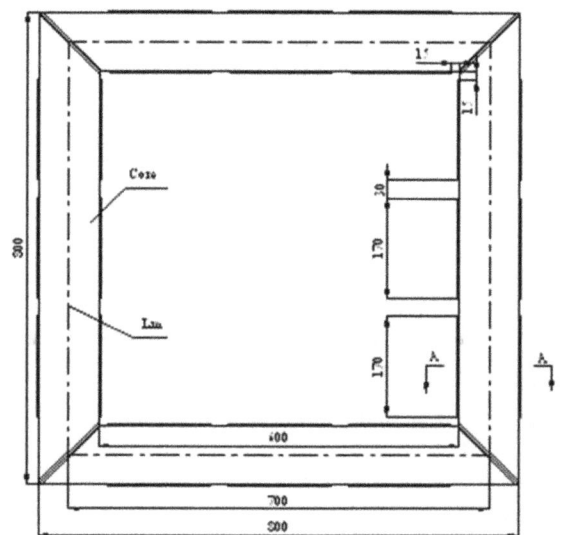

(a)

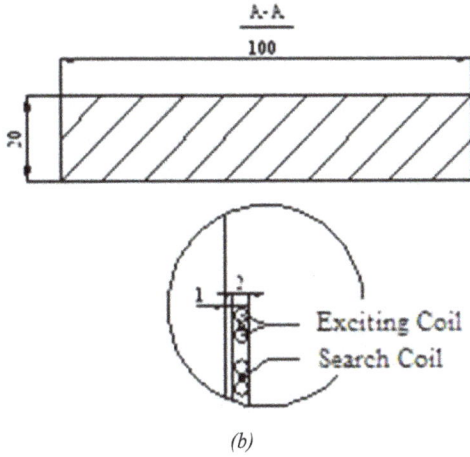

(b)

Figure 1. *Laminated core model. (a) structure of laminated core (b) view of cross section.*

Table 1. *Parameters of core and coils.*

Parameters (LCM)	
Number of turns of exciting coil	288
Number of turns of search coil	144
Mean length of magnetic path(m)	2.8
Length of silicon sheet (mm)	800
Width of silicon sheet (mm)	100
Thickness of core-leg (mm)	20.4(measured)
Net area of cross section(mm²)	1978.8(packing factor: 0.97)
Total weight of iron core(kg)	42.39

The well-established experiment system is shown in Fig. 2. The generator of arbitrary voltage waveforms composed of frequency converter with LC filter is employed as the exciting power source. The waveform and amplitude of flux density in the LCM can be controlled by adjusting the exciting power source.

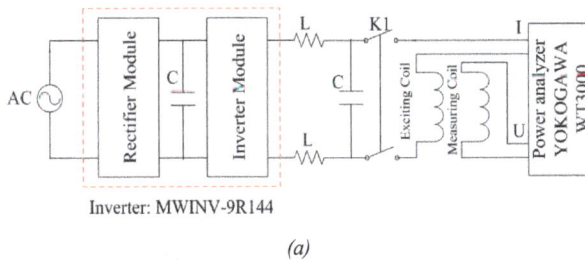

(a)

Inverter: MWINV-9R144

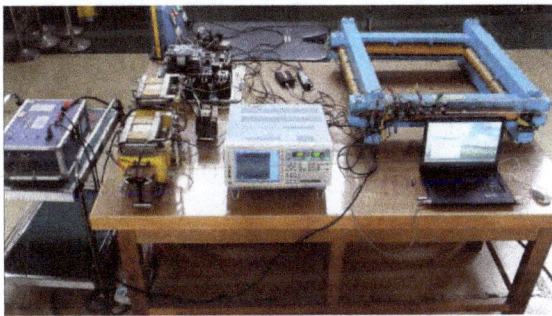

(b)

Figure 2. *Experiment system of magnetic property. (a) block diagram of the circuit; (b) experimental apparatus.*

3. Control Method of Flux Density in LCM

In the measurement, the flux density in LCM can be expressed as follows:

$$B = \sum_{n=1}^{j} B_n \sin(n\omega t + \varphi_n) \qquad (1)$$

where B_n and φ_n are the peak flux density and phase of the n-th harmonic. The percentage amplitude and the phase difference of the n-th harmonic can be defined as:

$$k_n = B_n / B_1 \times 100\% , \theta_n = \varphi_n - \varphi_1 \qquad (2)$$

where B_1 and φ_1 correspond to the peak flux density and phase of fundamental content.

In LCM, the induced voltage of exciting coil can be expressed as follows:

$$E(t) = -NS \frac{dB}{dt} \qquad (3)$$

where N is the number of turns of exciting coil and S is the effective cross area.

Substituting (1) and (2) into (3), the induced voltage of exciting coil can be given as

$$E(t) = -NS\omega B_1 \sum_{n=1}^{j} nk_n \cos(n\omega t + \varphi_n) \qquad (4)$$

As the resistance and leakage reactance of exciting coil can be neglected in no-load lamination core mode, the exciting voltage can be given as

$$U(t) = -E(t) = NS\omega B_1 \sum_{n=1}^{j} nk_n \cos(n\omega t + \varphi_n) \qquad (5)$$

In order to realize the control of the flux density in the LCM, the exciting power source is adjusted according to formulation (5).

4. Measurement Results

4.1. The Effects of Harmonic Phase on Magnetic Properties

The magnetic properties of LCM under distorted flux densities are measured using the above experiment system. It is worthwhile to note that all the distorted flux waveforms presented in the current version of this paper contain only a single harmonic content. Fig. 3 shows the measurement results of B-waveforms and hysteresis loops with the varying phase difference θ_5 of the 5-th harmonic. Fig.4 shows the corresponding specific total loss of the LCM. The specific total loss at $\theta_5 = 0°$ is considerably higher than those from 45° to 180°, because the larger minor loops occur in the major loop at $\theta_5 = 0°$ and the major loop is slightly larger than those from 45° to 180° (see Fig.3). Fig.5 and Fig.6 respectively shows the results of specific total loss at $k_3 = 20\%$ and $k_7 = 10\%$. Based on the measured results, it can be seen that the specific total loss shows a decreasing tread with the increase of harmonic phase difference at a given harmonic order and harmonic content, when the harmonic phase difference varied within the range of

0° and 180°.

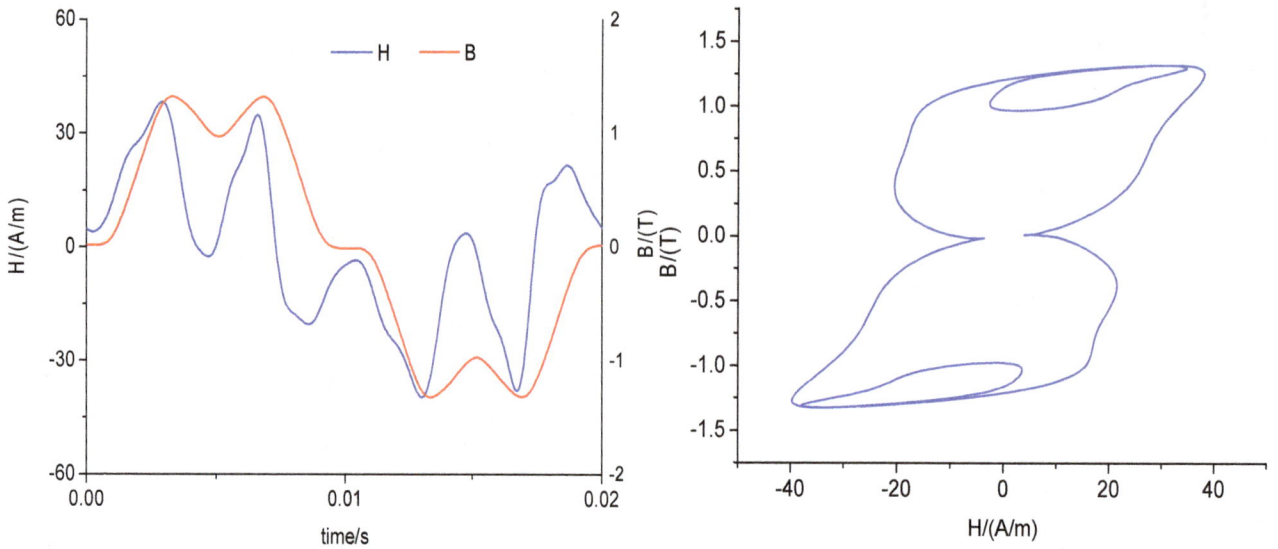

(a)

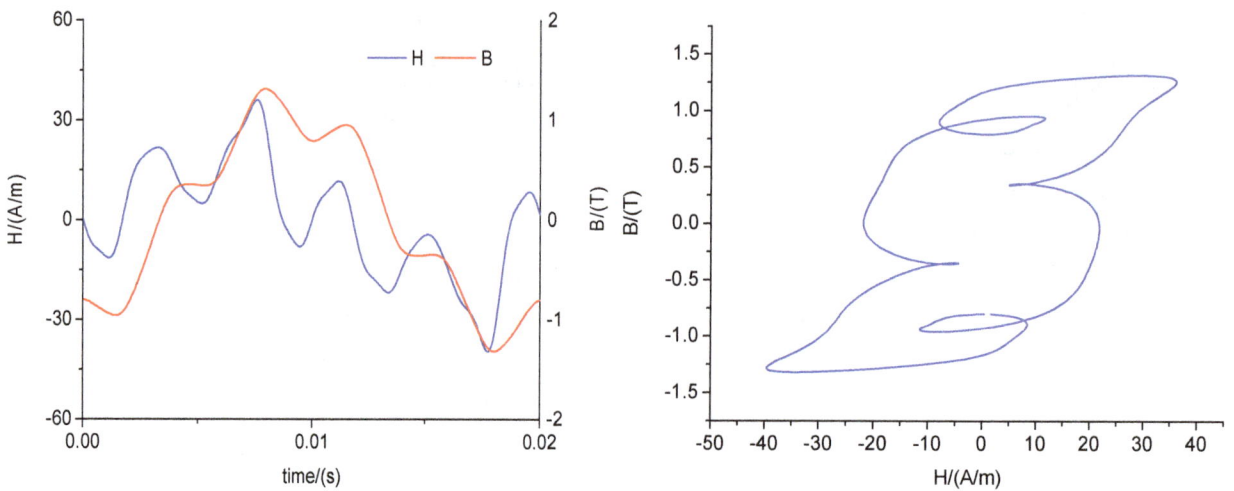

(b)

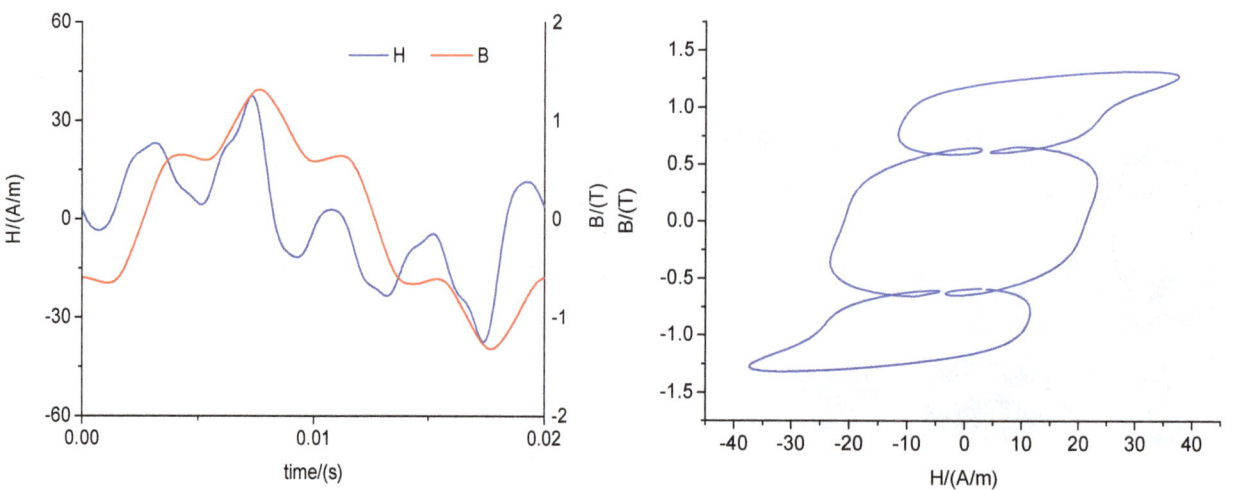

(c)

Figure 3. *Measurement results of B-waveforms and hysteresis loops at $k_5=20\%$, $B_m=1.32T$ (a) $\theta_5 =0°$(b) $\theta_5 =90°$(c) $\theta_5 =180°$.*

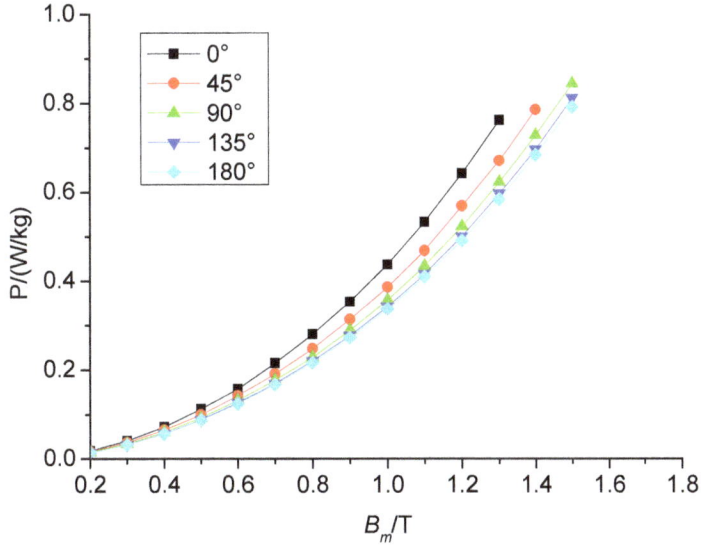

Figure 4. *Variation of the specific total loss with the phase difference of 5-th harmonic (k_5=15%).*

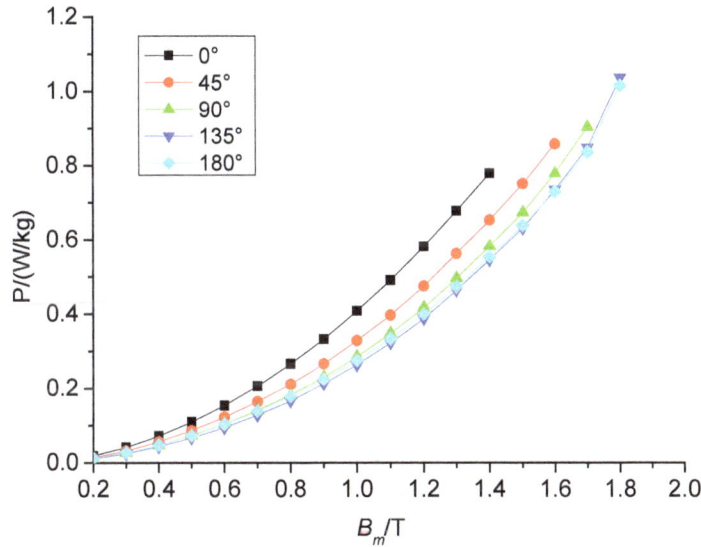

Figure 5. *Variation of the specific total loss with the phase difference of 5-th harmonic (k_3=20%).*

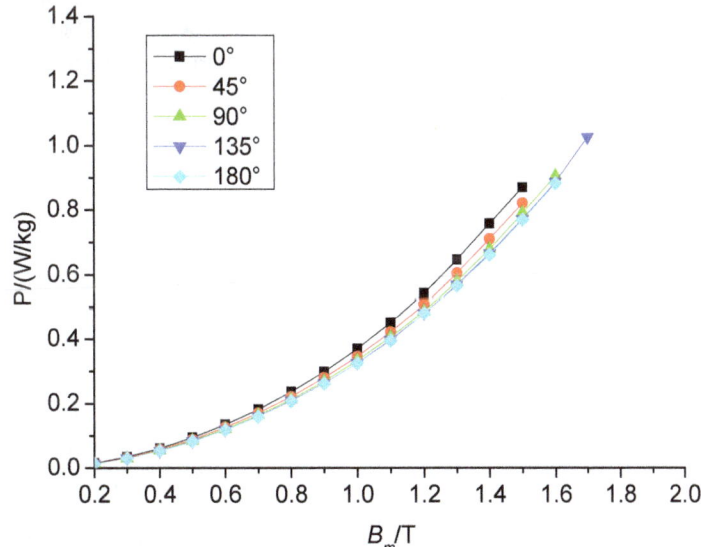

Figure 6. *Variation of the specific total loss with the phase difference of 5-th harmonic (k_7=10%).*

4.2. The Effects of Harmonic Content on Magnetic Properties

The Effects of the harmonic content on the specific total loss are shown in Fig.7 and Fig.8. It can be seen that the specific total loss increase with the variation of the harmonic content, as expected.

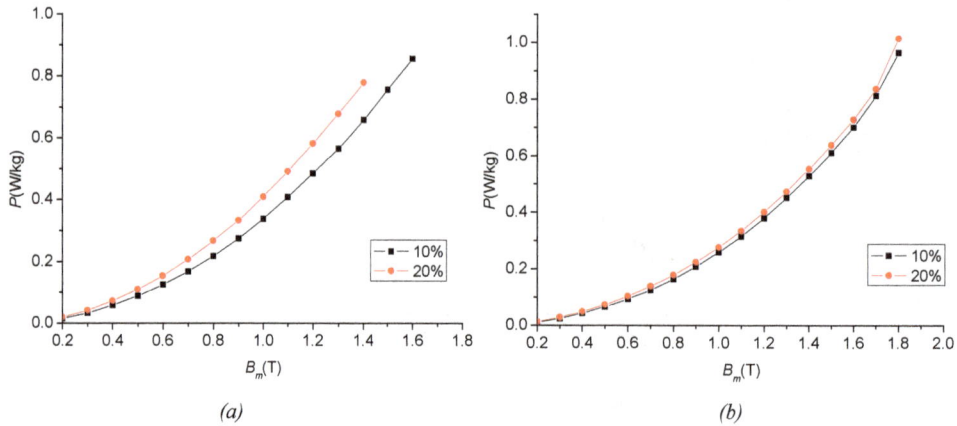

(a) *(b)*

Figure 7. *Variation of the specific total loss with the third harmonic content (a) 0° (b) 180°.*

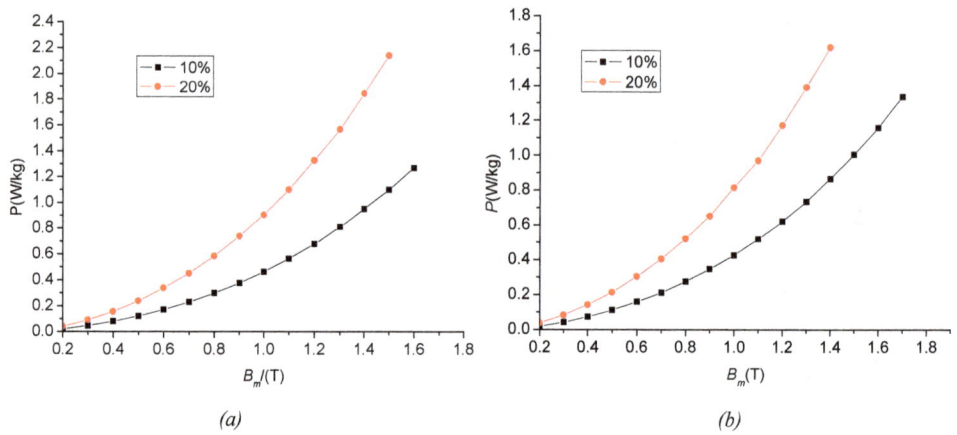

(a) *(b)*

Figure 8. *Variation of the specific total loss with the fifth harmonic content (a) 0° (b) 180°.*

4.3. The Effects of Harmonic Order on Magnetic Properties

The variation of the specific total loss with the harmonic orders is shown in Fig.9. The black line with the hollow box is the specific total loss under the sinusoidal flux condition. The specific total loss of 180° gradually approaches the one of 0° with the increment of the harmonic order at the same harmonic content. From this result, we can conclude that the specific total loss is likely to be less affected by the phase difference of harmonic when the flux waveforms include the higher harmonic order, which is more obviously affected by the harmonic content and harmonic order.

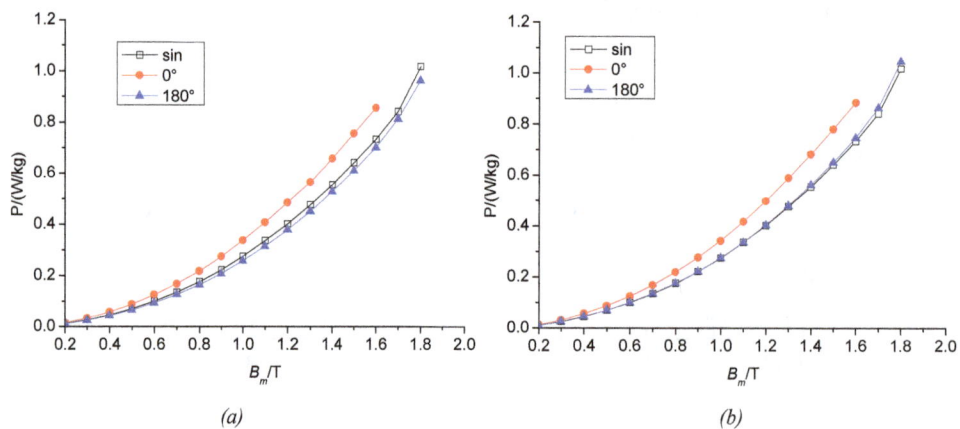

(a) *(b)*

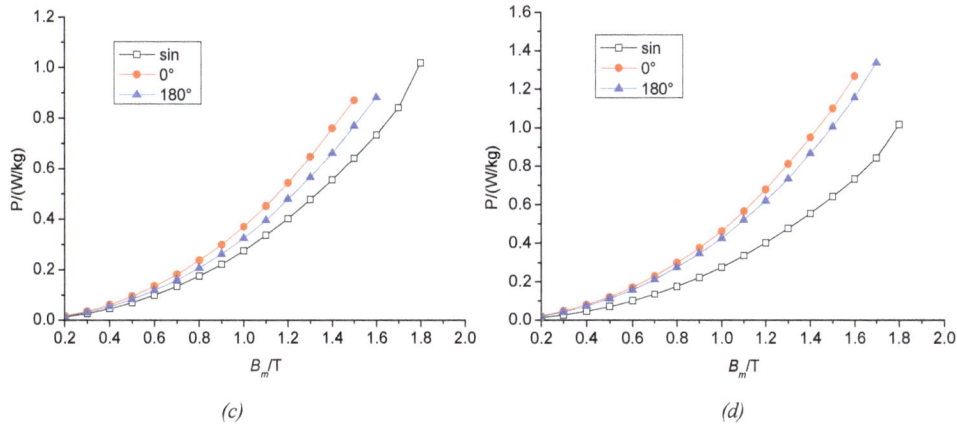

Figure 9. *Variation of the specific total loss with the harmonic order at 10% harmonic content. (a) the 3-th harmonic (b) the 5-th harmonic (c) the 7-th harmonic (d) the 9-th harmonic*

5. Summary

A measurement system of controllable flux density for the magnetic properties of GO silicon steel lamination is well established. The effects of harmonic phase difference, harmonic order and harmonic content on iron loss are examined, which can be briefly summarized as follows:

1) The specific total loss shows a decreasing tread with the increase of harmonic phase difference at a given harmonic order and harmonic content, when the harmonic phase difference varied within the range of 0° and 180°.

2) The effect of harmonic phase difference on specific total loss gradually decreases with the increasing of harmonic order. And the specific total loss is likely to be less affected by the phase difference of harmonic when the flux waveforms include the higher harmonic order.

3) The specific total loss also shows an increasing tread with the increase of harmonic order and harmonic content, as expected.

Acknowledgement

This project was supported in part by the State Grid Corporation of China under Grants sgri-wd-71-13-002, sgri-wd-71-14-002, and sgri-wd-71-14-009, and by the Youth Science Fund of Hebei Education Department, China, under Grant No. QN20131025.

References

[1]   IEC 60404-2 AMD 1-2008: Magnetic Materials - Part 2: Methods of measurement of the magnetic properties of electrical steel sheet and strip by means of an Epstein frame; Amendment 1.

[2]   IEC 60404-3-2010: Magnetic Materials – Part 3: Methods of measurement of the magnetic properties of electrical steel strip and sheet by means of a single sheet tester - Edition 2.2; Consolidated Reprint.

[3]   Z. Cheng, N. Takahashi, B. Forghani, et al, "Effect of variation of B-H properties on loss and flux inside silicon steel lamination," *IEEE Trans. Magn.*, vol.47, no.5, pp. 1346-1349, 2011.

[4]   A. J. Moses, "Characterisation and performance of electrical steels for power transformers operating under extremes of magnetization conditions," International Colloquium Transformer Research and Asset Management, Croatia, Nov. 12-14, 2009.

[5]   A. J. Moses and G. H. Shirkoohi, "Iron loss in non-oriented electrical steels under distorted flux condition," *IEEE Trans. Magn.*, vol. 23, no.5, pp. 3217-3220, 1987.

[6]   J. D. Lavers, P. P. Biringer, and H. Hollitscher. "The effect of third harmonic flux distortion on the core losses in thin magnetic steel laminations," *IEEE Trans. Magn.*, vol. PAS-96, no.6, pp.1856-1862, 1977.

22

Computation and Analysis of the DC-Biasing Magnetic Field by the Harmonic-Balanced Finite-Element Method

Xiaojun Zhao[1], Dawei Guan[1], Fanhui Meng[1], Yuting Zhong[1], and Zhiguang Cheng[2]

[1]Department of Electrical Engineering, North China Electric Power University, Baoding, China
[2]Institute of Power Transmission and Transformation Technology, Baobian Electric Co., Ltd, Baoding, China

Email address:
158748295@163.com (Xiaojun Zhao), zhynh123@163.com (Dawei Guan), mengfh1990@163.com (Fanhui Meng),
yuting315@yeah.net (Yuting Zhong), emlabzcheng@yahoo.com (Zhiguang Cheng)

Abstract: This paper sought to present harmonic-balanced method for finite element analysis of nonlinear eddy current field. The harmonic-balanced method can be used to compute the time-periodic electromagnetic field in harmonic domain, considering electric circuits coupled with the nonlinear magnetic field. 2-D and simplified 3-D model of laminated core is established and computed to prove the accuracy and validity of the proposed method. The calculated magnetizing current is compared with the measured results and the computed magnetic field is analyzed to investigate the effect of DC bias.

Keywords: Eddy Current, Finite Element Analysis, Harmonic Domain, Time-Periodic

1. Introduction

The DC bias phenomenon is an abnormal operation state of the power transformer. The direct current flows into the windings of transformers, which may lead to a series of problems in transformers and electric networks. Such problems include significant saturation of the ferromagnetic core, vibration and overheating of transformers [1]-[2], and reactive power demand [3] in the transmission system. The investigation of the DC biased problem originates from the effect of Geomagnetically Induced Current (GIC) [4]-[6] caused by solar magnetic disturbance on transformers and electric networks. With the development of high voltage direct current (HVDC) transmission, more attention has been paid to the DC biased problem in recent years. In the monopolar operating mode of the HVDC system, the earth is usually used as a return path [7]-[8]. In that case, there will be a large DC potential difference between the two converting plants. The electric potential difference generates a direct current that flows into the windings of the power transformer (through the earthed neutrals in the AC network).

Research on the mechanism of the DC biased problem contributes to important developments in transformer design. Many different methods have been used to investigate the electromagnetic field in transformers under the DC bias condition [9]-[12]. The circuit model [13]-[14] were proposed to calculate the excitation current in windings. The time-stepping finite element method [15] was also used by some researchers to compute the magnetizing current and magnetic field. However, it is difficult to obtain accurate results from the electric or magnetic circuit model, especially when magnetic field analysis is required to explore the mechanism of the DC biased problem. The time-stepping method is an alternative method to calculate the transient magnetic field, but accurate solutions of high order harmonic components in exciting current and magnetic induction require many more iterations in the time domain, which can reduce the effectiveness of this method.

In this paper the HBFEM is used to solve the nonlinear magnetic field under different DC bias conditions, considering the coupling of the electric circuit and the magnetic field. Harmonic solutions of the magnetizing current and the magnetic field can be calculated directly and quickly in the harmonic domain. The nonlinear magnetic field under DC bias condition is analyzed thoroughly through harmonic analysis in order to investigate the mechanism of the DC bias phenomenon in the power transformer. The calculated results and experimental data are compared on an Epstein frame-like core model, which proves that the HBFEM is an effective and efficient method in analyzing the DC biased problem of power transformers.

2. Harmonic-Balanced Finite-Element Method

2.1. Harmonic Balance Method in Finite Element Analysis of Nonlinear Magnetic Field

The nonlinear magnetic field can be described by using magnetic vector potential A,

$$\nabla \times \nu \nabla \times A = J \tag{1}$$

$$\nabla \cdot J = 0 \tag{2}$$

where v is the reluctivity, J is the current density including exciting current density J_0 and eddy current density J_e.

The finite element equation can be obtained by using Galerkin's method,

$$\int_{\Omega_n} N_i \cdot \left(\nabla \times \nu \nabla \times A\right) d\Omega + \int_{\Omega_e} N_i \cdot \sigma \left(\frac{\partial A}{\partial t} + \nabla \varphi\right) d\Omega = \int_{\Omega_n} N_i \cdot J_0 d\Omega \tag{3}$$

$$\int_{\Omega_e} N_i \cdot \sigma \left(\frac{\partial A}{\partial t} + \nabla \varphi\right) d\Omega = 0 \tag{4}$$

where N_i is the nodal shape function, φ is the electric scalar potential, Ω_n and Ω_c are the finite element in the whole domain and non-conducting domain, respectively. σ is the conductivity of conducting materials. The penalty function is always required in finite element analysis of eddy current problems based on nodal basis function [16].

The time-periodic solutions are focused in the DC bias phenomenon, since it is a harmonic problem with alternating and direct excitations. The magnetic flux density and magnetic vector potentials are both periodic functions in the time domain. Based on harmonic balance theory, all variables in (3) can be expressed by complex series as follows,

$$A^i(r,t) = \sum_{n=-N}^{N} A_k^i(r) e^{jn\omega t} \tag{5}$$

$$J(r,t) = \sum_{n=-N}^{N} J_k(r) e^{jn\omega t} \tag{6}$$

$$\nu(t) = \sum_{n=-N}^{N} \nu_k e^{jn\omega t} \tag{7}$$

where N is the truncated harmonic number.

Considering the orthogonal characteristics of trigonometric functions, the harmonic-balanced equation can be attained as follows,

$$\int_{\Omega_n} -\left(\nabla N_i \times \nabla \times N_j\right) D d\Omega A_j + \sigma \int_{\Omega_c} N_i \cdot N_j N d\Omega A_j$$
$$+ \sigma \int_{\Omega_c} N_i \cdot \nabla N_j N d\Omega \varphi_j = \int_{\Omega_n} N_i \cdot J_0 d\Omega \tag{8}$$

$$\sigma \int_{\Omega_c} \nabla N_i \cdot N_j N d\Omega A_j + \sigma \int_{\Omega_c} \nabla N_i \cdot \nabla N_j d\Omega \varphi_j = 0 \tag{9}$$

where A_j, φ_j are harmonic solutions, D and N are named as reluctivity matrix and harmonic matrix respectively,

$$D = \begin{bmatrix} v_0 & v_1 & v_{-1} & v_2 & v_{-2} & v_3 & v_{-3} & \cdots \\ v_{-1} & v_0 & v_{-2} & v_1 & v_{-3} & v_2 & v_{-4} & \cdots \\ v_1 & v_2 & v_0 & v_3 & v_{-1} & v_4 & v_{-2} & \cdots \\ v_{-2} & v_{-1} & v_{-3} & v_0 & v_{-4} & v_1 & v_{-5} & \cdots \\ v_2 & v_3 & v_1 & v_4 & v_0 & v_5 & v_{-1} & \cdots \\ v_{-3} & v_{-2} & v_{-4} & v_{-1} & v_{-5} & v_0 & v_{-6} & \cdots \\ v_3 & v_4 & v_2 & v_5 & v_1 & v_6 & v_0 & \cdots \\ \vdots & \vdots & \vdots & \vdots & \vdots & \vdots & \vdots & \ddots \end{bmatrix} \tag{10}$$

$$N = \omega \begin{bmatrix} 0 & 0 & 0 & 0 & 0 & 0 & 0 & \cdots \\ 0 & -j & 0 & 0 & 0 & 0 & 0 & \cdots \\ 0 & 0 & j & 0 & 0 & 0 & 0 & \cdots \\ 0 & 0 & 0 & -2j & 0 & 0 & 0 & \cdots \\ 0 & 0 & 0 & 0 & 2j & 0 & 0 & \cdots \\ 0 & 0 & 0 & 0 & 0 & -3j & 0 & \cdots \\ 0 & 0 & 0 & 0 & 0 & 0 & 3j & \cdots \\ \vdots & \vdots & \vdots & \vdots & \vdots & \vdots & \vdots & \ddots \end{bmatrix} \tag{11}$$

$$A_j = \begin{bmatrix} A_{j,0} & A_{j,-1} & A_{j,1} & A_{j,-2} & A_{j,2} & \cdots \end{bmatrix}^T \tag{12}$$

$$\varphi_j = \begin{bmatrix} \varphi_{j,0} & \varphi_{j,-1} & \varphi_{j,1} & \varphi_{j,-2} & \varphi_{j,2} & \cdots \end{bmatrix}^T \tag{13}$$

2.2. Harmonic-Balanced Method in Electric Circuits Coupled to Magnetic Field

Most of electromagnetic devices are connected to the voltage source and therefore electric circuits coupled with magnetic field should be considered.

According to Kirchhoff's Law the applied voltage on the external port of the electric circuit can be defined as follows,

$$U_{ink} = U_k + R_k I_k \tag{14}$$

where U_{ink} is the input voltage of circuit k, and U_k is the corresponding induced electromotive force.

The induced electromotive force can be obtained in (15) based on Faraday's Law,

$$U_k = \frac{N_k}{S_k} \frac{d}{dt} \iiint_\Omega A \cdot t d\Omega \tag{15}$$

where t is the current vector representing the flowing direction of the exciting current.

Consequently we can attain the harmonic-balanced equation of the electromagnetic coupling as follows,

$$\frac{N_c}{S_c} \iiint_{\Omega} tN d\Omega \cdot A_j + R_k I_k = U_{ink} \qquad (16)$$

where k indicate the circuit number, N_c is the number of turns and S_c is the cross-section area of the exciting coil. A_j and I_k are the harmonic solutions of the magnetic vector potential and exciting current, respectively.

Finally the harmonic-balanced system equation can be represented simply as follows,

$$\begin{bmatrix} H & G \\ C & Z \end{bmatrix} \begin{bmatrix} A \\ I \end{bmatrix} = \begin{bmatrix} 0 \\ U \end{bmatrix} \qquad (17)$$

3. Computational Results and Field Analysis

3.1. Laminated Core Under DC-Biased Condition

The Epstein frame-like core model for the DC biased test is shown in Fig. 1. The iron core is made up of silicon steel lamination of which the model number is 30Q140. There are two windings on the ferromagnetic core: the exciting coil (connected to alternating voltage source) and the measuring coil.

The peak value of excitation current without a DC bias is selected as a reference. This reference current causes the flux density in the silicon steel to reach the rated value (1.7T) in the transformer's no-load operation. The DC bias in the form of direct current is then applied in proportion to the reference current to the exciting coil.

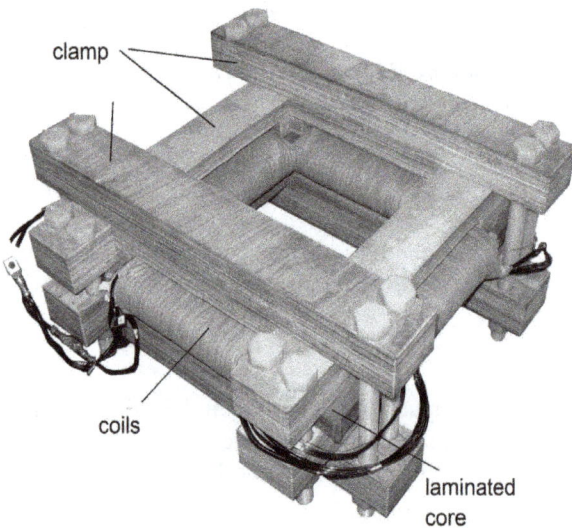

Figure 1. *Laminated Core Model Used in DC-Biased Experiment.*

The value of reference current I_0 measured on the square ferromagnetic core model is 1.68 ampere. The DC bias is

applied in incremental proportions of the reference current, which are represented by P_i ($i=1, 2, 3, 4$) in Table 1. The AC excitation is also applied in four different cases indicated by the subscript j ($j=1, 2, 3, 4$). I_{dc} represents the DC bias current that corresponds to different proportions of the reference current I_0, while H_{dc} is the subsequently generated magnetic intensity. The peak value of alternating flux density B_m in the magnetic core varies with the step-increase of alternating voltage U_m (peak value) [17], which is also shown in the same Table.

Table 1. *Different DC bias condition by quantity in the magnetic field.*

Cases (i/j)	DC bias			AC excitation	
	P_i ($\%I_0$)	$I_{dc,i}$(A)	$H_{dc,i}$(A/m)	$U_{m,j}$(V)	$B_{m,j}$(T)
1	25	0.4256	105.68	26	0.09
2	50	0.847	213.12	133	0.49
3	75	1.273	320.30	240	0.88
4	100	1.697	425.23	370	1.37
5	150	2.530	636.58	420	1.57
6				495	1.82

3.2. Computational Results and Analysis

3.2.1. Calculated Exciting Currents and Measured Results

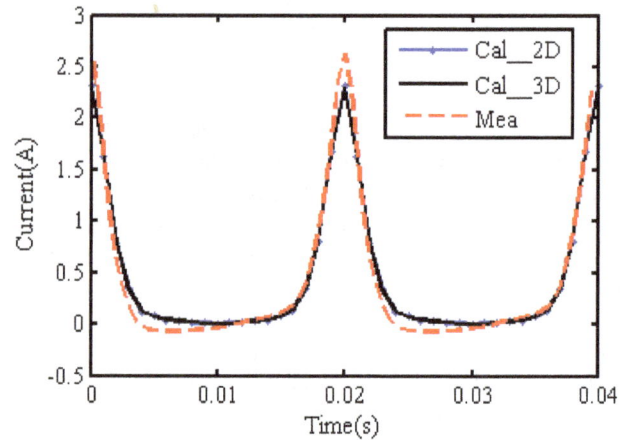

Figure 2. *Comparison of magnetizing currents between calculation and measurement ($U_m = U_{m,3} = 240V$, $B_m = B_{m,3} = 0.88T$).*

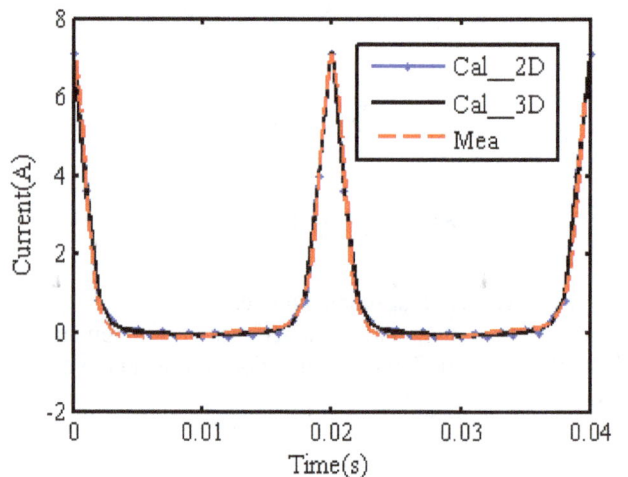

Figure 3. *Comparison of magnetizing currents between calculation and measurement ($U_m = U_{m,4} = 370V$, $B_m = B_{m,4} = 1.37T$).*

Harmonic solutions of the exciting currents and magnetic vector potential are computed simultaneously in (17). The calculated currents are compared with the measured data. As shown in Fig. 2 and Fig. 3, there is consistency in the computational and measured results obtained from the magnetizing current waveforms.

3.2.2. Harmonic Analysis of the Magnetizing Current

There are only odd harmonics in the magnetizing current when the transformer is fed by AC excitation. However, additional harmonics appear when the direct current invades the transformer windings. The generation of large harmonics results in significant saturation of the magnetic core and half-cycle saturation of the magnetizing current. Therefore, the relationship between the DC bias and harmonic components should be considered by using harmonic analysis [18].

Unlike the time-domain iterations and the Fourier transforming process of the solution in the time-stepping finite element method, all harmonic components in the magnetizing current can be obtained directly from the harmonic solution using the HBFEM. The histograms in Fig.4 and Fig.6 show the contribution of different harmonic components to the magnetizing current under different DC biases.

Fig. 4 shows that while the size of all harmonic components increases when additional DC bias is applied, the growth rate varies in different components. The growth tendency of each harmonic is shown in Fig. 5. The numbers 1, 2, 3, 4 in the horizontal coordinate represent different proportions (25%, 50%, 75%, 100%) of the DC bias reference current respectively. It is obvious that the fundamental and second harmonic components increase near-linearly, while higher order harmonics (the third and fourth) grow faster rather than linearly.

The contribution of each harmonic component is different when the peak value of alternating voltage is increased up to 495 volts, which is given in Fig. 6. Odd harmonics are greater than even order components under 25% and 50% DC bias respectively. It is implied that the growth of odd harmonic components is related to the increased AC excitation.

Curves in Fig. 7 display a relationship between odd harmonics and AC excitation. With the increased alternating voltage odd harmonics grow faster (and are greater in size) than the even harmonics. On the other hand, the negative influence of DC bias on each harmonic is analysed in Fig. 8 when the ferromagnetic core is significantly saturated as a consequence of high alternating voltage. Even harmonics increase faster than the odd harmonics with the increased DC bias and constant AC excitation.

It can be concluded that the appearance of DC bias in exciting current leads to the generation of even harmonics in the DC biased problem, and each harmonic component in the exciting current is affected by DC and AC excitation simultaneously. The applied alternating voltage makes the main contribution to the growth of odd harmonics while the DC bias plays a more important role in the variation of even harmonics, especially when the ferromagnetic core is

significantly saturated.

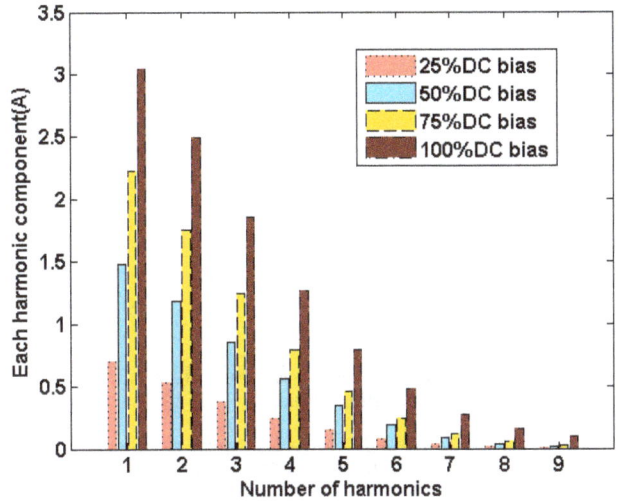

Figure 4. Each harmonic component of exciting current under different DC bias ($U_m = U_{m,3} = 240V$; $B_m = B_{m,3} = 0.88T$).

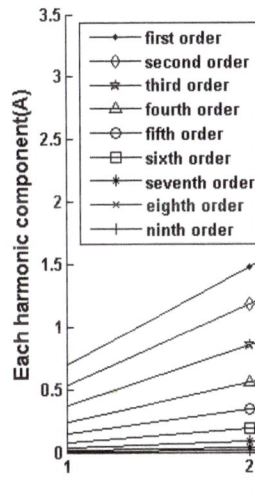

Figure 5. DC bias effect on different harmonics ($U_m = U_{m,3} = 240V$; $B_m = B_{m,3} = 0.88T$).

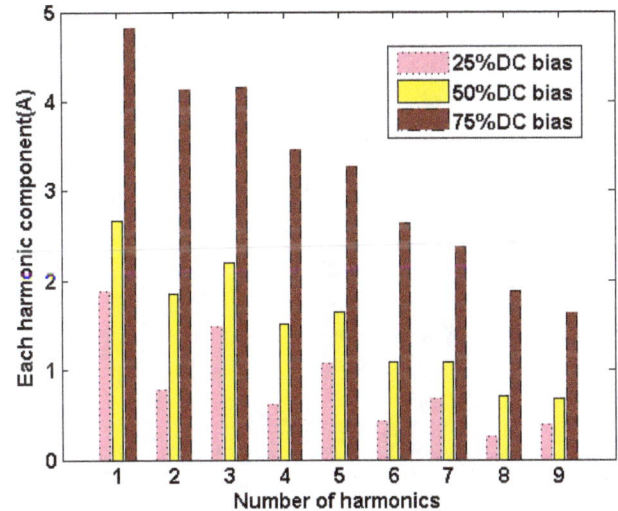

Figure 6. Each harmonic component of exciting current under different DC bias ($U_m = U_{m,6} = 495V$; $B_m = B_{m,6} = 1.82T$).

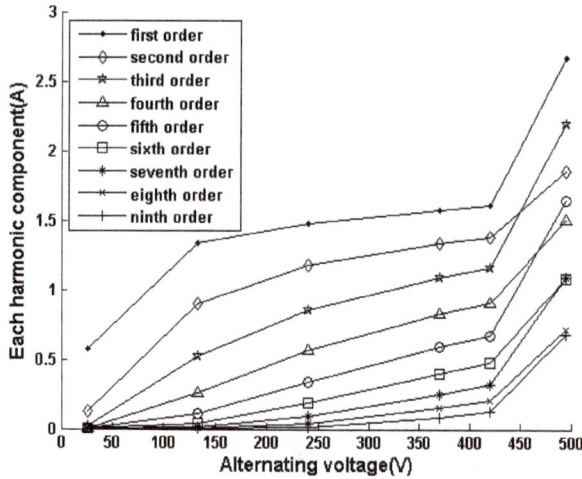

Figure 7. AC voltage (peak value) effect on each harmonic component under 50% DC bias ($I_{dc} = I_{dc,2} = 0.847A$; $H_{dc} = H_{dc,2} = 213.12A/m$).

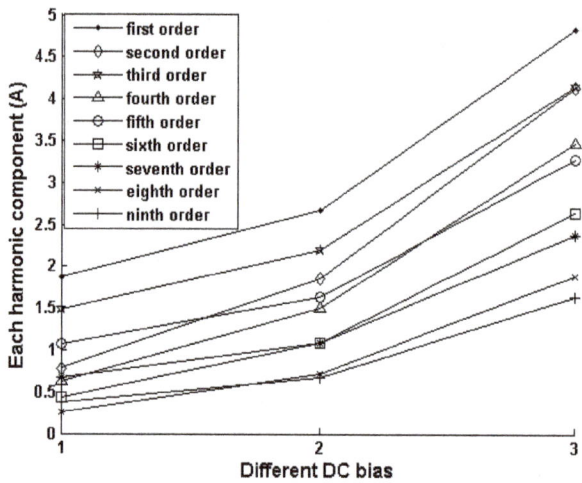

Figure 8. DC bias effect on different harmonics ($U_m = U_{m,6} = 495V$; $B_m = B_{m,6} = 1.82T$).

3.2.3. Calculated Magnetic Field

The averaged flux density in the laminated core is attained from the harmonic solutions of vector potential. As shown in Fig. 9, flux density in the laminated core calculated in 2-D model agrees well with that in 3-D model.

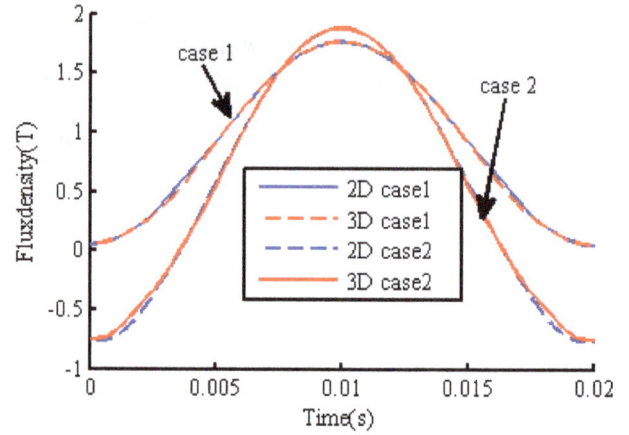

Figure 9. Flux density in the laminated core computed in 2-D and 3-D model (case 1:$U_m = U_{m,3} = 240V$, $B_m = B_{m,3} = 0.88T$; case 2:$U_m = U_{m,4} = 370V$, $B_m = B_{m,4} = 1.37T$;).

Harmonic flux distributions under DC bias condition are presented in Fig.10. The harmonic flux distributions vary with time (related to phase angle) and excitations (related to DC bias and alternating voltage) [19].

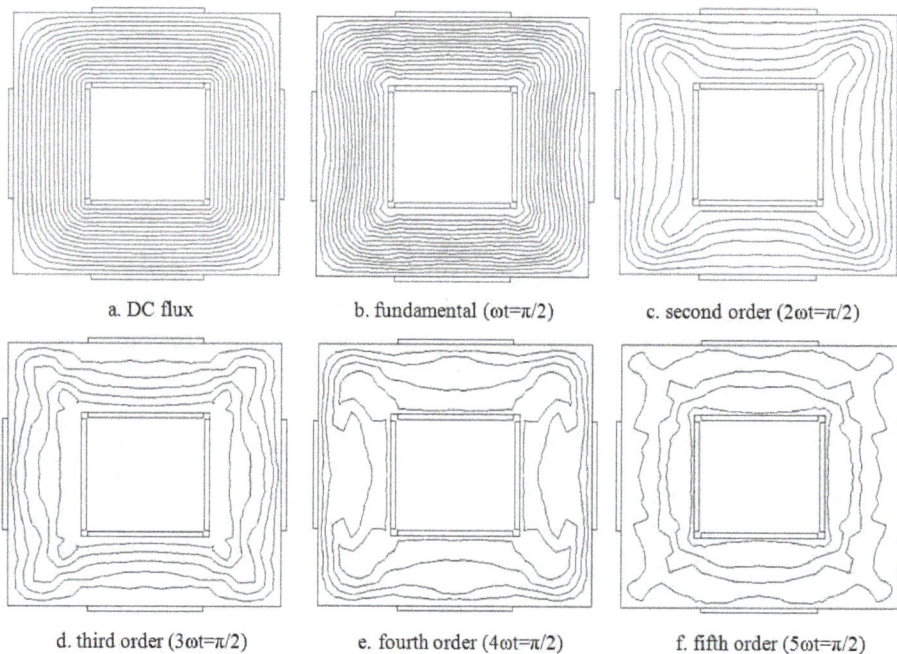

a. DC flux   b. fundamental ($\omega t=\pi/2$)   c. second order ($2\omega t=\pi/2$)

d. third order ($3\omega t=\pi/2$)   e. fourth order ($4\omega t=\pi/2$)   f. fifth order ($5\omega t=\pi/2$)

Figure 10. Harmonic flux distributions ($I_{dc}=1.27A$, $U_m= U_{m,4} =370V$).

4. Conclusion

The time-periodic magnetic field can be computed in harmonic domain by the harmonic-balanced finite-element method. The time-stepping method requires many periods to approach the steady state solution in time domain, instead, the harmonic-balanced method computes nonlinear magnetic field directly in frequency domain. Harmonic solutions can be decoupled by the decomposed algorithm [19] and the system equation is suitable for concurrent computation. It can be concluded that the harmonic-balanced method requires less memory for the solution since the number of harmonics is usually less than the number of time steps in computation.

Acknowledgement

This work is supported by the National Natural Science Foundation of China (Grant No. 51307057), Hebei Province Natural Science Foundation (Grant No. E2013502323), Research Fund for the Doctoral Program of Higher Education of China (Grant No. 20130036120011), and the Fundamental Research Funds for the Central Universities (Grant No. 2015MS82).

References

[1] P. Picher, L. Bolduc, A. Dutil, and V. Q. Pham, "Study of the acceptable DC current limit in core-form power transformers," *IEEE Trans. Power Del.*, vol. 12, no. 1, pp. 257–265, Jan. 1997.

[2] H. C. Tay and G. W. Swift, "On the problem of transformer overheating due to geomagnetically induced current," *IEEE Trans. Power App. Syst.*, vol. PAS-104, no. 1, pp. 212–219, Jan. 1985.

[3] Y. You, E. F. Fuchs, D. Lin, and P. R. Barnes, "Reactive power demand of transformers with DC bias," *IEEE Ind. Appl. Mag.*, vol. 2, no. 4, pp. 45–52, Jul. 1996.

[4] V. D. Albertson, B. Bozoki, W. E. Feero, J. G. Kappenman, E. V. Larsen, and D. E. Nordell *et al.*, "Geomagnetic disturbance effects on power systems," *IEEE Trans. Power Del.*, vol. 8, no. 3, pp. 1206–1216, Jul. 1993.

[5] L. Bolduc, A. Gaudreau, and A. Dutil, "Saturation time of transformers under DC excitation," *Elsevier, Elect. Power Syst. Res.*, vol. 56, no. 2, pp. 95–102, Sep. 2000.

[6] R. Pirjola, "Geomagnetically induced current during magnetic storms," *IEEE Trans. Plasma Sci.*, vol. 28, no. 6, pp. 1867–1873, Dec. 2000.

[7] Y. Yao, "Research on the DC bias phenomena of large power transformers," Ph.D. dissertation, Shenyang Univ. Technol., Shenyang, China, 2000

[8] B. Zhang, X. Cui, R. Zeng, and J. He, "Calculation of DC current distribution in AC power system near HVDC system by using moment method coupled to circuit equations," *IEEE Trans. Magn.*, vol. 42, no. 4, pp. 703–706, Nov. 2006.

[9] N. Takasu, T. Oshi, F. Miyawaki, and S. Saito, "An experimental analysis of DC excitation of transformers by geomagnetically induced current," *IEEE Trans. Power Del.*, vol. 9, no. 2, pp. 1173–1182, Apr. 1994.

[10] P. R. Price, "Geomagnetically induced current effects on transformers," *IEEE Trans. Magn.*, vol. 17, no. 4, pp. 1002–1008, Oct. 2002.

[11] O. Biro, G. Buchgraber, G. Leber, and K. Preis, "Prediction of magnetizing current wave-forms in a three-phase power transformer under DC bias," *IEEE Trans. Magn.*, vol. 44, no. 6, pp. 1554–1557, Jun. 2008.

[12] O. Biro, S. Auberhofer, G. Buchgraber, K. Preis, and W. Seitlinger, "Prediction of magnetizing current waveform in a single-phase power transformer under DC bias," *Inst. Eng. Technol. Sci., Meas. Technol.*, vol. 1, no. 1, pp. 2–5, 2007.

[13] E. F. Fuchs, Y. You, and D. J. Roesler, "Modeling and simulation, and their validation of three-phase transformers with three legs under DC bias," *IEEE Trans. Power Del.*, vol. 14, no. 2, pp. 443–449, Apr. 1999.

[14] L. Cao, J. Zhao, and J. He, "Improved power transformer model for DC biasing analysis considering transient leakage reluctance," in *Proc. Int. Conf. Power System Technol.*, 2006, pp. 1–5.

[15] Y. Yao, C. S. Koh, G. Ni, and D. Xie, "3-D nonlinear transient eddy current calculation of online power transformer under DC bias," *IEEE Trans. Magn.*, vol. 41, no. 5, pp. 1840–1843, May 2005.

[16] Z. Cheng, S. Gao, and L. Li, *Eddy Current Analysis and Validation in Electrical Engineering*. Beijing, China: Higher Education Press, ISBN 7-04-009888-1, pp. 70–83, 2001.

[17] Z. Cheng, N. Takahashi, and B. Forghani, *Electromagnetic and Thermal Field Modeling and Application in Electrical Engineering*. Beijing, China: Science Press, ISBN 978-7-03-023561-9, pp. 386–393, 2009.

[18] X. Zhao, J. Lu, L. Li, Z. Cheng and T. Lu, "Analysis of the dc bias phenomenon by the harmonic balance finite element method," *IEEE Trans. Power Delivery*, vol. 26, no. 1, pp. 475–487, 2011.

[19] X. Zhao, J. Lu, L. Li, Z. Cheng and T. Lu, "Analysis of the saturated electromagnetic devices under DC bias condition by the decomposed harmonic balance finite element method", *COMPEL.*, vol. 31, no. 2, pp. 498-513, 2012.

23

Development of Transesterification System with Acid and Base Homogeneous Catalysts For Mangifera Indica Seed Oil to Mangifera Indica Methyl Ester (MOME Biodiesel)

Shubhangi S. Nigade[1], Sangram D. Jadhav[2, *], Abhimanyu K. Chandgude[1]

[1]Department of Mechanical Engineering, KJEI's Trinity college of Engineering and Research Pune, Maharashtra, India
[2]Department of Mechanical Engineering, Government of Maharashtra Dr. B. A. Technological University Mangaon, Maharashtra, India

Email address:

shubhangisnigade@gmail.com (S. S. Nigade), jdsangram@gmail.com (S. D. Jadhav), akc0107@gmail.com (A. K. Chandgude)

Abstract: The depletion of resources, increased cost of fossil fuel and increased environmental awareness reaching the critical condition. Development of viable alternative fuels from renewable resources is gaining the international attention and acceptance. The vegetable oils have the potential of alternative fuel for compression ignition engines by converting it into biodiesel. The mangifera indica oil is a nonedible vegetable oil, available in large quantities in mangifera indica cultivating countries including India. Very little research has been done on utilization of oil in general and optimization of transesterification process for biodiesel production. However, direct base catalyzed transesterification produced no biodiesel due to the high Free Fatty Acid (FFA) value of the oil. Hence, acid pretreatment was preferred prior to base transesterification which afforded a significant reduction of the FFA value from 3.3% to 0.9% . Various input parameters like oil-to-methanol molar ratio (1:08, 1:12 and 1:16), catalyst type (NaOH, KOH and NaOCH3), catalyst concentration (0.5, 1 and 1.5 wt %) and reaction temperature (59, 64 and 69°C) were studied. The optimum conditions for transesterification process are: 1:12 oil-to-methanol molar ratio, 1.0 wt.% catalyst concentration, KOH catalyst, & 64°C reaction temperature. The optimum yield of MOME was 89.8%. The biodiesel produced (MOME) is within the limits prescribed by EN-14214 standard.

Keywords: Biodiesel, Extraction, Mangifera Indica, Pretreatment, Transesterification

1. Introduction

India currently ranks as the worlds 11th greatest energy producer accounting for about 2.4% of the world's total annual energy production, while it ranks as the 6th largest energy consumer accounting for about 3.3 % of the world's total annual energy consumption. Currently India is such a country where present level of energy consumption by world standard is very low, with per capita energy consumption is less than 500 Kgoe (Kilogram oil equivalent) compared to Global average of nearly 1800 Kgoe. The major part of all energy consumed in most parts of the world comes from fossil sources such as petroleum, coal and natural gas. However, these non-renewable sources will be exhausted in near future. Recent assessments of remaining petroleum reserves show the world will soon face a relentless oil-supply conventional crude oil is projected to peak and decline irreversibly. Alternative sources for petroleum products will then be critical. In India 95 % energy need of transportation sector are provided by the Diesel and the demand for diesel is five times higher than the diesel it was estimated that for sustaining India's 8% average annual economic growth and to support its growing population. India needs to generate 2 to 3 fold more energy than present It is estimated that India has only 0.4% of the world's proven reserves of crude oil .India meets about 70 % of its petroleum requirement through import which are expected to expand in the coming years. In India volume of crude oil imported increased 14 fold from 11.66 million tons during 1970-71 to 163.59 million tons by 2010 to 2011.During last 7 years; India's foreign exchange outflow due to this purpose has increased 5 fold because of escalation of international oil prices [1-2].

If the nation's source of petroleum products continues to be limited to conventional crude oil, this situation is certain to become worse. Widely acknowledged estimates of remaining recoverable reserves of conventional crude oil worldwide total 1 trillion barrels. Thus, the search for alternative sources of renewable and sustainable energy has gained importance with the potential to solve many current social issues such as the rising price of petroleum crude and environmental concerns like air pollution and global warming caused by combustion of fossil fuels [3-4]. The term biofuel or renewable fuel is referred to as solid, liquid or gaseous fuels that are predominantly produced from biomass. Liquid and gaseous biofuels have become more attractive recently because of its environmental benefits. Biofuels are non-polluting, locally available, accessible, sustainable and reliable fuel obtained from renewable sources [5-6]. Among other reasons why biofuels are considered as relevant technologies by both developing and industrialized countries are: energy security, environmental concerns, foreign exchange savings, and socio-economic issues related to the rural sectors of all countries in the world [7]. In recent years, many studies have investigated the economic and environmental impacts of biofuels, especially bioethanol, biodiesel, biogas, and biohydrogen [6]. Biodiesel (fatty-acid alkyl esters) is a renewable and environmentally friendly energy source. It can be produced from plant oils and animal fats. Several techniques are available for biodiesel production. The most commonly used technique is transesterification in which triglycerides are reacted with alcohol, usually methanol, in the presence of a catalyst, usually potassium or sodium hydroxide (KOH or NaOH), to produce mono alkyl esters. Many factors affect the biodiesel yield and process economics. The most important factors are alcohol type, alcohol/oil molar ratio, reaction temperature and time, catalyst type and amount and water content of the reactants [8]. Mangifera indicas belong to the genus Mangifera of the family Anacardiaceae. The genus Mangifera contains several species that bear edible fruit. Most of the fruit trees that are commonly known as mangifera indicas belong to the species Mangifera indica. The other edible Mangifera species generally have lower quality fruit and are commonly referred to as wild mangifera indicas. Mangifera indica fruit is classed as a drupe (fleshy with a single seed enclosed in a leathery endocarp).

Fruits from different varieties can be highly variable in shape, color, taste and flesh texture. Fruit shapes vary from round to ovate to oblong and long with variable lateral compression. Fruits can weigh from less than 50 g (0.35 lb) to over 2 kg (4.4 lb). The fruit has a dark green background color when developing on the tree that turns lighter green to yellow as it ripens [9]. Currently, Nigerian government has shown great interest in Jatropha and other biofuel plants. The aim of the government is to gradually reduce the nation's dependency on gasoline, reduce environmental pollutions as well as create commercially viable industry that can precipitate domestic job [10-11,26-30].

India is the largest producer of mangifera indica in the world, with the annual production about 15.19 million tonnes.

Indian share in production of mangifera indica is 42.04% of world production. Mangifera indica is by-product of mangifera indica, which contain approximately 25.6 to 32.6% oil("UN FAOSTAT 2012.pdf", S. S. Raju et al. 2012, Thammarat 2013). Such a huge amount of feedstock volume, high oil contents & low cost favors mangifera indica oil for biodiesel production in India (Tapasvi et al. 2005). A large green tree mainly valued for its fruits, raw and ripe. It can grow up to 15-30 m tall and its yield in kg per tree is given in table no. 1.

Table 1. Yield of Mangifera indica as per age.

Sr. No	Age (Years)	Fruit no./ tree	Yield (kg/yr/tree)
1	5-8	450	292
2	9-10	800	657
3	11-25	1250	892

In view of this, it was decided to optimize the transesterification process for production of biodiesel using homogeneous catalysts. The main objective of research was to maximize yield with respect to input variables; methanol to oil molar ratio, catalyst type, catalyst amount and reaction temperature.

2. Experimentation

2.1. Materials

Methanol (Purity 99.8%, IR spectrum), potassium hydroxide pellets (Purity 98.0%), sodium hydroxide pellets (Purity 98.0%), N-hexane (Purity 99.0%, IR spectrum), Acetone hexane (Purity 99.0%, IR spectrum), Silicon oil (oil bath upto 250°C), Sulfuric acid (Purity 98.0%), Stearic acid (Purity 98.0%), Palmitic acid (Purity 98.0%), Oleic acid, Linoleic acid(Purity 90.0%), Linolenic acid (Purity 98.0%), Arachidic acid (Purity 90.0%), 1250 Grade1 filter paper (10-13μm, <0.06 Cenizas), phenolphthalein pH indicator were purchased Thomas baker Chemicals pvt. Ltd. India. The kernels of Mangifera indica were collected at canning industries around coastal area konkan, India.

2.2. Extraction of the Oil

Before the extraction process, mangifera indica kernels were dried overnight at 58°C in an oven to remove the excess moisture. The dried seeds were then weighted and crushed into fine particles of 0.5 to 10 mm in size. The oil was then extracted using Soxhelet extractor with N-hexane as the solvent. The duration for each batch of extraction was fixed at 4h; while the volume of solvent per kilogram of seed was varied from 4 liter to 6 liter for maximization of oil yield. The oil was recovered at the RBF which put into the heating mental at the bottom. The spiral coil condenser is used for recovery of solvent which get collected in RBF and re-used again in the process till completion of extraction. The extracted oil was then measured to calculate the content of oil in the kernel of mangifera indica. The physiochemical properties and fatty acid composition of mangifera indica oil (MIO) are shown in Table 2 and 3.

Table 2. Physiochemical properties of mangifera indica oil.

Sr. No	Physical character	Value
1	Refractive Index at 40°C	1.4560
2	Iodine Value	47.3
3	Saponification value	192.4
4	Unsaponificable matter	1.2%
5	Specific gravity	0.998
6	colour	Dark Yellow

Table 3. Fatty acid composition of the mangifera indica oil.

Fatty acid	Chemical Structure	Percentage (%)
Palmitic	$C_{16}H_{32}O_2$	5.6
Stearic	$C_{18}H_{36}O_2$	40.3
Oleic	$C_{18}H_{34}O_2$	46.6
Linoleic	$C_{18}H_{32}O_2$	5.1
Arachidic	$C_{20}H_{40}O_2$	3.2
Linolenic	$C_{18}H36O_2$	0.3

2.3. Acid Pretreatment Process

The pretreatment was conducted in a corked 250ml flat bottom flask, placed on a hot plate magnetic stirrer preset at the required temperature. In the experiments, flasks loaded with Mangifera indica oil samples was first heated to the designated temperature of 500C[17, 24, 29-31]. This was followed by the addition of the methanol and sulfuric acid (The solution of concentrated H_2SO_4 acid 1.0% based on the weight of oil, and the oil to methanol ratio of 1:6 by volume and a reaction time of 70min). The Transesterification pretreated products oil was separated in a tap funnel to obtain the upper layer, which was then washed with water several times until the pH of the washing water was close to 7.0 [14]. The resultant pretreated oil was dried in an oven before the subsequent transesterification process [18, 26-30].

2.4. Transesterification

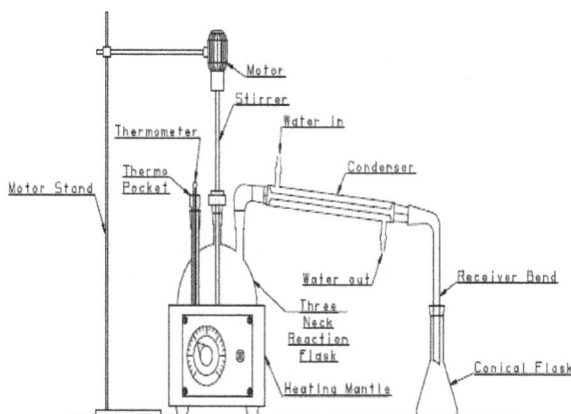

Fig. 1. Experimental setup for Pretreatment and Transesterification.

Base catalyzed transesterification was carried out according to the Ireland method 1. In a 250 ml conical flask equipped with a magnetic stirrer. 30ml of the extracted oil was taken in flask and potassium hydroxide (1 percent of oil's weight) dissolved in methanol (22.5 percent of oil's weight) was added to flask (as shown figure 1). Stirring was continued for 90mins at 600C, the mixture was transferred to a separatory funnel and glycerol was allowed to separate, leaving the upper layer biodiesel and the lower layer glycerin [19, 30-31].

2.5. Experimental Conditions

Experiments were planned to ascertain the oil/methanol ratio (w/w), catalyst concentration, reaction temperature and agitation intensity on transesterification reaction. The ratio of Oil/Methanol was varied as per w/w 4:1, 5:1, 6:1, the catalysts was sodium hydroxide and its concentrations were varied as 0.25, 0.50, 0.75, 1.00 and 1.50% of the oil. Reaction temperatures considered were 50, 55, 60 and 65°C and the agitation intensity were varied through 150, 300, 450, 600 and 700 rpm.

2.6. Biodiesel Separation and Washing

After obtaining the biodiesel phase, methyl ester was washed with hot water three times until the residual catalyst is finally off the solution. Warm water at temperature of about 500C, usually ratio 1:1 to the biodiesel was used in each washing step to clean up the esters. Finally, the biodiesel was dried in an oven at 105 degree for 30mins [19-23,30].

2.7. Fuel Properties

The following properties of the biodiesel produced were determined: density and specific gravity [11], kinematic viscosity 40°C (ASTM D 445), flash point (ASTM D 93), Sulfated Ash (ASTM D847) carbon residue (ASTM D524).

3. Results and Discussion

3.1. Effect of Molar Ratio

In stoichiometric transesterification reaction, each mol of triglyceride requires three moles of alcohol to produce three moles of fatty acid alkyl ester and one mole of glycerol, whereas esterification requires one mole of FFA and one mole of alcohol to generate one mole of ester and one mole of water. Since these reactions are reversible, excess alcohol is required to drive the reaction toward the product side for increasing and completing the conversion [22-27].

The average molecular weight of mangifera indica oil was calculated from its composition and accordingly the amount of methanol was taken in the reaction. The methanol to oil molar ratio varied from 1:8 to 1:16 to study its effect on yield of conversion process. The mean yield of MOME at different molar ratio 1:8, 1:12 and 1:16 are shown in Fig. 2. The effect of methanol in the range of 1:8 to 1:16 (molar ratio) was investigated. It was found that the ester yield increases with increase in molar ratio of methanol to vegetable oil. Molar ratio is in between 1:08 to 1:12 shows faster conversion rate compared to molar ratio after 1:12.

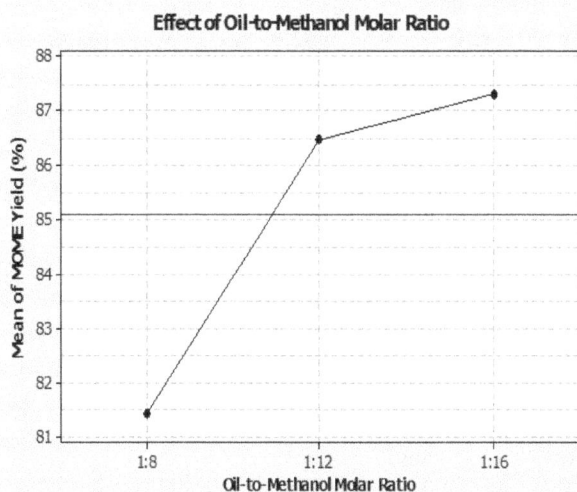

Fig. 2. *Effect of Oil-to-Methanol Molar Ratio.*

At low molar ratio, low proportion of methanol reduces the probability of braking bonds between glycerol and tryglyceriods reducing the yield. Though the stoichiometric molar ratio is 3:1, the general trend is increase in yield of reaction with molar ratio. Therefore molar ratio at 1:12 shows the optimum molar ratio for the yield of mangifera indica oil. However, when the ratio of oil to alcohol is too high, it could give adverse effect on the yield of fatty acid alkyl esters eg. phase separation of ester and glycerol, mass transfer problem between triglycerides etc.

3.2. Effect of Catalyst Type

The type of catalyst required in the transesterification process usually depends on the quality of the feedstock and method applied for the transesterification process. For a purified feedstock, any type of catalyst could be used for the transesterification process.

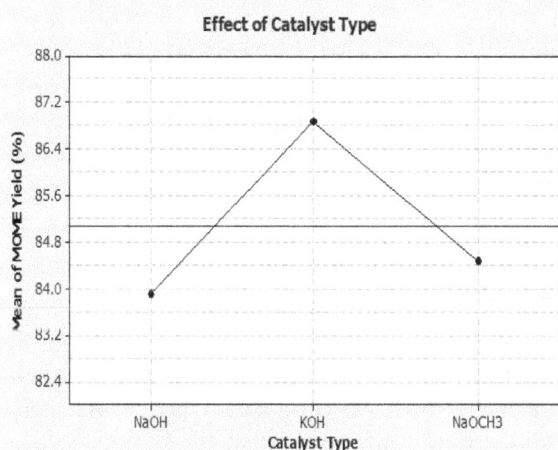

Fig. 3. *Effect of Catalyst Type.*

However, for feedstock with high moisture and free fatty acids contents, homogenous transesterification process is unsuitable due to high possibility of saponification process instead of transesterification process to occur. Two step

transesterification processes had been suggested by several researchers (13-18,21). At present, homogeneous catalysts such as NaOH and KOH are primarily used by biodiesel industry due to their simple usage and short time required for conversion of oils to ester. The homogeneous catalysts (e.g NaOH, KOH and $NaOCH_3$) catalyst forms sodium and potassium methoxide due to its solubility in methanol which augment the completion of reaction. The main advantage homogenous acid and alkali catalysts are high yield and low cost. Transesterification process was carried out by using three homogeneous catalysts (NaOH, KOH, $NaOCH_3$). The results obtained are presented in Fig. 3. The output values represented in figure are mean percentage values of yield. There is significant effect of type of catalyst on yield. Amongst the homogeneous catalysts the output with KOH (86.6%) catalyst is greater than NaOH, $NaOCH_3$ catalyst.

3.3. Effect of Catalyst Concentration

Tests were conducted to study the effect of amount of catalyst used (concentration) on conversion of mangifera indica oil to ester. The results of the tests are presented in Fig. 4. The amount of catalyst used during the tests was varied from 0.5 to 1.5 % of weight of the oil in a step of 0.5. The yield of MOME increases with increase of catalyst concentration. If we see the rate of yield conversion it is faster upto 1.0 wt.% catalyst concentration after it there is slightly lower the yield conversion rate. As increase in catalyst concentration no doubt yield increases, this is due to availability of more active sites by additions of larger amount of catalyst in the transesterification process.

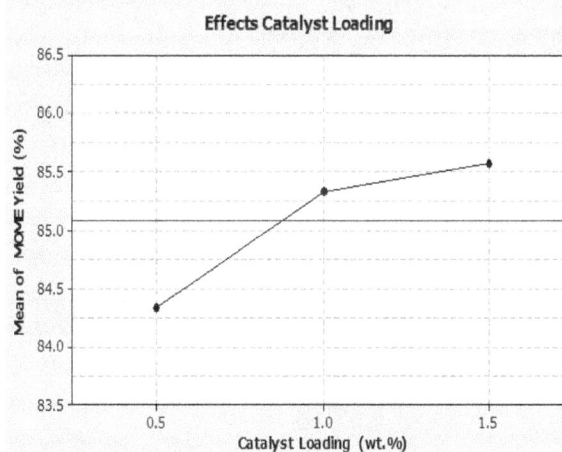

Fig. 4. *Effect of Catalyst Loading.*

However, on economic perspective, larger amount of catalyst may not be profitable due to cost of the catalysts itself. Hence for economic and performance points of view, 1.0 wt.% catalyst loading is optimum for mangifera indica oil.

3.4. Effect of Reaction Temperature

The reaction temperature was varied from 59 to 69 °C. The results of the test are presented in Fig. 5. The yield of mangifera indica oil increases with increase in temperature upto 64 °C and it drops beyond this value of temperature. The maximum value of mean yield (85.82%) was observed at 64 °C. It is interesting to note that maximum yield was observed at temperature 64 °C which is very close to boiling point of methanol (64.6 °C). The formation of vapour causes the volatilization and the momentum between monoglycerid, diglyceride, triglyceride and glycerol which helps for breaking the bond.

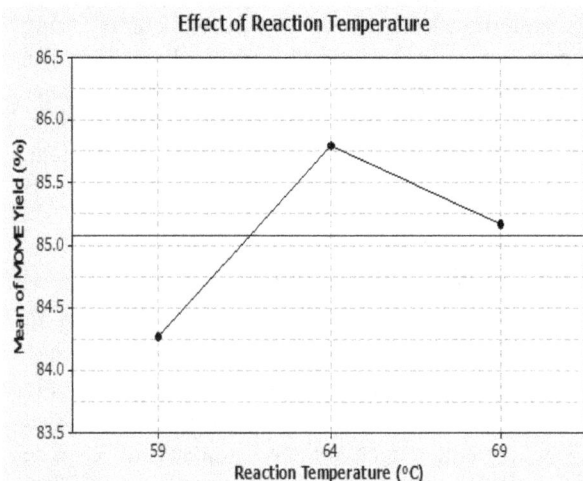

Fig. 5. Effect of reaction Temperature.

3.5. Biodiesel Characterization

The fuel properties of the biodiesel (Mangifera indica methyl ester) and mineral Diesel were determined using standard test procedures (are given in Table 4). The calorific value is a measure of the energy content of the fuel and is a very important property of biodiesel, which determines its suitability as an alternative to mineral Diesel. The calorific value of Mangifera indica methyl ester (MOME) and Mangifera indica oil is 43.1 and 42.3 MJ/kg, which is almost 96% and 93% of the diesel calorific value (44.8 MJ/kg), respectively.

Table 4. *Properties of diesel, mangifera indica oil and mangifera indica methyl ester (MOME).*

Property	Std.	Diesel	MVO	MOME
Specific gravity	---	0.839	0.912	0.872
Kinematic viscosity @ 40°C (cSt)	ASTM D445	3.18	38.46	4.62
Cloud Point (°C)	ASTM D2500	6	11	8
Pour Point(°C)	ASTM D2500	-7	1	-2
Flash Point(°C)	ASTM D93	68	310	176
Fire point(°C)	ASTM D93	103	332	179
Carbon Residue (%, w/w)	ASTM D189	0.1	0.8	0.38
Calorific Value (MJ/kg)	ASTM D240	44.8	42.3	43.1

The lower calorific value of MOME is because of the presence of oxygen in the molecular structure, which is confirmed by elemental analysis also. The flash point and fire point were tested with a closed cup Pensky Marten's apparatus. The flash point is the measure of the tendency of a substance to form flammable mixtures when exposed to air. This parameter is considered in the handling, storage and safety of fuels. The high value of flash point and fire point in the case of MOME represents it is a safer fuel to handle.

4. Conclusions

The present study was aimed to optimize transesterification process for mangifera indica oil using different types of homogeneous catalysts. The study involved following processes:
- Extraction of mangifera indica oil,
- Transesterification with different catalysts
- Catalyst removal from biodiesel
- Measurement of properties and quality of ester as product of the process.

Extensive work has been done mainly on jatropha and karanja oil. This may be partly due to a) the policy of Indian government to promote use of nonedible feedstock for alternative fuel b) large scale potential for their growth. Varity of crops are grown in different regions of the country. Hence regional feedstocks like mahua, simaruba, mangifera indica, neem, kokam, mahua etc., apart from jatropha and karanja, are also promoted for biodiesel production. Literature review on mangifera indica oil revealed that very little work has been done on this oil. This was one of the motivating factors to carry out the present study on mangifera indica oil. In west coast of India huge quantity (15.19 million of tonnes) of mangifera indica kernel is available. Hence there is good opportunity for obtaining biodiesel from this oil in India. High oil content, large availability, low cost of feedstock, huge amount of mangifera indica kernel per tree per year, low initial investment are the key features in favors of its use for biodiesel production.

Alkali catalyzed transesterification was studied with NaOH, KOH, and $NaOCH_3$ as catalyst for biodiesel production from Mangifera indica oil. Apart from optimization of transesterification process, catalyst removal from biodiesel also has been studied in detail. Transesterification reaction parameters decide the yield of the ester, while catalyst removal decides the quality of ester.

5. Overall Conclusion

The effect of the parameters, namely molar ratio of oil-to-alcohol, catalyst type, catalyst concentration and reaction temperature on the yield of ester (biodiesel) was investigated. The optimum values of input parameters for maximum yield of biodiesel (MOME) were obtained as: 1:12 oil-to-alcohol molar ratio, KOH catalyst, 1.0 wt.% catalysts concentration and 64°C reaction temperature. The biodiesel produced is within the limits prescribed by ASTM standard. The results of the study revealed that KOH catalyst (homogeneous) is best for mangifera indica oil, which is one of the promising resources for biodiesel production and possible substitute for diesel.

References

[1] Abbaszaadeh Ahmad, Barat Ghobadian, Mohammad Reza Omidkhah, and Gholamhassan Najafi. 2012. "Current Biodiesel Production Technologies: A Comparative Review." Energy Conversion and Management 63:138–48.

[2] Bagby M, and B Freedman. 1987. "Preparation and Properties from Vegetable Oils" 66: 1372–78.

[3] Balat Mustafa, and Havva Balat. 2010. "Progress in Biodiesel Processing." Applied Energy 87 (6): 1815–35.

[4] Dorado M.P., Ballesteros, E., Almeida, J.A., Schellert, C., Lohrlein, H.P. and Krause, R., "An alkali-catalyzed transesterification process for high free fatty acid waste oils". ASAE 45(2002) 525-29.

[5] Barnwal B.K., and M.P. Sharma. 2005. "Prospects of Biodiesel Production from Vegetable Oils in India." Renewable and Sustainable Energy Reviews 9 (4): 363–78.

[6] Tapasvi D., Wisenborn D. and Gustafson C., "Process model for biodiesel production from various feedstocks". ASAE, 48 (2005) 2215-21

[7] Bozbas Kahraman. 2008. "Biodiesel as an Alternative Motor Fuel: Production and Policies in the European Union." Renewable and Sustainable Energy Reviews 12 (2): 542–52.

[8] "BP Statistical Review of World Energy June 2012, 2013, 2014. 68th edition.

[9] Government of India Planning Commission, 2011. "Faster , Sustainable and More Inclusive Growth An Approach to the Twelfth five year plan 2012-17 ."

[10] Demirbas Ayhan. 2009. "Progress and Recent Trends in Biodiesel Fuels." Energy Conversion and Management 50 (1): 14–34.

[11] Formo Marvin W, and Archer-daniels-midland Company. 1953. "Ester Reactions of Fatty Materials" 31 (130): 548–59.

[12] Fukuda Hidekl, Akihiko Kond, and Hide Noda. 2001. "Biodiesel Fuel Production by Transesterification" 92 (5) 405-416.

[13] Haas Michael J, Andrew J McAloon, Winnie C Yee, and Thomas a Foglia. 2006. "A Process Model to Estimate Biodiesel Production Costs." Bioresource Technology 97 (4): 671–78.

[14] Jadhav Sangram D, and Madhukar S Tandale. 2014. "Production, Performance and Emission Study of Mahua (Madhuca Indica) Methyl Ester in Multicylinder 4-Stroke Petrol Engine" 3 (1): 2582–86.

[15] Karaosmanog Filiz. 2004. "Optimization of Base-Catalyzed Transesterification" 5 (2): 1888–95.

[16] W. Kdrbltz, 1999. "Biodiesel production in Europe and North America, An Encourging prospects" Renewable Energy 16, 1078–83.

[17] Knothe Gerhard. 2005. "Dependence of Biodiesel Fuel Properties on the Structure of Fatty Acid Alkyl Esters" 86: 1059–70.

[18] Marchetti J M A, V U Miguel, and A F Errazu. 2007. "Possible Methods for Biodiesel Production" 11: 1300–1311.

[19] Meher L, D Vidyasagar, and S Naik. 2006. "Technical Aspects of Biodiesel Production by Transesterification—a Review." Renewable and Sustainable Energy Reviews 10 (3): 248–68.

[20] S. S. Raju, R. Chand, P. Kumar, and Siwa Msangi. 2012. " Biofuels in India: Potential , Policy and Emerging Paradigms."NCAP New Delhi 14: 1–18.

[21] Olutoye M a, and B H Hameed. 2013. "Production of Biodiesel Fuel by Transesterification of Different Vegetable Oils with Methanol Using Al2O3 Modified MgZnO Catalyst." Bioresource Technology 132: 103–8.

[22] Pinzi S., I. L. Garcia, F. J. Lopez-Gimenez, M. D. Luque de Castro, G. Dorado, and M. P. Dorado. 2009. "The Ideal Vegetable Oil-Based Biodiesel Composition: A Review of Social, Economical and Technical Implications." Energy & Fuels 23 (5): 2325–41.

[23] Saravanan N., Sukumar Puhan, G. Nagarajan, and N. Vedaraman. 2010. "An Experimental Comparison of Transesterification Process with Different Alcohols Using Acid Catalysts." Biomass and Bioenergy 34 (7): 999–1005.

[24] Sharma Y C, and B Singh. 2009. "Development of Biodiesel : Current Scenario" 13: 1646–51.

[25] Shay E.Griffin. 1993. "Diesel Fuel from Vegetable Oils: Status and Opportunities." Biomass and Bioenergy 4 (4): 227–42.

[26] Shiu Pei-jing, Setiyo Gunawan, Wen-hao Hsieh, Novy S Kasim, and Yi-hsu Ju. 2010. "Bioresource Technology Biodiesel Production from Rice Bran by a Two-Step in-Situ Process." Bioresource Technology 101 (3): 984–89.

[27] Srivastava Anjana, and Ram Prasad. 2000. "Triglycerides-Based Diesel Fuels." Renewable and Sustainable Energy Reviews 4 (2): 111–33.

[28] Thammarat Nakhon et al. 2013. "Mangifera indica Seed Kernel Oil and Its Physicochemical Properties" 20 (3): 1145–49.

[29] Tu Le, Kenji Okitsu, Yasuhiro Sadanaga, Norimichi Takenaka, and Yasuaki Maeda. 2010. "Bioresource Technology A Two-Step Continuous Ultrasound Assisted Production of Biodiesel Fuel from Waste Cooking Oils: A Practical and Economical Approach to Produce High Quality Biodiesel Fuel." Bioresource Technology 101 (14): 5394–5401.

[30] "Statistical Division Food and Agricultural Organisation of United Nations: Economic and social Department. UN Food and Agricultural Organisation Corporate statistical Database 2011-12."

[31] Zelenka Paul, Wolfgang Cartellieri, and Peter Herzog. 1996. "Worldwide Diesel Emission Standards , Current Experiences and Future Needs" 10: 3–28.

24

Effect of Tail Shapes on Yawing Performance of Micro Wind Turbine

Nikhil C. Raikar, Sandip A. Kale

Mechanical Engineering Department, Trinity College of Engineering and Research, Pune, India

Email address:

nikhilraikar90@gmail.com (N. C. Raikar), sakale2050@gmail.com (S. A. Kale)

Abstract: Wind energy is one of the most widely used renewable energy resources. Large amounts of research and resources are being spent today in order to harness the energy from wind effectively. Large wind turbines are erected at windy site and delivering satisfactory performance. Small wind turbines are erected in low wind regions and are in development stage. Research activities are increasing at a significant rate in the field of small wind turbines. World wind energy association forecast considerable development in this field. To face the wind directions, small wind turbines are using mechanical systems in the form of tail. The conventional tail changes the direction of the wind turbine to accommodate the variation of the incoming direction of winds. Quick and steady response is important as per change in wind directions. It has considerable effect on the wind turbine performance. The tails are having different shapes, but literature is not available about it. Different manufacturers used different tail shapes for their model, but the effect of tail shapes has not been studied yet. Hence, it is necessary to study the effect of tail shapes on the performance of wind turbine. This paper presents the effect of tail shapes on the performance of wind turbine. In this work three different tail shapes; rectangular, trapezoidal and triangular are considered. The yawing performance for these tail shapes using CFD analysis is carried out and presented in this paper.

Keywords: CFD, Small Wind Turbine, Tail Shape, Yawing

1. Introduction

Due to the increase in energy consumption and depletion of non renewable sources; there is need of clean renewable energy sources. Renewable energy devices such as wind turbines are playing a tremendous and significantly increasing role in the generation of electric power, in worldwide [1-4]. Countries in Europe and Asia are putting massive effort in the development of wind energy technology. At the end of 2013; the recorded small wind capacity installed worldwide has reached more than 755MW. This is nothing but the growth of more than 12% compared with 2012, when 678MW were registered. The technology of large wind turbines has been developed well; while small wind turbine technology is in development stage. The market for small wind turbine technology is encouraging in India also [2].

Generally, wind turbines, which are producing power in the range of 1-100kW, are called as small wind turbines. This definition of small wind turbines is found different for different countries [2]. Small wind turbines operating at low wind speeds regularly face the problem of yawing performance due to the uneven nature of wind. Because of this uneven nature of wind, the rotor axis of a wind turbine rotor is usually not aligned with the wind [11]. Due to which the rotor is not capable of following the variability of wind and so spends most of its time in a yawed condition. The yawed rotor is less efficient than the non-yawed rotor and so it is important to assess the efficiency for purposes of energy production estimation. In order to ensure the extraction of maximum wind potential even at lower wind speeds, most of the small wind turbines point into the wind using a tail assembly [5-6].

Presently, most of the leading small wind turbine manufacturers use different kind of tail shapes in their assembly [2]. The sufficient information about tails shapes and their effect on yawing performance of wind turbine is not available. This paper presents an investigation on the effect of different tail shapes on yawing performance of a micro wind turbine. Also, different pressure response and forces acting on the front side of different tail shape at different angle is observed and evaluated. In order to investigate this effect, three different tailshapes; rectangular, trapezoidal and triangular are considered and yawing performance for these shapes is recorded computationally.

2. Yawing in Wind Turbine

In small wind turbine, alignment of rotor surface area facing the wind direction is nothing but the yawing. In other words, rotation of the rotor axis about a vertical axis (for horizontal axis wind turbine only) is called yawing [7]. Horizontal-axis wind turbine yaw system is mainly divided into two types: passive yaw and active yaw. Mostly the active yaw system is used in large wind turbines; while the passive yaw system is adopted by small wind turbines. The active yaw systems are equipped with some of torque producing device able to rotate the nacelle of the wind turbine against the stationary tower based on automatic signals from wind direction sensors or mutual actuation. The passive yaw systems utilize the wind force in order to adjust the orientation of the wind turbine rotor into the wind. In their simplest form these systems comprise a simple roller bearing connections between the tower and the nacelle and tail fin mounted on the nacelle and designed in such way that it turns the wind turbine rotor into the wind by exerting a corrective torque to the nacelle. Therefore, the power of the wind is responsible for the rotor rotation and the nacelle orientation. Alternatively in case of downwind turbines the tail fin is not necessary since the rotor itself is able to yaw the nacelle into the wind. Yaw system used in downwind turbines sometime called as semiactive yaw system [8]. Types of yaw system used in wind turbine are shown in Figure 1.

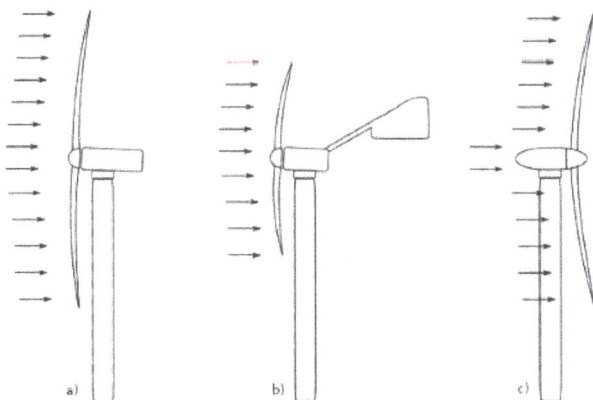

Figure 1. a) active yaw system in large WT; b) passive yaw system used in upwind turbine; c) semiactive yaw system in upwind turbine.

3. Computational Analysis

In order to know yawing performance of a micro wind turbine, it is necessary to determine the pressure distribution and forces acting on the surface of tail shape. The pressure and forces acting on the vane area were determined using Computational Fluid Dynamics. During this analysis, wind having 7m/s velocity and rotor with 350 rpm was considered.

Pressure plots for 0,10,20,30 degree angle of inclination were obtained. This section presents various pressure plots of trapezoidal, rectangular, triangular tail shapes at different inclinations. The computational results of these tail shapes are presented in Figure. 2 to Figure 13.

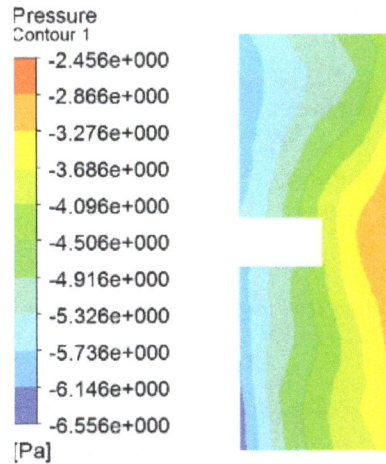

Figure 2. Pressure Plot for rectangular tail at 0 degree.

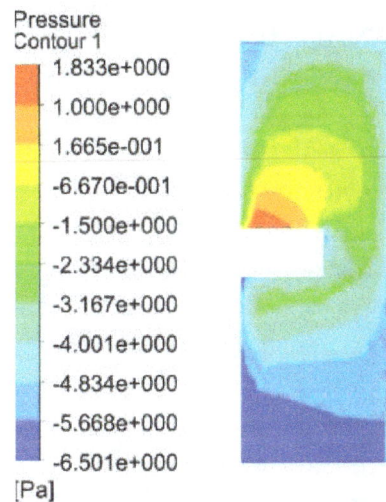

Figure 3. Pressure Plot for rectangular tail at 10 degree.

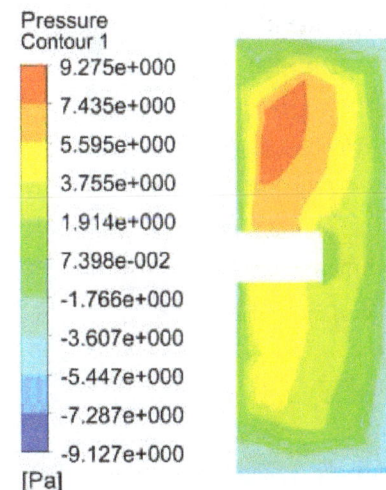

Figure 4. Pressure Plot for rectangular tail at 20 degree.

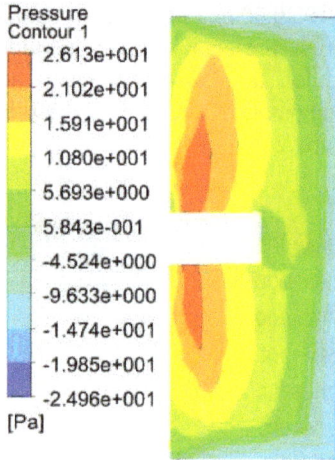

Figure 5. *Pressure Plot for rectangular tail at 30 degree.*

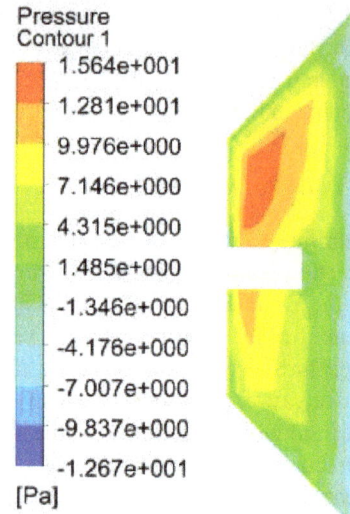

Figure 6. *Pressure Plot for trapezoidal tail at 0 degree.*

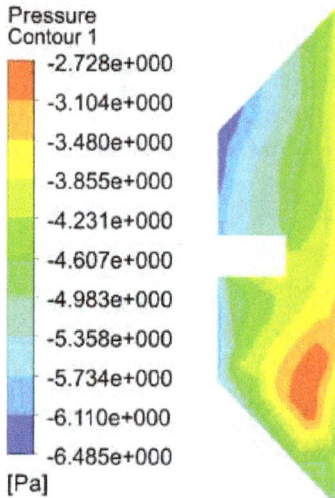

Figure 7. *Pressure Plot for trapezoidal tail at 10 degree.*

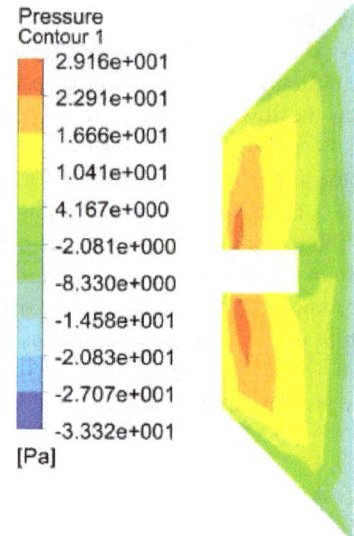

Figure 8. *Pressure Plot for trapezoidal tail at 20 degree.*

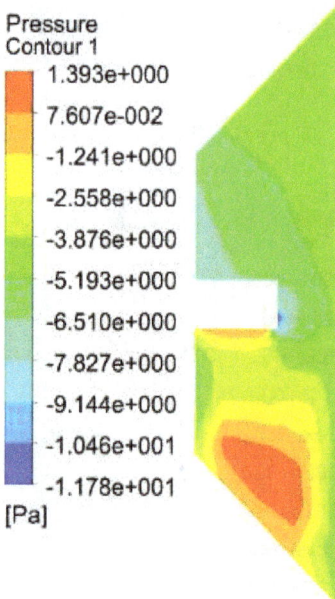

Figure 9. *Pressure Plot for trapezoidal tail at 30 degree.*

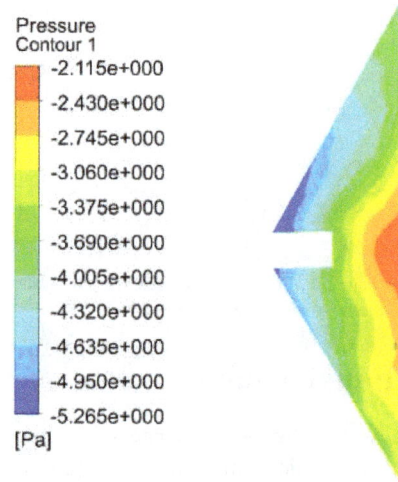

Figure 10. *Pressure Plot for triangular tail at 0 degree.*

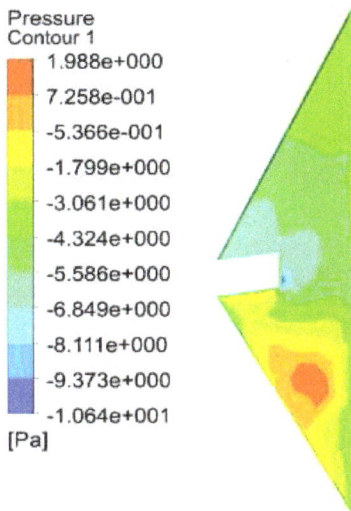

Figure 11. Pressure Plot for triangular tail at 10 degree.

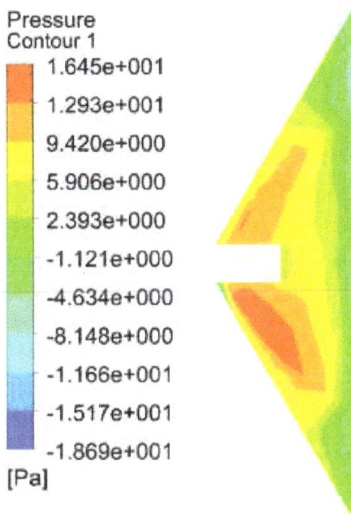

Figure 12. Pressure Plot for triangular tail at 20 degree.

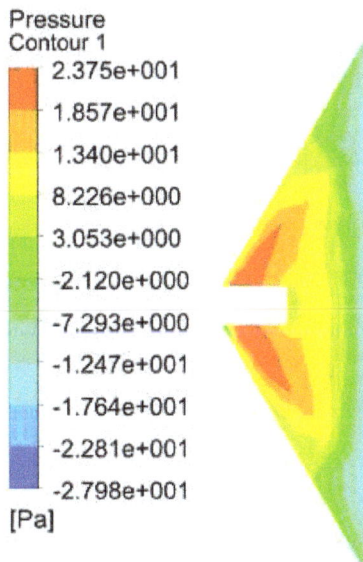

Figure 13. Pressure Plot for triangular tail at 30 degree.

4. Results and Discussion

The results obtained through computational analysis were significant. Table.1 shows the forces acting on trapezoidal, rectangular, triangular tail shapes at different angle of inclination. These results show that the pressure and forces induced on triangular tail shape are slightly higher than that of trapezoidal and rectangular tail.

Table 1. Force Acting on Different Tail Shapes.

Angle of Inclination (degree)	Forces(N) acting on		
	Rectangular Tail	Trapezoidal Tail	Triangular Tail
0	0.00106	0.00113	0.00196
10	0.00557	0.00574	0.00660
20	0.02797	0.04420	0.05280
30	0.07152	0.07551	0.07759

5. Conclusion

Three uniform micro wind turbines having different tail shapes of same area are modeled and analyzed successfully using CFD software. The results obtained from computational analysis are favorable for triangular shape. From which it is clear that yawing performance of triangular shape is better followed by trapezoidal and then rectangular shape. These results show that use of triangular tail shape will slightly improves performance of wind turbine compared to other two tail shape. It is also observed that as angle of inclination increases forces on tail shapes are also increases.

References

[1]   U.S. Gov, Dept. of Energy. Energy Information Administration - Official Energy Statistics from the U.S. Government.

[2]   http://www.eia.doe.gov/, 2009.

[3]   Stefan Gsanger and Jean-Daniel Pitteloud, World wind energy association 2015 report. 1-39p.www.wwindea.org

[4]   Khan B H. Non-Conventional Energy Resources, Tata McGraw-Hill, New Delhi. 2006.

[5]   Rakesh Das. Energy Conversion Systems. New Age International Publishers, Delhi. 2006.

[6]   Burton T, Jenkins N, Sharpe D and Bossanyi E Wind Energy Handbook (Wiley) ISBN 0470699752, 2011

[7]   Clausen P. D. and Wood D. H. Renewable Energy Journal 16.1999. 922-927p.

[8]   Wood. D. Small Wind Turbine Analysis, Design and Application Springer-Verlag London Limited, ISSN 1865-3529,2011

[9]   Nielsen E. Yawing device and method of controlling it. USPatent 4966525A, 1990

[10] Mayer C, Bechly M E and Wood DH. The starting behaviour of a small horizontal-axis wind turbine. Renewable Energy 2001.

[11] Hugh Piggott. Wind power workshop, Springer

[12] A.K. Wright and D.H. Wood, "The starting and low wind speed behaviour of a small horizontal axis wind turbine", Journal of Wind Engineering and Industrial Aerodynamics. September 2004.

25

Modeling and Analysis of Leakage Flux and Iron Loss Inside Silicon Steel Laminations

Yong Du[*]**, Wanqun Zheng, Jingjun Zhang**

Department of Electrical Engineering, Hebei University of Engineering, Handan, Hebei Province, China

Email address:

duyong_1970@126.com (Yong Du)

Abstract: This paper investigates the leakage flux and the iron loss generated in the laminated silicon sheets of the core or the magnetic shields of large power transformers. A verification model is well established, and proposed parabolic model (non-saturated region) and hybrid model (saturation region) to simulate the magnetic properties of the silicon steel with different angles to the rolling direction. An efficient analysis method is implemented and validated. The calculated and measured results with respect to the test models are in good agreement.

Keywords: Benchmark Problem 21, Modeling of Silicon Steel Laminations, Arbitrary Anisotropy, Parabolic Model, Hybrid Model

1. Introduction

The iron losses caused in grain-oriented silicon steel sheets of both laminated core and magnetic shields due to the strong leakage magnetic fluxes are highly concerned in very large power transformers. Under operating conditions, the silicon steel sheets performance exhibit non-linearity and anisotropy so that the distributions of the flux density and the iron loss inside each lamination become non-uniform and complex, this usually causes a local overheating. So an exact and quick calculation using three-dimensional finite element method is important [1,2,3]. But if each lamination of core is subdivided into a fine mesh, the number of unknown variables becomes very huge and the calculation is also difficult within an acceptable CPU time. Therefore, a practical modeling method of laminations should be investigated.

Some techniques have been proposed to deal with the iron loss problems about the laminations [4, 5, 6], however, the distributions of the iron loss and the magnetic flux inside the laminations are usually not known clearly. In this paper, based on a benchmark Problem 21—P21^c-M1[7,8], a verification model is proposed, an efficient and practical nonlinear analysis is performed by subdividing only the region near the surface of lamination into a fine mesh, and by modeling the inner part of lamination as a bulk core having anisotropic conductivity. and then proposed parabolic model

(non-saturated region) and hybrid model (saturation region) to simulate the magnetic properties of the silicon steel with different angles to the rolling direction. The distributions of the iron loss and the magnetic flux density inside silicon steel laminations are analyzed in details under the different exciting source, and some calculated and measured results of both the magnetic flux density and the iron loss distribution inside the silicon steel sheets are shown.

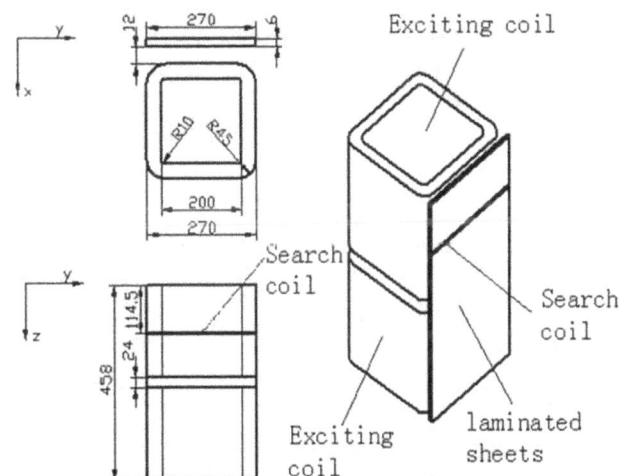

Figure 1. *Model P21^c-M1.*

2. Model Configuration

2.1. Verification Model

In order to detail the electromagnetic behavior of the grain-oriented silicon steel lamination excited by different applied fields, the original benchmark model of P21^c-M1 is simplified, i.e., the solid magnetic steel plate of 10 mm thick is removed so that only the laminated sheets are driven by the exciting source of twin coils (coil 1 and coil 2), which is called Model P21^c-M1$^-$ [9, 10], as shown in Fig.1.

2.2. Search Coils and Excitations

In the verification model, the grain-oriented silicon steel sheets (30RGH120) are used (total 20 sheets, each sheet is 0.3mm thick). 4 search coils (20 turns for each) are located at the specified positions on the laminated sheets of interest, as shown in Fig.2. The magnetic flux at those specified positions can be obtained by integrating the emf measured from the search coils.

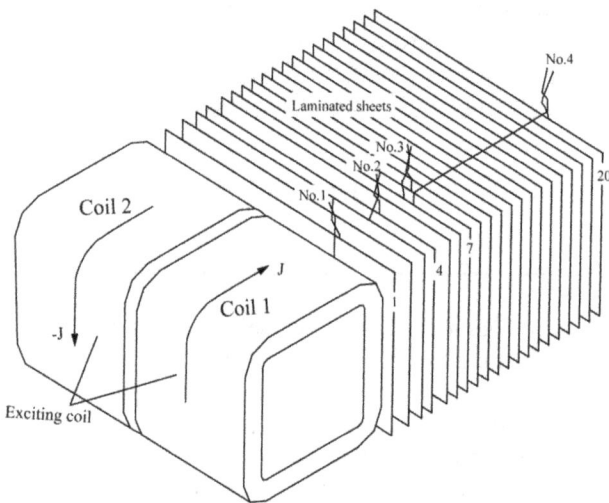

Figure 2. *Located search coils and excitations.*

In order to investigate the magnetic flux and the loss inside the laminations of the verification model under different excitation conditions. There are three testing cases depended on the exciting source, i.e., the exciting current, J, is different in two exciting coils, coil 1 and coil 2 for three cases, as shown in Table1.

Table 1. *Different excitation conditions.*

Cases	Exciting currents (A, rms, 50Hz)	
	in Coil 1	in Coil 2
I	J	-J
II	J	J
III	J	0

Note that the exciting current, J, is ranged from 0A to 25A (rms, 50Hz). The search coils are located at different layers of 20 sheets, and the number of the sheets included in each search coil (e.g., no.1 to no.4) is also different. See Fig.2.

3. Experiments

To simplify the physical model and establish a reasonable computational model, some advance measurements have been done to observe the electromagnetic behavior. The experiments upon the magnetic flux and the loss inside the laminations of the verification model under different excitation conditions have been carried out.

3.1. Average Flux Density Inside Laminated Sheets

The measured results of magnetic flux (or averaged magnetic flux density) passing through the different search coil under the different exciting conditions are shown in Fig. 3.

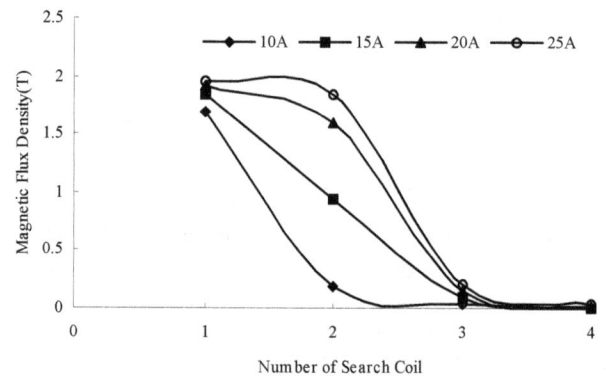

(a) Case I

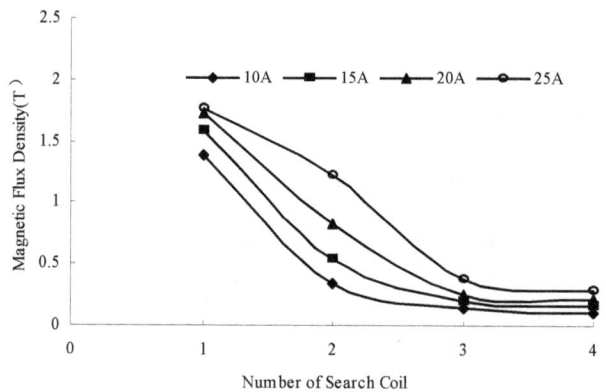

(b) Case II

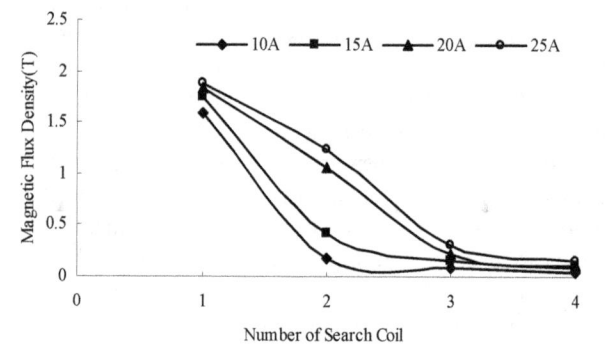

(c) Case III

Figure 3. *Measured average flux density inside laminated sheets under different exciting conditions.*

It can be seen that the averaged magnetic flux density of the laminations closest to the exciting source is considerably high than that far away from the exciting source. And the averaged magnetic flux density will be reduced quickly with the increase of the number of the laminated sheets.

3.2. Total Iron Loss Inside All the Laminated Sheets

The total iron loss inside all the laminated sheets (20 sheets) are also measured at different exciting currents, the measured results are shown in Fig.4.

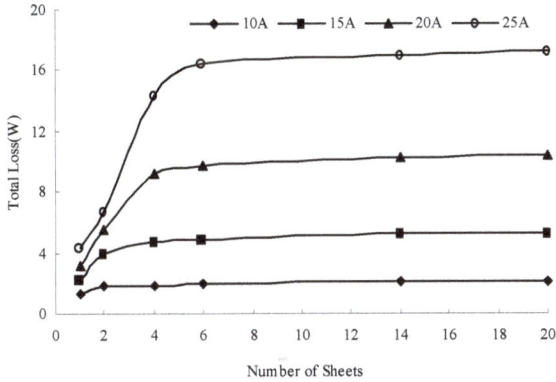

(a) Case I

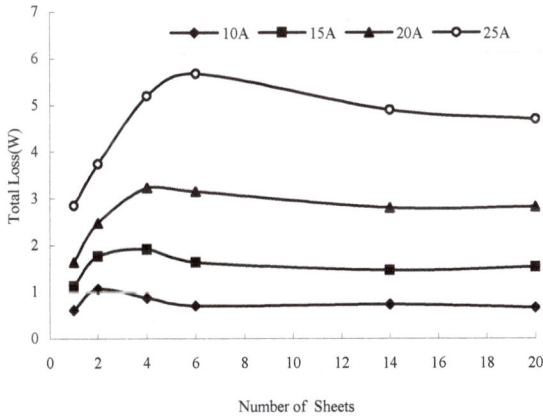

(b) Case II

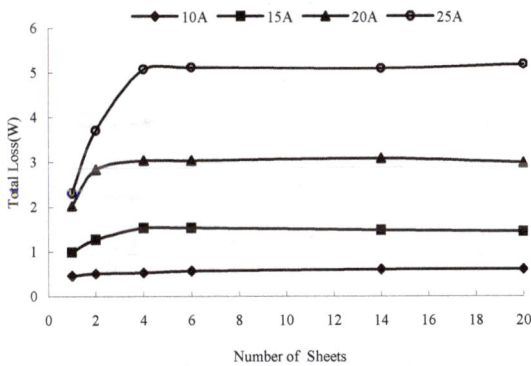

(c) Case III

Figure 4. *Measured iron loss inside laminated sheets under different exciting conditions.*

It can be seen from Fig.4. The total iron loss inside the laminated sheets is also focused on the laminations that closest to the exciting source, and it will be constant with the increase of the number of the laminated sheets.

4. Finite Element Formulation

4.1. The Simplified Computation Model

The experiment results show that the averaged magnetic flux density and the total iron loss within the laminations are quite non-uniform. The eddy currents induced in the laminations by 3-D leakage flux become a real 3-D distribution, especially in the first few laminations closest to the exciting source. In that case the detailed 3-D eddy currents must be investigated.

On the other hand, the magnetic flux density and then the iron loss inside the remained laminations, far away from the exciting source is reduced very quickly.

To simplify the analysis model, the actual silicon steel laminations of the iron core must be simplified on computation, the electromagnetic non-linearity and anisotropy of the laminations should be taken into account. The simplified computation model is shown in Fig.5. In order to simulate the effects of the eddy currents, the first few laminations closest to the exciting source are necessary to model the individual laminations and divided by 0.1 mm thick mesh layers，and the number of the individual lamination models increases with the exciting current increasing. The remaining laminations are modeled as a bulk with the anisotropic material property and divided by 0.9mm thick mesh layers.

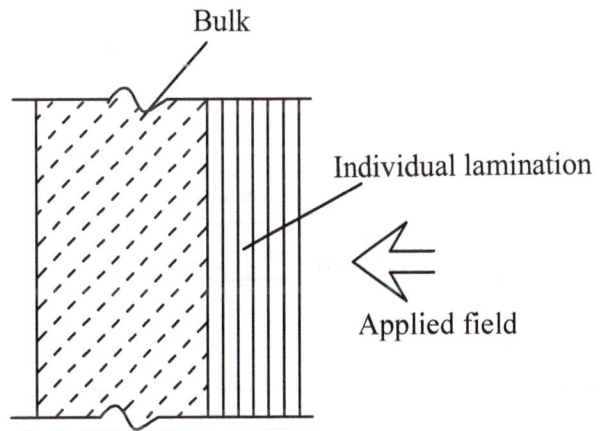

Figure 5. *Simplified computation model.*

4.2. Formulation

The well established T-ψ (or expressed as T-Ω) method is applied in this simplification of model[11,12], which features the hierarchical element method (based on polynomial orders 1 to 3) and in which the magnetic field is represented as the sum of two parts: the gradient of a scalar potential in non-conducting medium, and an additional vector field represented with vector-edge elements in conducting medium, described as follows:

1) In the conducting medium, the governing equation is:

$$\nabla \times \left(\frac{1}{\sigma} \nabla \times H \right) + \mu \frac{\partial H}{\partial t} = 0 \qquad (1)$$

2) In the non-conducting medium, the governing equation is:

$$\nabla \cdot \left[\mu \left(-\nabla \psi + H_s \right) \right] = 0 \qquad (2)$$

$$H = -\nabla \psi + H_s \qquad (3)$$

Where H_s represents the contribution from the exciting source.

4.2. B-H Modeling

The simplest non-linear and anisotropic material model is the elliptical model that uses the properties in the rolling and transverse directions [13]. This model approximately models the material properties in any directions other than the major axes of the material. it derives the permeability in a principal direction directly from the relevant B–H curve based on the magnitude of the applied magnetic field. If a two-dimensional magnetic field with constant magnitude is applied in different directions, the permeability of each principal direction is constant. and, therefore, the vector of the flux density traces an ellipse. However, the new experiment results show that the elliptical model does not provide good accuracy.

For more accurate modeling of the non-linear anisotropic B-H property has been proposed, they are parabolic model (non-saturated region) and hybrid model (saturation region), Through comparing the simulation results that obtained form the elliptic model and the new model with the experiment results, verified that the new model has higher simulated accuracy. In this paper, the new model can be used. They can be basically formulated by (4) and (5) respectively.

$$\frac{B_y^2}{\mu_{y0}^2} + H \frac{B_x}{\mu_{x0}} = H^2 \qquad (4)$$

$$\frac{B_y^2}{\mu_{y0}^2} + \frac{1}{2} \left(\frac{B_x^2}{\mu_{x0}^2} + H \frac{B_x}{\mu_{x0}} \right) = H^2 \qquad (5)$$

Where parameters μ_{x0} and μ_{y0} are the longitudinal and transverse permeability. The global flux density B is then defined by the following equations:

$$B = H / \sqrt{\left(\cos\phi / \mu_{x0} \right)^2 + \left(\sin\phi / \mu_{y0} \right)^2} \qquad (6)$$

$$\phi = \arctan \left(\mu_{y0} / \mu_{x0} \tan \tau \right) \qquad (7)$$

$$B_Y = B \cdot \sin \phi \qquad (8)$$

$$B_x = B \cdot \cos\phi \qquad (9)$$

Where ϕ is the angle between the rolling direction and flux density, τ is the angle between the rolling direction and the magnetic field.

The new models require only two curves in the principal directions when we perform the analysis in finite element formulation. In addition to its simplicity, an advantage of the models is that these B–H curves are available directly from the manufacturers.

4.3. Results and Discussion

The measured and calculated total iron loss and flux results are shown in Table 2 and Table 3 respectively. Table 2 and Table 3 show a good agreement between the measured and calculated results for each test case at different exciting currents.

In the further 3-D eddy current analysis, the effect of the different exciting conditions on the total iron loss generated in the laminated sheets is taken into account.

Table 2. Measured and calculated total loss under different cases.

Exciting currents (A, rms, 50Hz)	Case I(W)		Case II(W)		Case III(W)	
	Measured	Calculated	Measured	Calculated	Measured	Calculated
10	2.20	2.17	0.66	0.64	0.59	0.58
15	5.30	5.25	1.43	1.39	1.39	1.34
20	10.20	9.96	2.71	2.65	2.99	2.91
25	16.80	15.73	4.72	4.68	5.19	4.97

Table 3. Measured and calculated total flux inside twenty sheets under different cases.

Exciting current (A, rms, 50Hz)	CaseI/ (mWb)		CaseII/ (mWb)		CaseIII/ (mWb)	
	Measured	Calculated	Measured	Calculated	Measured	Calculated
10	0.297	0.311	0.357	0.381	0.329	0.323
15	0.444	0.447	0.532	0.569	0.490	0.501
20	0.589	0.594	0.708	0.707	0.652	0.672
25	0.738	0.702	0.886	0.893	0.817	0.832

From Table 2 it can be also seen that the total iron loss of the Case I is about three times of that of the Case III, while, the total iron loss of the Case II is approximately equal to that of Case III. However, the fluxes within 20 sheets for the three cases do not have the same relationship (Table 3). So, there is a complex function relationship between the loss, the fluxes and the exciting sources.

5. Conclusion

A simplified benchmark model involving lamination structures is well established to investigate the iron loss caused by the normal leakage flux, and examine the effects of the eddy currents on the total iron loss. A practical approach, in which the whole solved region is divided into the 3-D and 2-D eddy current sub-regions, is implemented to deal with the lamination configuration and the additional iron loss problems. The electric and magnetic anisotropic properties of the grain-oriented silicon steel are taken into account in the FEM analysis, as well as the magnetic nonlinearity of it according to the parabolic model (non-saturated region) and hybrid model (saturation region).

The distributions of the magnetic flux and the iron loss inside silicon steel laminations are analyzed in detail under different excitation conditions. The simplified analysis method is validated based on the proposed model by comparing the calculated results of the iron losses and the fluxes in sheets.

Acknowledgement

This work was supported by the Youth Science Fund of Hebei Education Department, China, under Grant No. QN20131025 and the National Natural Science Foundation of China, under Grant No. 11272112. This work was also supported by R & D Center, Baoding Electric Co., LTD, China. The Author thanks the all colleagues of R & D Center for their energetic supports and helpful discussions.

References

[1] Du Y, Cheng Z, Zhang J, Liu L, Fan Y, Wu W, Zhai Z, and Wang J, "Additional iron loss modeling inside silicon steel laminations," IEEE International Electric Machines and Drives Conference, pp.826-831, 2009.

[2] Cheng Z, Takahashi N, Forghani B, Du Y, Zhang J, Liu L, Fan Y, Hu Q, Jiao C, and Wang J, "Large power transformer-based stray-field loss modeling and validation," IEEE International Electric Machines and Drives Conference, pp.548-555, 2009.

[3] E. teNyenhuis, R. Girgis, and G. Mechler, "Other factors contributing to the core loss performance of power and distribution transformers," IEEE Trans. on Power Delivery, vol., no.4, pp.648-653, October 2001.

[4] L. Krähenbühl, P. Dular, T. Zeidan, and F. Buret, "Homogenization of lamination stacks in linear magnetodynamics," IEEE Trans. Magn., vol. 40, no.2, pp.912-915, March 2004.

[5] K. Muramatsu, T. Shimizu, A. Kameari, I. Yanagisawa, S. Tokura, O. Saito, and C. Kaido, "Analysis of eddy currents in surface layer of laminated core in magnetic bearing system using leaf edge elements," IEEE Trans. Magn., vol. 42, no.4, pp. 883-886, April 2006.

[6] Z. Cheng, N. Takahashi, S. Yang, T. Asano, Q. Hu, S. Gao, X. Ren, H. Yang, L. Liu, and L. Gou, "Loss spectrum and electromagnetic behavior of problem 21 family," IEEE Trans. Magn., vol.42, no.4, pp.1467-1470, 2006.

[7] Z. Cheng, N. Takahashi, B. Forghani, G. Gilbert, J. Zhang, L.Liu, Y. Fan, X. Zhang, Y. Du, J. Wang, and C. Jiao, "Analysis and measurements of iron loss and flux inside silicon steel laminations," IEEE Trans. Magn.,45(3): 1222-1225, 2009.

[8] Z. Cheng, N. Takahashi, B. Forghani, L. Liu, Y. Fan, T. Liu, J. Zhang, and X. Wang, "3-D finite element modeling and validation of power frequency multi-shielding effect," IEEE Trans. Magn., vol.48, 243-246, 2012.

[9] Z. Cheng, N. Takahashi, B. Forghani, Y. Du, Y. Fan, L. Liu, and H. Wang, "Effect of variation of B-H properties on both iron loss and flux in silicon steel lamination," IEEE Trans. Magn., vol.47,1346-1349, 2011.

[10] W. Zheng, and Z. Cheng, "An inner-constrained separation technique for 3-D finite-element modeling of grain-oriented silicon steel laminations," IEEE Trans. Magn., vol.48, no.8, pp. 2277-2283, 2012.

[11] Yong Du, Zhiguang Cheng, Zhigang Zhao, Yana Fan, Lanrong Liu, Junjie Zhang, and Jianmin Wang, "Magnetic Flux and Iron Loss Modeling at Laminated Core Joints in Power Transformers," IEEE Trans. on Applied Superconductivity, 20(3):1878-1882, 2010.

[12] J.P. Webb and B. Forghani, "A T-Omega method using hierarchal edge elements," IEE Proc.-Sci. Meas. Technol., vol. 142, no. 2, pp.133-141, 1995.

[13] A Di Napoli, and R Paggi, "A model of anisotropic grain-oriented steel," IEEE Trans. Magn., vol. 23, no.5, pp. 1557-1561, July 1983.

26

Fixed-Point Harmonic-Balanced Method for Nonlinear Eddy Current Problems

Xiaojun Zhao[1], Yuting Zhong[1], Dawei Guan[1], Fanhui Meng[1], Zhiguang Cheng[2]

[1]Department of Electrical Engineering, North China Electric Power University, Baoding, China
[2]Institute of Power Transmission and Transformation Technology, Baobian Electric Co., Ltd, Baoding, China

Email address:

158748295@163.com (Xiaojun Zhao), yuting315@yeah.net (Yuting Zhong), zhynh123@163.com (Dawei Guan),
mengfh1990@163.com (Fanhui Meng), emlabzcheng@yahoo.com (Zhiguang Cheng)

Abstract: A new method to optimally determine the fixed-point reluctivity is presented to ensure the stable and fast convergence of harmonic solutions. Nonlinear system matrix is linearized by using the fixed-point technique, and harmonic solutions can be decoupled by the diagonal reluctivity matrix. The 1-D and 2-D non-linear eddy current problems under DC-biased magnetization are computed by the proposed method. The computational performance of the new algorithm proves the validity and efficiency of the new algorithm. The corresponding decomposed method is proposed to solve the nonlinear differential equation, in which harmonic solutions of magnetic field and exciting current are decoupled in harmonic domain.

Keywords: Eddy Current, Fixed-Point, Harmonic solutions, Reluctivity

1. Introduction

Non-linear eddy current problems can be solved by the time-stepping method [1] or the harmonic-balanced method [2]. The time-stepping method requires many periods to approach the accurate steady-state solution, while the harmonic-balanced method computes the magnetic field directly in the frequency domain. Compared with the frequency-domain method, the so called brute force method using time-stepping technique spends much more time on the transient process.

A relationship between magnetic field intensity H and the magnetic induction B is represented by introducing the fixed-point reluctivity v_{FP}[3]. The fixed-point reluctivity v_{FP} can be regarded as a periodic variable or a constant in harmonic-balanced method. The convergent speed of the solutions depends mainly on the strategies to determine the value of v_{FP} in non-linear iterations. Different methods to optimally determine v_{FP} have been presented and investigated in order to ensure the stable and fast convergence of solutions [4-6].

When the power transformer works under DC-biased magnetization, the ferromagnetic core will be significantly saturated. Owing to the nonlinearity of magnetic material in electromagnetic devices, there often are high-order harmonics in the exciting current and magnetic field. Therefore, electrical devices such as power transformers and reactors may work abnormally due to the magnetic storm and the transmission of high voltage direct current [7]. In that case, the DC-biased eddy current problem should be solved efficiently and accurately. Furthermore, electromagnetic coupling should always be considered in numerical computation when the strand coils or solid conductors are connected to the voltage source.

In this paper an efficient algorithm to determine the fixed-point reluctivity v_{FP} is proposed. It is aimed at efficiently computing the non-linear eddy current problem under DC-biased magnetization. The nonlinear magnetic field is computed in harmonic domain. The corresponding decomposed algorithm is presented to solve the nonlinear differential equation sequentially or concurrently, which decreases the memory cost of harmonic-balanced computation of large scale problems.

2. Fixed-Point Harmonic-Balanced Method

2.1. Fixed-Point Method

Maxwell's equations hold in non-linear eddy current

problems,

$$\nabla \times \boldsymbol{H} = \boldsymbol{J}_0 + \sigma \boldsymbol{E} \tag{1}$$

$$\nabla \times \boldsymbol{E} = -\frac{\partial \boldsymbol{B}}{\partial t} \tag{2}$$

$$\nabla \cdot \boldsymbol{B} = 0 \tag{3}$$

where E is the electric field intensity, J_0 is the impressed current density and σ is the conductivity.

A relationship between magnetic field intensity H and magnetic flux density B is represented by introducing the fixed-point reluctivity ν_{FP} [8],

$$\boldsymbol{H}\left(\boldsymbol{B}\right) = \nu_{\text{FP}} \boldsymbol{B} - \boldsymbol{M}\left(\boldsymbol{B}\right) \tag{4}$$

where M is a magnetization-like quantity which varies nonlinearly with B. Therefore the magnetic field intensity H is split into two parts: the linear part, which is related to the ν_{FP}, and the non-linear part, which varies with the magnetic induction B [3].

2.2. Harmonic-Balanced Method

The periodic variables in the electromagnetic field under DC-biased excitation can be approximated by the Fourier-series with a finite number of harmonics [9],

$$W(t) = W_0 + \sum_{n=1}^{\infty} \left(W_{2n-1} \sin n\omega t + W_{2n} \cos n\omega t \right) \tag{5}$$

where $W(t)$ can be replaced by current density J, vector potential A, magnetic flux density B, magnetic field intensity H, and the magnetization-like quantity M.

Equation (1) can be rewritten in isotropic material by means of harmonic vector in 2-D problems,

$$\begin{cases} \boldsymbol{H}_x = \nu_{\text{FP}} \boldsymbol{B}_x - \boldsymbol{M}_x \\ \boldsymbol{H}_y = \nu_{\text{FP}} \boldsymbol{B}_y - \boldsymbol{M}_y \end{cases} \tag{6}$$

in which

$$\begin{cases} \boldsymbol{H}_x = \begin{bmatrix} H_{x,0} & H_{x,1} & H_{x,2} & H_{x,3} & H_{x,4} & \cdots \end{bmatrix}^T \\ \boldsymbol{H}_y = \begin{bmatrix} H_{y,0} & H_{y,1} & H_{y,2} & H_{y,3} & H_{y,4} & \cdots \end{bmatrix}^T \end{cases} \tag{7}$$

Each of harmonic vectors $\boldsymbol{B}_x$, $\boldsymbol{B}_y$, $\boldsymbol{M}_x$ and $\boldsymbol{M}_y$ has a similar expression with (7).

2.3. 2-D Eddy Current Problems

Non-linear eddy current problems can be solved directly with the prescribed impressed current density. However, the impressed current density is unknown when the solid conductor or strand coil is connected to the voltage source. Therefore, the coupling between the magnetic field and electric circuits should be investigated if the non-linear eddy

current problem is solved in the harmonic domain.

2.3.1. Solid Conductor Connected to Voltage Sources

When the solid conductor is fed by the voltage source, the eddy current exists in the solid conductor and the other conducting materials. The non-linear problem can be described by the following equation,

$$\nabla \times \nu_{FP} \left(\nabla \times \boldsymbol{A} \right) = \boldsymbol{J} - \nabla \times \boldsymbol{M} \tag{8}$$

where J is the current density.

The magnetic vector potential A and the scalar potential V can be linked to the current density J by the equation as follows,

$$\boldsymbol{J} = -\sigma \frac{\partial \boldsymbol{A}}{\partial t} - \sigma \nabla V \tag{9}$$

The non-linear equation, including the 2-D magnetic and electric fields, can be presented as follows [10],

$$\nabla \times \nu_{FP} \left(\nabla \times \boldsymbol{A} \right) + \sigma \frac{\partial \boldsymbol{A}}{\partial t} + \sigma \left(\nabla V \right) = -\nabla \times \boldsymbol{M} \tag{10}$$

The fixed-point harmonic-balanced equation can be established by applying the finite element method on the whole problem domain,

$$\sum_{i=1}^{m} \left(S_{ij}^e \boldsymbol{D}_{FP}^e + T_{ij}^e \boldsymbol{N} \right) \boldsymbol{A}_j^e = \boldsymbol{K}_j^e + \boldsymbol{P}_j^e \quad (j = 1, 2, \dots, m) \tag{11}$$

$$S_{ij}^e = \int_{\Omega_{ew}} \nabla N_i \cdot \nabla N_j d\Omega \tag{12}$$

$$T_{ij}^e = \int_{\Omega_{ed}} \omega \sigma N_i N_j d\Omega \tag{13}$$

$$\boldsymbol{P}_j^e = -\int_{\Omega_{en}} \left(\nabla \times \boldsymbol{M}^e \right) N_j d\Omega \tag{14}$$

$$\boldsymbol{K}_j^e = \int_{\Omega_{ec}} \sigma N_j U d\Omega \tag{15}$$

where Ω_{ew}, Ω_{ed} and Ω_{en} represent the finite element in the whole problem domain, eddy current region and the non-linear region, respectively. N_i is the shape function on node i in each finite element, and m is the total number of nodes in one element. D_{FP} and N are the square matrices related to the fixed-point reluctivity and harmonic number [2], respectively. A_j is the harmonic vector of the magnetic vector potential on node j and P_j is the harmonic vector obtained from M. U is the voltage in harmonic domain per unit length.

By integrating (13) on the solid conductor, we can obtain

$$\sum_{\Omega_{ec}} \boldsymbol{C}^e \boldsymbol{A}^e + \boldsymbol{ZI} = \boldsymbol{U} \tag{16}$$

since

$$C_i^e = N \frac{\omega}{S_{cd}} \int_{\Omega_{ec}} N_i d\Omega \qquad (17)$$

$$Z = \begin{bmatrix} R & 0 & \cdots \\ 0 & R & \cdots \\ \vdots & \vdots & \ddots \end{bmatrix} \qquad (18)$$

where Ω_{ec} represents the finite element in the conducting region, R is the conductor's resistance per unit length, S_{cd} is the cross-sectional area of the solid conductor.

2.3.2. Strand Coil Connected to Voltage Sources

The strand coil consists of fine wires where the eddy current is generally too small to be considered for computation. The supplied voltage U and the exciting current I in the coil can be linked by Kirchhoff's Law and Faraday's Law [11],

$$N_{coil} \int_{\Omega_c} \frac{\partial A}{\partial t} d\Omega + RI = U \qquad (19)$$

where N_{coil} is the turn number of the strand coil.

The frequency-domain system equation considering electromagnetic coupling can be obtained according to the harmonic-balanced theory,

$$\sum_{\Omega_{ew}} S^e A^e + \sum_{\Omega_{ed}} T^e A^e + \sum_{\Omega_{ec}} G^e I = \sum_{\Omega_{en}} P^e \qquad (20)$$

$$\sum_{\Omega_{ec}} C^e A^e + ZI = U \qquad (21)$$

since

$$G_i^e = -I_u \frac{N_{coil}}{S_{coil}} \int_{\Omega_{ec}} N_i d\Omega \qquad (22)$$

where I_u is a unit matrix of the same size with D_{FP} and N.

Consequently, the harmonic solutions of the magnetic field and magnetizing current can be computed simultaneously by solving (21) when the solid conductor and strand coil are both connected to the voltage source,

$$\begin{bmatrix} S+T & G \\ C & Z \end{bmatrix} \begin{bmatrix} A \\ I \end{bmatrix} = \begin{bmatrix} K+P \\ U \end{bmatrix} \qquad (23)$$

where G is related to the spatial distribution of the magnetizing current when the strand coil is fed by the voltage source, while K appears on the right side of the equation when the solid conductor is connected to the voltage source.

2.4. Determination of Fixed-Point Reluctivity

The fixed-point reluctivity v_{FP} can be regarded as a periodic variable when it is determined in each time step. Consequently, the harmonic coefficients of v_{FP} can be used to calculate D_{FP} in the harmonic-balanced method [2]. All elements in the square

matrix D_{FP} are non-zero, which indicates harmonic solutions are coupled with each other. In that case the memory demand will increase significantly in the large-scale computation, although fast convergence is achieved. In fact the v_{FP} can be a constant in the harmonic domain, and is determined as follows,

$$\nu_{FP} = \partial H \left(B_{max} \right) / \partial B_{max} \qquad (24)$$

where B_{max} represents the maximum value of the magnetic induction in one period.

The mean (e_{mean}) and maximum (e_{max}) variation of the reluctivity defined by $v = H/B$ can be observed to check the convergence of the harmonic solutions. In this paper the stopping criterions are set to $e_{mean} = 0.1\%$ and $e_{max} = 1\%$ in the numerical computation of the one-dimensional and two-dimensional eddy current problems.

3. Computational Results and Analysis

3.1. Laminated Steel Sheet

As shown in Fig.1, A thin electrical steel sheet carrying eddy current is modelled and computed to observe variation of the magnetic induction when the lamination operates under different types of magnetization.

The 30Q140 oriented steel sheet of 0.3 mm thickness is first tested under sinusoidal flux in 50 Hz, and then the DC flux is provided. The conductivity of the sheet is $\sigma = 2.22 \times 10^6$ S/m.

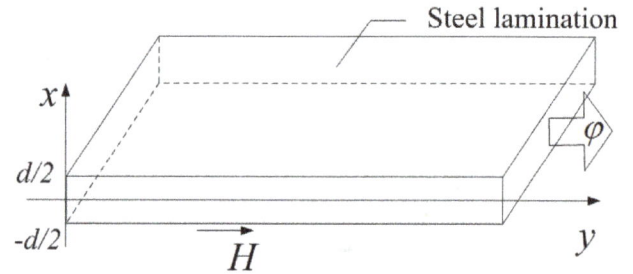

Figure 1. *Electrical steel sheet.*

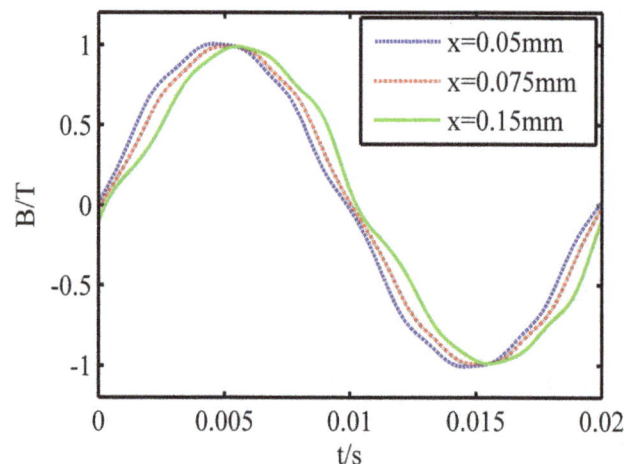

Figure 2. *Magnetic induction in different depths of the steel sheet.*

The spatial distribution of magnetic induction in the thin sheet is well clarified in Fig. 2. The magnetic flux for the boundary condition is $B_{av,ac} = 0.994$T and $B_{av,dc} = 0.7036$T. Fig. 3 compares the waveforms of magnetic induction under sinusoidal (indicated by "ac") and dc-biased (indicated by "dc") magnetizations. Notice that the eddy current in the sheet leads to the non-sinusoidal waveform of the magnetic induction. Furthermore, the distribution of the magnetic induction varies with the depth of the sheet in the x-direction.

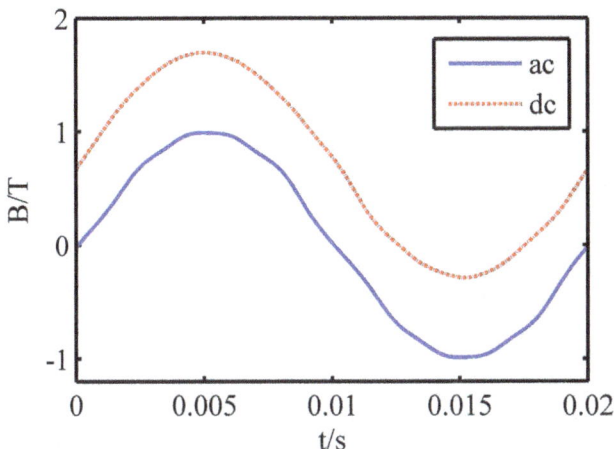

Figure 3. Comparison of the magnetic induction under sinusoidal and dc-biased magnetizations.

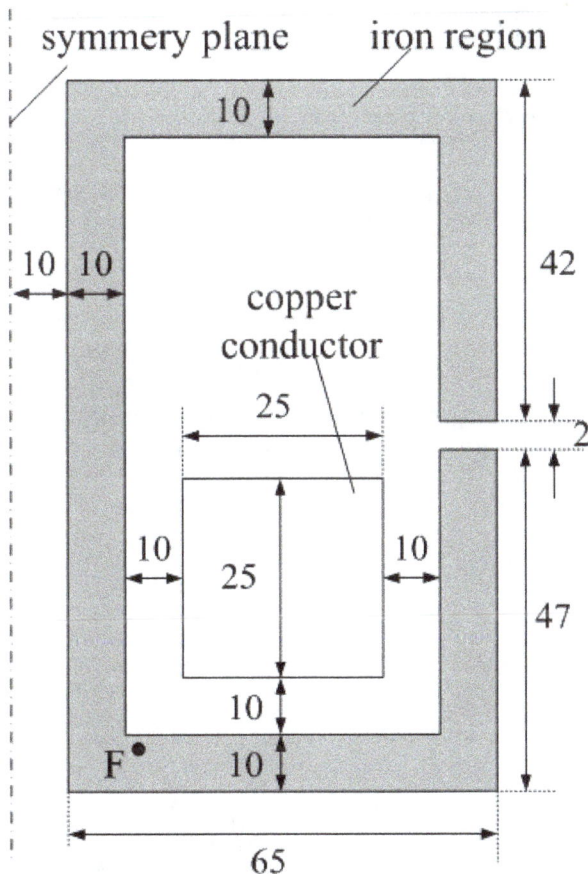

Figure 4. Geometric structure of the 2-D model.

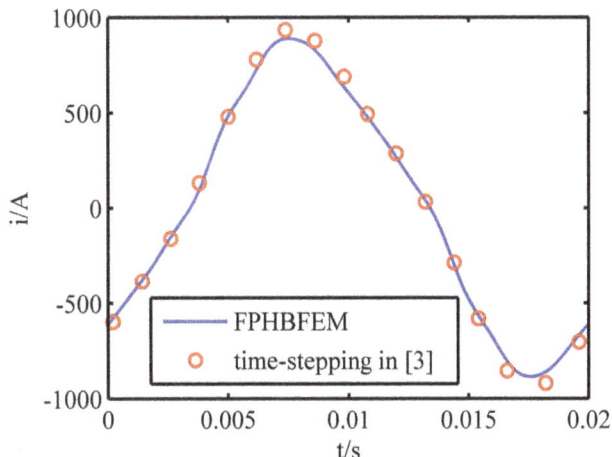

Figure 5. Calculated current in the copper conductor.

3.2. Copper Conductor Surrounded by Ferromagnetic Screen

The two-dimensional problem consists of a solid copper conductor and an iron screen with an air gap. As shown in Fig. 4, the iron screen surrounds the conductor. The eddy current exists in both the copper conductor and iron screen. The conductivities of the copper and iron are $\sigma = 5.7 \times 10^7$ S/m and $\sigma = 1.0 \times 10^6$ S/m, respectively. The copper conductor is connected to a voltage source of 50 Hz. The B-H curve is detailed in [3]. 892 second-order elements with 2781 nodes are used in the numerical computation. Computational costs of the proposed method and the traditional method [2] are compared in Table I. M_c and T_c represent the memory demand and computational time, respectively. N_h is the truncated harmonic number. Compared to the traditional method, the proposed method significantly reduces memory requirements with a slight increase in computational time due to a few more non-linear iterations. The calculated magnetizing current is compared with that obtained by using time-stepping method, and the good congruency proves the validity of the proposed method.

Table 1. Comparison between two different methods.

Method	Traditional	Proposed
Memory/Mb	17.51	7.92
Time/s	951.31	812.86
N_h	11	11

4. Decomposed Algorithm

When the fixed-point reluctivity is computed according to (24), the nonlinear equation in (23) can be linearized by v_{FP} which is space-dependent and time-independent. Therefore harmonic solutions can be decoupled and calculated separately. Equation (11) can be decomposed as follows,

$$\sum_{i=1}^{m} \left(S_{ij}^e \boldsymbol{d}_{k,FP}^e + T_{ij}^e \boldsymbol{N}_k \right) \boldsymbol{A}_{k,j}^e$$
$$= \boldsymbol{K}_{k,j}^e + \boldsymbol{P}_{k,j}^e \quad (j=1,2,...,m; \ k=0,1,2,...,N_h) \tag{25}$$

$$d_k^e = \begin{cases} \nu_{FP}(i=1) \\ \begin{bmatrix} \nu_{FP} & 0 \\ 0 & \nu_{FP} \end{bmatrix} & (k>1) \end{cases} \qquad (26)$$

$$h_k = \begin{cases} 0 & (k=1) \\ \omega \begin{bmatrix} 0 & -k \\ k & 0 \end{bmatrix} & (k \geq 1) \end{cases} \qquad (27)$$

where k is the harmonic number.

The decoupled equation system can be solved sequentially and concurrently, updating harmonic solutions by Gauss-Seidel and Jacobi iterative method respectively.

5. Conclusion

The convergent speed of harmonic solutions highly depends on the determination of optimal fixed point reluctivity in the fixed-point harmonic-balanced method. Differential reluctivity can be used to guarantee the stable and fast convergence of harmonic solutions. Due to the linearized system equation, harmonic solutions can be decoupled and computed in parallel, which can improve the computational efficiency and reduce the memory cost greatly. The proposed algorithm is more efficient than the traditional harmonic-balanced method.

Acknowledgement

This work is supported by the National Natural Science Foundation of China (Grant No. 51307057), Hebei Province Natural Science Foundation (Grant No. E2013502323), Research Fund for the Doctoral Program of Higher Education of China (Grant No. 20130036120011), and the Fundamental Research Funds for the Central Universities (Grant No. 2015MS82).

References

[1]   E. Dlala, A. Belahcen, and A. Arkkio, "Locally convergent fixed-point method for solving time-stepping nonlinear field problems," *IEEE Trans. Magn.*, vol.43, pp. 3969-3975, 2007.

[2]   X. Zhao, L. Li, J. Lu, Z. Cheng and T. Lu, "Characteristic analysis of the square laminated core under dc-biased magnetization by the fix-point harmonic-balanced mehtod," *IEEE Trans. Magn.*, vol. 48, no. 2, pp. 747-750, 2012.

[3]   O. Biro and K. Preis, "An efficient time domain method for nonlinear periodic eddy current problems," *IEEE Trans. Magn.*, vol. 42, no. 4, pp. 695-698, 2006.

[4]   E. Dlala and A. Arkkio, "Analysis of the convergence of the fixed-point method used for solving nonlinear rotational magnetic field problems," *IEEE Trans. Magn.*, vol. 44, no. 4, pp. 473-478, 2008.

[5]   S. Ausserhofer, O. Biro, and K. Preis, "A strategy to improve the convergence of the fixed-point method for nonlinear eddy current problmes," *IEEE Trans. Magn.*, vol. 44, no. 6, pp. 1282-1285, 2008.

[6]   G. Koczka, S. Auberhofer, O. Biro and K. Preis, "Optimal convergence of the fixed point method for nonlinear eddy current problmes," *IEEE Trans. Magn.*, vol. 45, no. 3, pp. 948-951, 2009.

[7]   X. Zhao, J. Lu, L. Li, Z. Cheng and T. Lu, "Analysis of the saturated electromagnetic devices under DC bias condition by the decomposed harmonic balance finite element method", *COMPEL.*, vol. 31, no. 2, pp. 498-513, 2012.

[8]   F. I. Hantila, G. Preda and M. Vasiliu, "Polarization method for static field" *IEEE Trans. Magn.*, vol.36, no.4, pp. 672-675, 2000.

[9]   X. Zhao, J. Lu, L. Li, Z. Cheng and T. Lu, "Analysis of the DC Bias phenomenon by the harmonic balance finite-element method," *IEEE Trans. on Power Delivery.*, vol.26, no.1, pp. 475-485, 2011.

[10] I. Ciric, and F. Hantila, "An efficient harmonic method for solving nonlinear time-periodic eddy-current problmes," *IEEE Trans. Magn.*, vol. 43, no. 4, pp. 1185-1188, 2007.

[11] P. Zhou, W. N. Fu, D. Lin, and Z. J. Cendes, "Numerical modeling of magnetic devices," *IEEE Trans. Magn.*, vol. 40, no. 4, pp. 1803-1809, 2004.

27

A Review of Technical Issues for Grid Connected Renewable Energy Sources

S. Yasmeena*, **G. Tulasiram Das**

EEE Department Vishnu Institute of Technology, Bhimvaram, A. P., India

Email address:

yasminab4u@gmail.com (S. Yasmeena), das_tulasiram@yahoo.co.in (G. T. Das)

Abstract: Renewable energy in recent years become more and more common, due to the large increase in generation from renewable energy sources such as small hydropower stations, wind turbines, photovoltaic's (PV) etc. This paper gives the report on two forms of renewable energy wind and solar energy, and on the role of smart grids in addressing the problems associated with the efficient and reliable delivery and use of electricity and with the integration of renewable sources. In this paper different power quality issues are addressed and a FACTS device STATIC COMPENSATOR (STATCOM) is connected at a point of common coupling for grid connected wind turbine to reduce the power quality problems likeharmonics in the grid current, by injecting superior reactive power in to the grid of wind turbine. And also an active power filter implemented with a four leg voltage-source inverter using DQ (Synchronous Reference Frame) based Current Reference Generator scheme is presented for renewable based distributed generation system of PV cell.

Keywords: PV Cell, Wind Turbine, STATCOM, Power Quality

1. Introduction

As the worlds electricity demand increases, more environmental constraints is given to conventional energy sources such as fossil or nuclear energy. This comes as a direct result of the problem with global warming where the emissions from energy production from fossil fuels are a big contributor. Which is clearly emphasized by the fact that in 2008, 81% of the worlds energy was produced by fossil fuels. Reasons for increasing Renewable energy sources: Declining of fossil fuel supplies

Environmental issues, Increasing cost of fossil fuels, business opportunities, Energy security, Energy independence

Wind energy: Wind turbines extract the kinetic energy from the wind and converts into generator torque. Generator converts this torque into electricity and feeds in to the grid. 1 MW of wind plant in one year can displace 1500 tons of CO_2, 6.5 tons of SO_2 and 3.2 tons of NOx. (REPP report, Washington July 2003)

Photovoltaic (PV) cells: PV generation is the technique which uses photovoltaic cell to convert solar energy into electrical energy. Now a days, PV generation is developing increasingly fast as a renewable energy source.

Large scale power generations are connected to transmission systems where as small scale distributed power generation is connected to distribution systems. There are certain challenges in the integration of both types of systems directly. Due to this, wind energy has gained a lot of investments from all over the world. However, due to the wind speed's uncertain behavior it is difficult to obtain good quality power, since wind speed fluctuations reflect on the voltage and active power output of the electric machine connected to the wind turbine. Table 1 shows the transmittable power of grid connected wind turbines

Table 1. Transmittable power of grid connected wind turbine.

Rated voltage of the system	Size of wind turbine or wind farm	Transmittable power
Low voltage system less than 600V.	For small to medium wind turbines	Upto =300kW
Medium voltage systems (600V -35KV)	For medium to large wind turbines and small wind forms	Upto =10-40 MW
High Voltage (35 kV -132 KV)	For medium to large onshore wind forms	upto 100MW
Extra high Voltage (132 KV and above)	Large offshore wind forms	Greater than 0.5GW

Solar penetration also changes the voltage profile and frequency response of the system and affects the transmission and distribution systems of utility grid.

2. Power Quality Problems IN Grid Connected Renewable Energy Sources

There are certain challenges in the integration of wind and solar systems with grid directly. For grid connection of renewable energy sources we use Grid Integration – Grid-tie

Inverter. The use of Inverter is to take energy from grid when renewable energy is insufficient. And supply energy when more power is generated. The connection of grid with renewable energy and disconnection is done in 100ms.The block diagram for grid connected PV array is shown in figure 1 (a), Fig.1(b) shows the grid connected wind energy system.

The main function of converter in PV array connected grid system is to correct the magnitude and phase of the output of PV system by taking the feedback from utility grid. And in case of wind turbine connected grid system it works as isolation of mechanical and electrical frequencies.

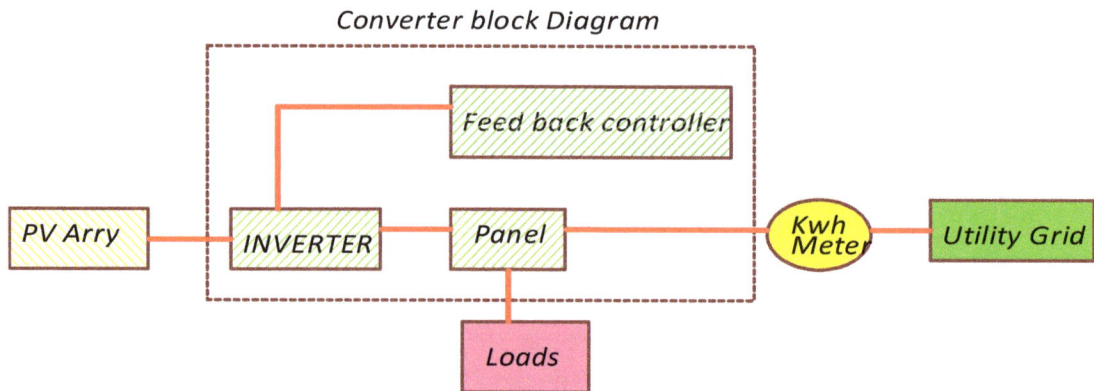

Fig. 1(a). Block Diagram for Gridconnected PV Array.

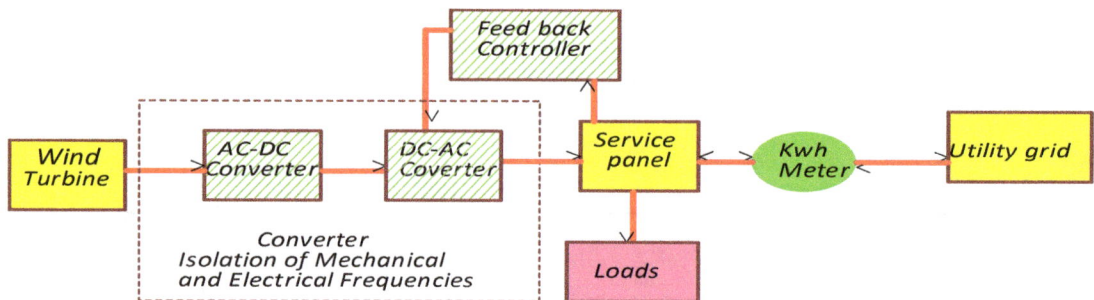

Fig. 1(b). Block Diagram for Grid connected Wind Turbine.

There are several technical issues associated with grid connected systems like Power Quality Issues, Power and voltage fluctuations, Storage, Protection issues, Islanding.

Power Quality issues are harmonics and voltage and frequency fluctuations.

2.1. Harmonics

Harmonics are currents or voltages with frequencies that are integer multiples of the fundamental power frequency. Electrical appliances and generators all produce harmonics and in large volumes (eg. computers and compact fluorescent lamps), can cause interference that results in a number of power quality problems.

Most grid-connected inverters for DG applications put out very low levels of harmonic current, and because of their distribution on the network are unlikely to cause harmonic issues, even at high penetration levels.

While the most common type of inverters (current-source) can not provide the harmonic support required by the grid,

voltage-source inverters can, but do so at an energy cost and there are a variety of harmonic compensators that are likely to be cheaper. Labeling that identifies the type of inverter (voltage or current source) would help purchase of voltage source or current source inverters as required, as would financial compensation for reducing energy losses if voltage source inverters are installed. Note that, unless specially configured, PV inverters disconnect from the grid when there is insufficient sunlight to cover the switching losses, meaning that no harmonic support would be provided outside daylight hours.

2.2 Frequency and Voltage Fluctuations

Frequency and voltage fluctuation again classified as
1. Grid-derived voltage fluctuations
2. Voltage imbalance
3. Voltage rise and reverse power flow
4. Power factor Correction

(i). Grid-Derived Voltage Fluctuations

Inverters are generally configured to operate in grid 'voltage-following' mode and to disconnect DG when the grid voltage moves outside set parameters, This is both to help ensure they contribute suitable power quality as well as help to protect against unintentional islanding. Where there are large numbers of DG systems or large DG systems on a particular feeder, their automatic disconnection due to the grid voltage being out of range can be problematic because other generators on the network will suddenly have to provide additional power[9,10].

To avoid this happening, voltage sag tolerances could be broadened and where possible, Low Voltage Ride-through Techniques (LVRT) could be incorporated into inverter design. LVRT allows inverters to continue to operate for a defined period if the grid voltage is moderately low but they will still disconnect rapidly if the grid voltage drops below a set level. Inverters can also be configured to operate in 'voltage-regulating' mode, where they actively attempt to influence the network voltage. Inverters operating in voltage-regulating mode help boost network voltage by injecting reactive power during voltage sags, as well as reduce network voltage by drawing reactive power during voltage rise.Thus, connection standards need to be developed to incorporate and allow inverters to provide reactive power where appropriate, in a manner that did not interfere with any islanding detection systems. Utility staff may also need to be trained regarding integration of such inverters with other options used to provide voltage regulation - such as SVCs (Static VAr Compensator) or STATCOMS (static synchronous compensators)[1].

(ii). Voltage Imbalance

Voltage imbalance is when the amplitude of each phase voltage is different in a three-phase system or the phase difference is not exactly 120°. Single phase systems installed disproportionately on a single phase may cause severely unbalanced networks leading to damage to controls, transformers, DG, motors and power electronic devices. Thus, at high PV penetrations, the cumulative size of all systems connected to each phase should be as equal as possible. All systems above a minimum power output level of between 5-10kW typically should have a balanced three phase output.

(iii). Voltage Rise and Reverse Power Flow

Traditional centralized power networks involve power flow in one direction only: from power plant to transmission network, to distribution network, to load. In order to accommodate line losses, voltage is usually supplied at 5-10% higher than the nominal end use voltage. Voltage regulators are also used to compensate for voltage drop and maintain the voltage in the designated range along the line.

(iv). Power Factor Correction

Because of poor power factor line losses increases and voltage regulation become difficult.Poor power factor on the grid increases line losses and makes voltage regulation more difficult. Inverters configured to be voltage-following have unity power factor, while inverters in voltage-regulating mode provide current that is out of phase with the grid voltage and so provide power factor correction. This can be either a simple fixed power factor or one that is automatically controlled by, for example, the power system voltage.[8,9]

A number of factors need to be taken into consideration when using inverters to provide power factor correction

- To provide reactive power injection while supplying maximum active power, the inverter size must be increased.
- The provision of reactive power support comes at an energy cost, and how the VAr compensation is valued and who pays for the energy has generally not been addressed.
- Simple reactive power support can probably be provided more cost-effectively by SVCs or STATCOMS[1,2], which have lower energy losses, however inverter VAr compensation is infinitely variable and has very fast response times. In areas where rapid changes in voltage are experienced due to large load transients (eg. motor starts) then an inverter VAr compensator may be justified.
- While this sort of reactive power compensation is effective for voltage control on most networks, in fringe of grid locations system impedances seen at the point of connection are considerably more resistive, and so VAr compensation is less effective for voltage control. In these situations, real power injection is more effective for voltage regulation.

Studies into the use of inverters to regulate network voltage at high PV penetrations have found that in order to achieve optimal operation of the network as a whole, some form of centralized control was also required. In addition, reactive power injection by inverters may be limited by the feeder voltage limits, and so coordinated control of utility equipment and inverters, as well as additional utility equipment, may be required.

3. Grid Connected PVGeneration System

Figure 2shows the configuration of the grid-connected PV /Battery generation system. PV array and battery are connected to the common dc bus via a DCIDC converter respectively, and then interconnected to the ac grid via a common DC/AC inverter. Battery energy storage can charge and discharge to help balance the power between PV generation and loads demand. When the generation exceeds the demand, PV array will charge the battery to store the extra power, meanwhile, when the generation is less than the demand, the battery will discharge the stored power to supply loads. Each of PV system, battery energy storage system and the inverter has its independent control objective, and by controlling each part, the entire system is operating safely.

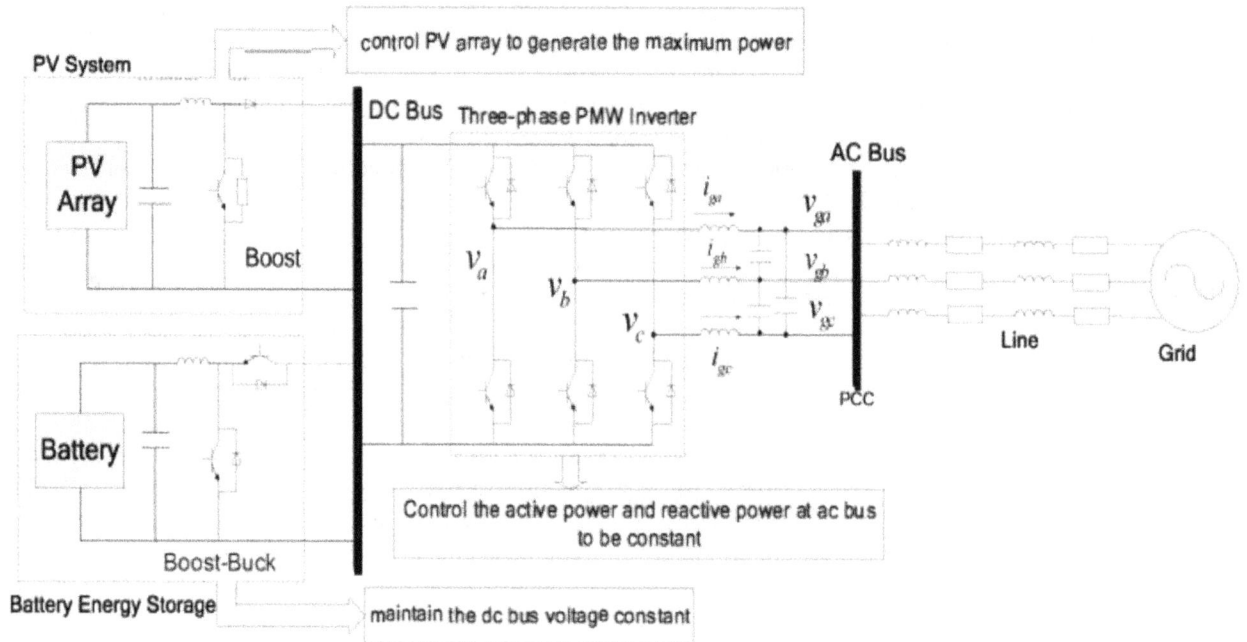

Fig. 2. *Block diagram for Grid conncted PV system.*

3.1. Boost Circuit and Its Control

For two-stage PV generation system, boost chopper circuit is always used as the DC/DC converter. Since the output voltage of PV cell is low, the use of boost circuit will enable low-voltage PV array to be used, as a result, the total cost will be reduced. A capacitor is generally connected between PV array and the boost circuit, which is used to reduce high frequency harmonics. Figure 3(a)is the configuration of the boost circuit and its control system.

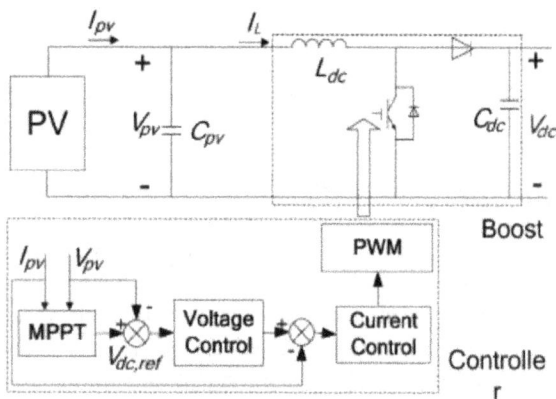

Fig. 3 (a). *Boost circuit of PV system.*

Fig.3(b) shows the Battery energy storage system (BESS).Battery energy storage system (BESS) is composed of a battery bank, a bi-directional DC/DC converter and control system [10]. The system should be able to operating in two directions: the battery can be charged to store the extra energy and also can discharge the energy to loads[4].

The utility grid is considered as a backup source and the

battery bank serves as a short-duration power source to meet the load demands which cannot be fully met by the PV system, particularly during fluctuations of the solar or transient periods.

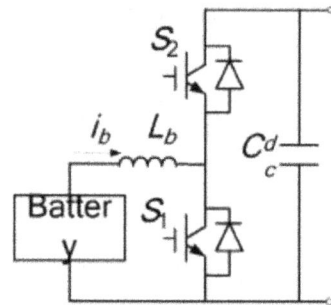

Fig. 3 (b). *BESS for PV arry.*

The primary objective of the battery converter is to maintain the common dc link voltage constant. Tn this way, no matter the battery is charging or discharging, the voltage of the dc bus can be stable and thus the ripple in the capacitor voltage is much less. When charging, switch Sf is activated and the converter works as a boost circuit. otherwise. Whendischarging, switch S2 is activated and the converter works as a buck circuit.

3.2. Control of Grid-Connected Inverter

PV array and the battery are connected to the ac grid via a common DC/AC inverter.Fig 4 Shows the control block diagram for the inverter. The inverter is used in current control method with PWM switching mechanism to make the inductance current track the sinusoidal reference current command closely and obtain a low THD injected current. The control strategy mainly consists of two cascaded loops, namely a fast internal current loop and an external voltage loop. The proposed multi-level control scheme is based on the concept of instantaneous power on the synchronous-rotating dq reference frame.

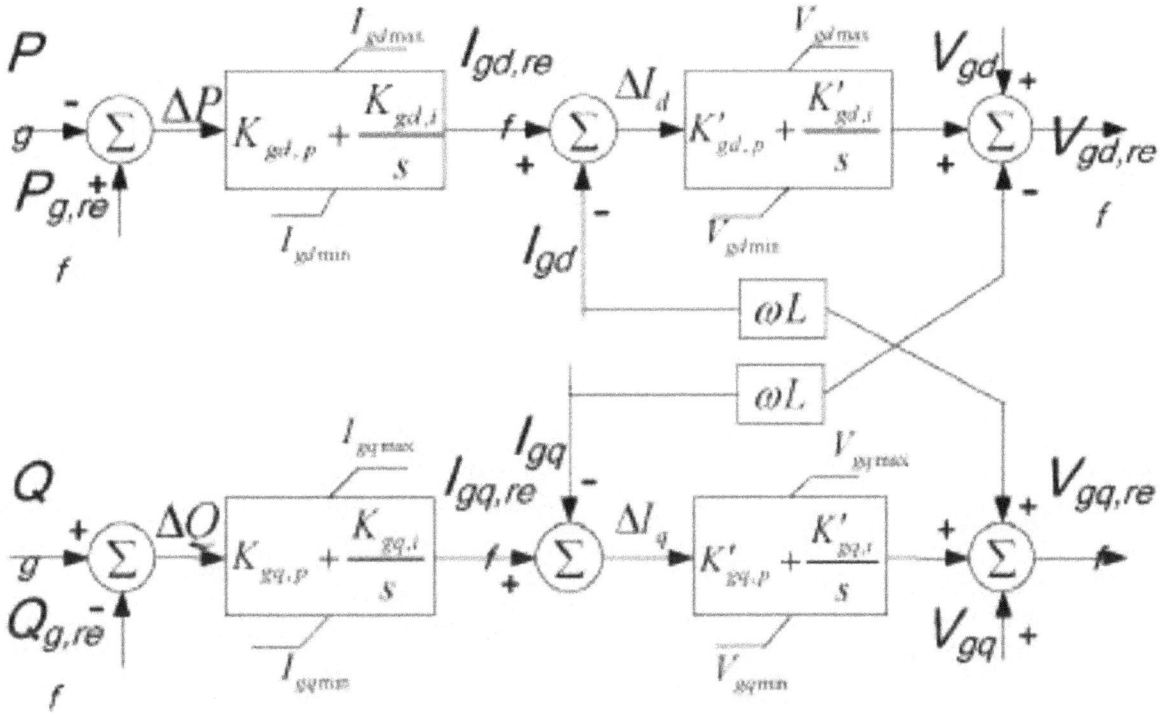

Fig. 4. *Control circuit block diagram of PV connected Inverter.*

3.3. Grid Connected Solar Energy System with Shunt APF

Renewable energy source (RES) integrated at distribution level is termed as distributed generation (DG). The utility is concerned due to the high penetration level of intermittent RES in distribution systems as it may pose a threat to network in terms of stability, voltage regulation and power-quality (PQ) issues. Therefore, the DG systems are required to comply with strict technical and regulatory frameworks to ensure safe, reliable and efficient operation of overall network. With the advancement in power electronics and digital control technology, the DG systems can now be actively controlled to enhance the system operation with improved PQ at PCC[3]. However, the extensive use of power electronics based equipment and non-linear loads at PCC generate harmonic currents, which may deteriorate the quality of power.

3.4. Control Circuit for the Four Leg VSI

A dq-based current reference generator scheme is used to obtain the active power filter current reference signals. Four leg VSI Schematic Diagram is shown in Fig. 5. The current reference signals are obtained from the corresponding load currents as shown in Fig 6. The dq-based scheme operated in a rotating reference frame. Therefore, the measured currents must be multiplied by the sin ωt and cos ωt signals. By using dq transformation, the d current component is synchronized with the corresponding phase-to-neutral system voltage and the q current components are phase-shifted by 90°. The sin ωt and cosωt synchronized reference signals are obtained from a Synchronous Reference Frame (SRF) PLL. The SRF-PLL generates a pure sinusoidal waveform even when the system voltage is severely distorted.

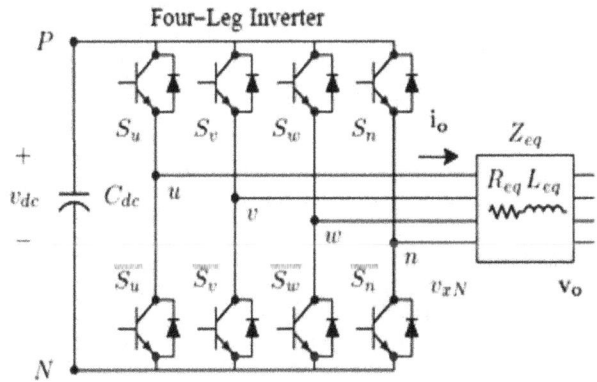

Fig. 5. *Four leg VSI Schematic Diagram.*

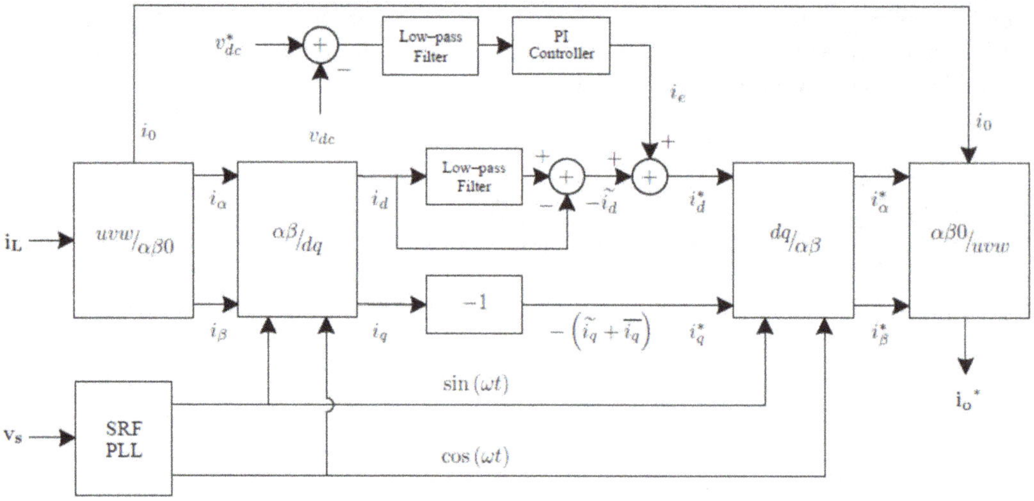

Fig. 6. *Block Diagram for control circuit of four Leg based VSI.*

3.5. Simulation Results of PV Based System

A simulation model for the three-phase four-leg PWM converter with the source voltage of 55V, System frequency of 50 Hz, dc capacitor of 2200μF and filter inductor of 5.0 mH with a sampling time of 20 micro seconds has been developed using MATLAB-Simulink. The objective is to verify the current harmonic compensation effectiveness of the proposed control scheme under different operating conditions. A six pulse rectifier was used as a non-linear load.

In the simulated results shown in Figures 7-16, the active filter starts to compensate at t =0.2. At this time, the active power filter injects an output current i to compensate oucurrent harmonic components, current unbalanced, and neutral current simultaneously. During compensation, the system currents (is) show sinusoidal waveform, with low total harmonic distortion. At t =0.4, a three-phase balanced load step change is generated from 0.6 to 1.0 p.u. The compensated system currents remain sinusoidal despite the change in the load current magnitude. Finally, at t=0.6, a single-phase load step change is introduced in phase u from 1.0 to 1.3 p.u., which is equivalent to an 11% current imbalance. As expected on the load side, a neutralcurrent flow through the neutral conductor (iLn), but on the source side, no neutral current is observed (isn). Simulated results show that the proposed control scheme effectively eliminates unbalanced currents. Additionally, Results show that the dc-voltage remains stable throughout the whole active power filter operation.

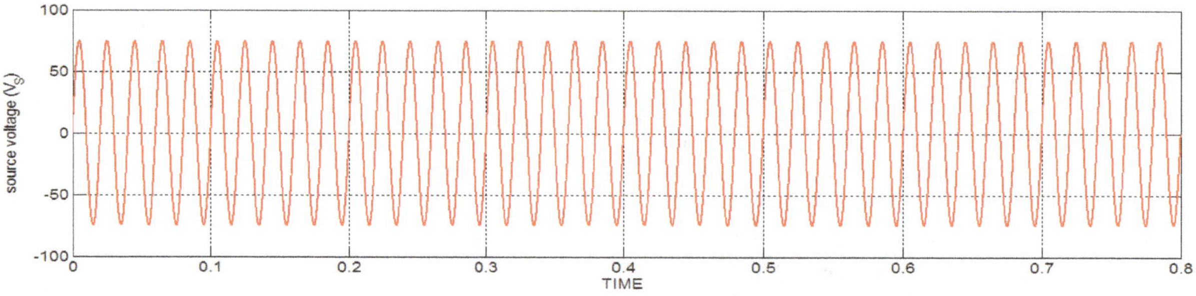

Fig. 7(a). *Phase to neutral Source voltages.*

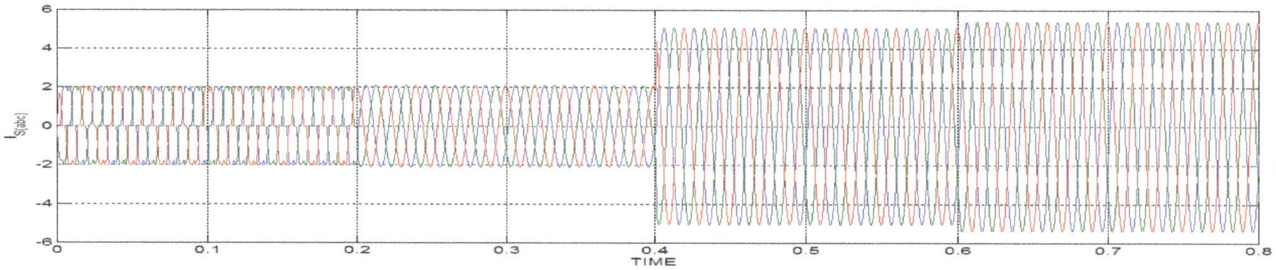

Fig. 7(b). *Source Currents.*

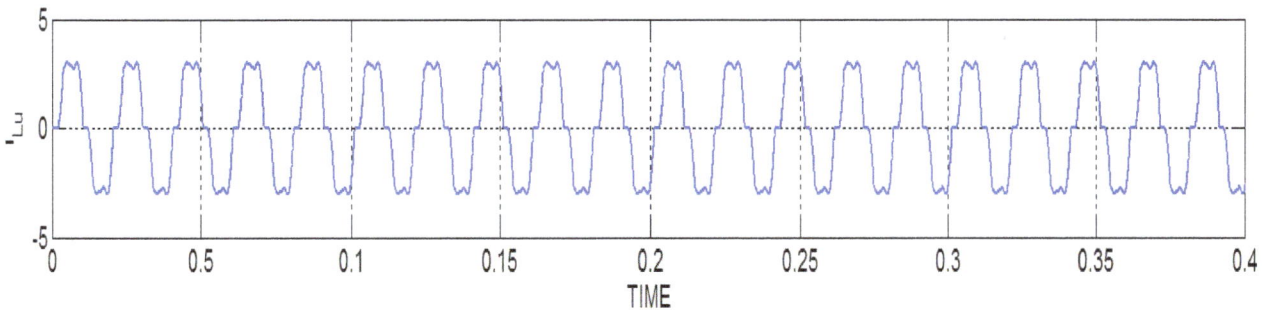

Fig. 8. *Load current at 0<t<0.4sec.*

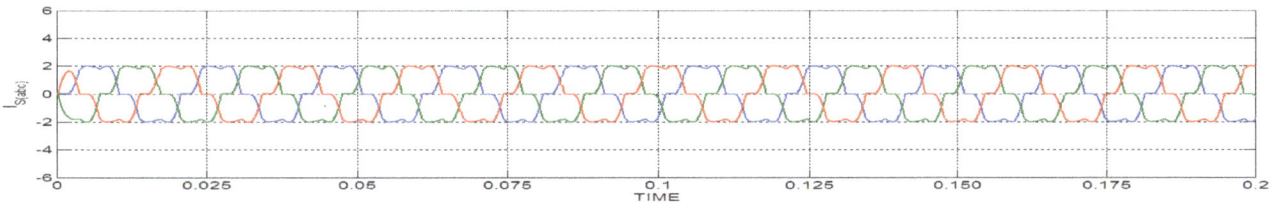

Fig. 9. *Source current at 0<t<0.2 Sec.*

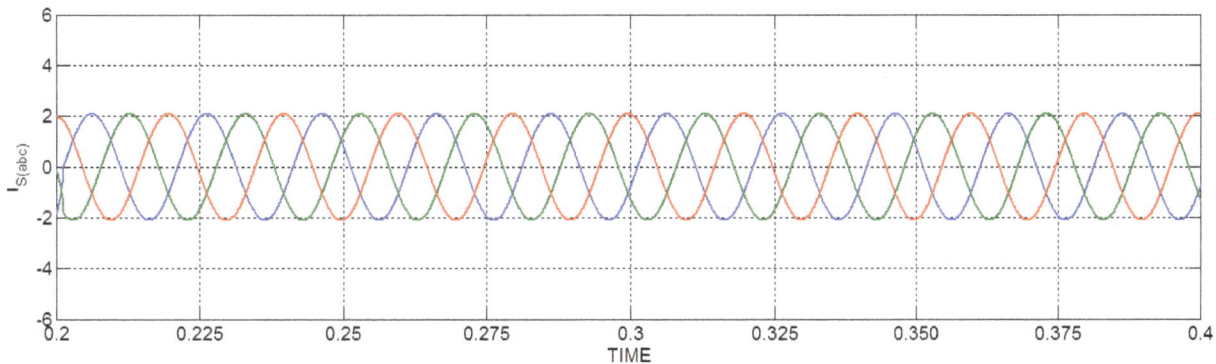

Fig. 10. *Source current at 0.2 <t<0.4sec.*

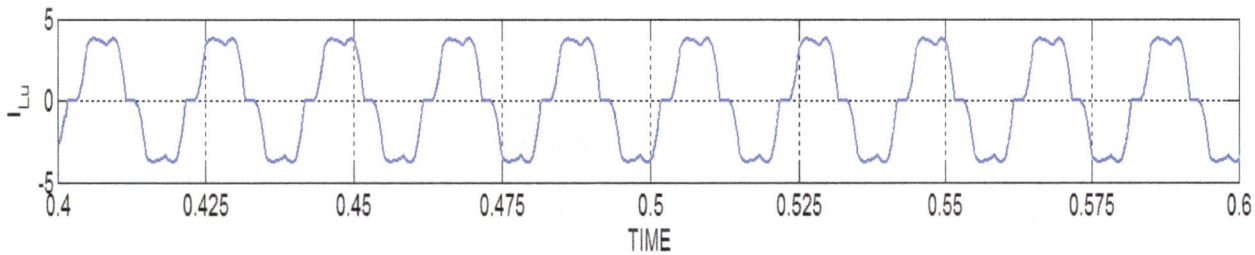

Fig. 11. *Load current due to step change 0.4 <t<0.6 sec.*

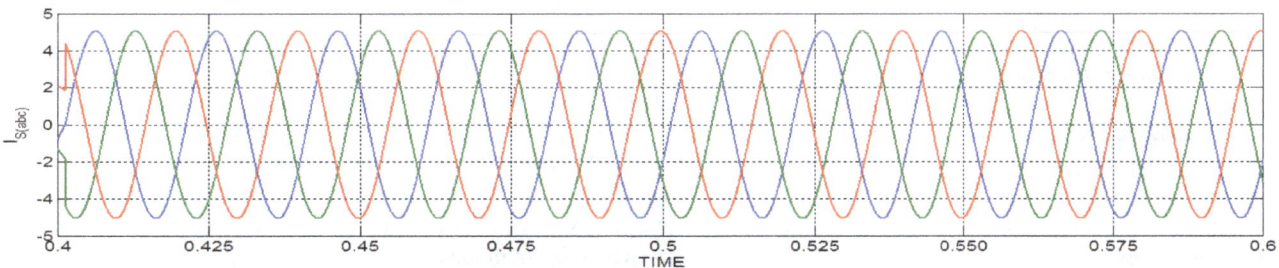

Fig. 12. *Compensated load current 0.4<t<0.6 sec.*

4. Grid Connected Wind Energy System

The three main components for energy conversion in WT are rotor, gear box and generator. The rotor converts the fluctuating wind energy into mechanical energy and is thus the driving component in the conversion system. The block diagram for wind energy system with grid connection is shown in Fig 13.

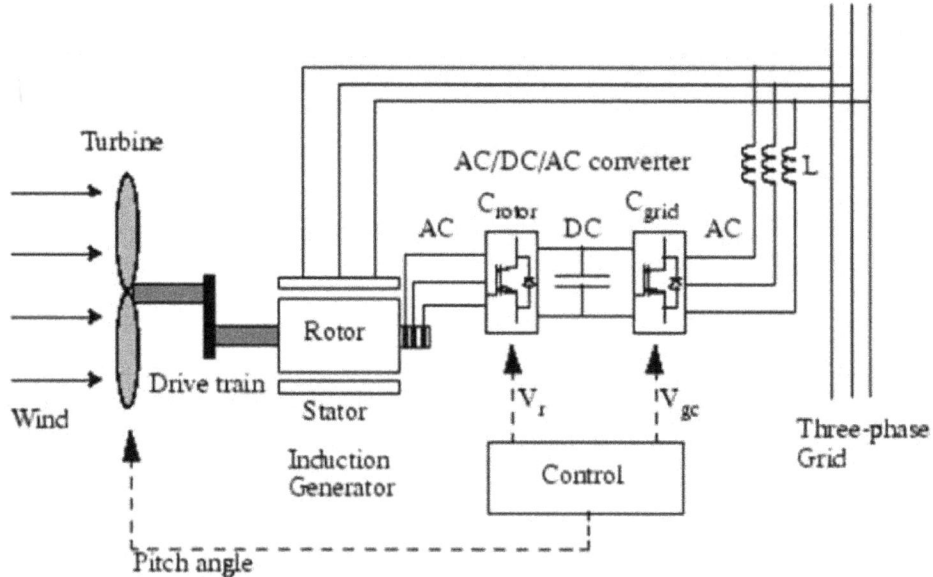

***Fig. 13.** Schematic diagram of Gidconncted wind turbine.*

At the point of common coupling (PCC) betweenthe single WT or the wind farm and the grid a circuitbreaker for the disconnection of the whole windfarm or of the WT must exist. In general this circuitbreaker is located at the medium voltage systeminside a substation, where also the electricity meterfor the settlement purposes is installed. This usuallyhas its own voltage and current transformers. The medium voltage connection to the grid can beperformed as a radial feeder or as a ring feeder,depending on the individual conditions of theexisting supply system.

4.1. Classification of Induction Generators for Wind Turbines

Induction generators can be classified by different ways as rotorconstruction, excitation process, and prime movers.

4.1.1. Classification on the Basis of Their Rotor Construction
- Squirrel cage induction generator
- Wound rotor induction generator

4.1.2. Classification on the Basis of Their Excitement Process
- Grid connected induction generator
- Self-excited induction generator

4.1.3. Classification on the Basis of Prime Movers Used, and Their Locations
- Fixed speed concept using a multistage gearbox
- Limited Variable speed concept using a multistage

Gearbox
- Variable speed concept with a partial scale power Converter
- Variable speed direct drive concept with a full-scalepower converter

4.2. Grid Connected Wind Energy System with STATCOM Control

The wind energy generating system is connected with grid having the nonlinear load. It is observed that the source current on the grid is affected due to the effects of nonlinear load and wind generator, thus purity of waveform may be lost on both sides in the system. The three phase injected current into the grid from STATCOM will cancel out the distortion caused by the non-linear load and wind generator.

The Fig.14shows the complete simulation diagram for the grid connected wind energy conversion system using hybrid fuzzy controller. Here the grid voltage 415 volts and frequency 50 Hz is maintained continuously and a nonlinear load is connected to it and it is represented by the subsystem. A constant speed (10 m/s) wind turbine, with asynchronous generator is connected to the grid.

Induction generator is connected to the distribution network; it requires an external reactive source connected to its stator winding to provide an output voltage control. This reactive support is given by the STATCOM since STATCOM operates in two different modes. One is voltage regulation and the other is VAR control mode. In voltage regulation mode the STATCOM regulates at its connection point by controlling the amount of reactive power that is absorbed from or injecting into the power system through VSC.

When the system voltage is high the STATCOM will absorb the reactive power (inductive behavior). When the system voltage is low the STATCOM will generate and inject reactive power into the system. That's how it will give reactive support to the induction generator for its excitation.

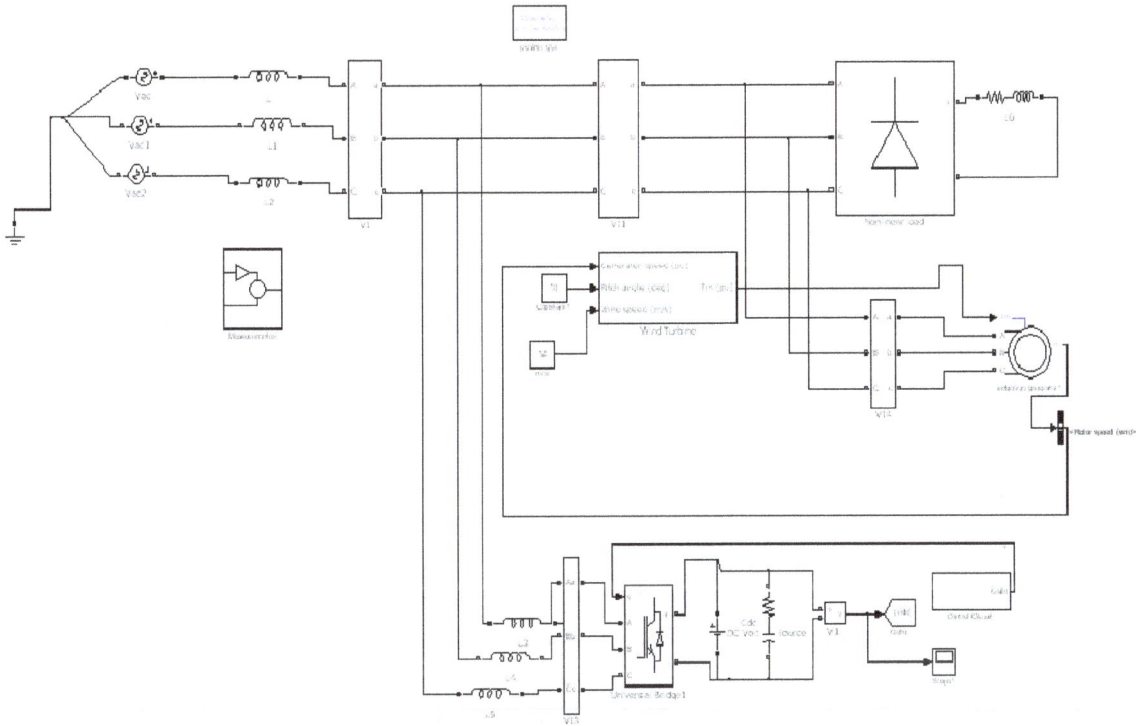

Fig. 14. Simulation diagram for grid connected Wind turbine with STATCOM Control.

To control the distortions caused by the nonlinear load and wind turbine a battery energy storage system with STATCOM is also connected at the point of common coupling. The battery energy storage system (BESS) is used as an energy storage element for the purpose of voltage regulation. The BESS will naturally maintain dc capacitor voltage constant and is best suited in STATCOM since it rapidly injects or absorbed reactive power to stabilize the grid system. It also controls the distribution and transmission system in a very fast rate. When power fluctuation occurs in the system, the BESS can be used to level the power fluctuation by charging and discharging operation. The battery is connected in parallel to the dc capacitor of STATCOM.

4.3. Simulation Results of Wind System

The wind energy generating system is connected with grid having the nonlinear load. It is observed that the source current on the grid is affected due to the effects of nonlinear load and wind generator, thus purity of waveform may be lost on both sides in the system. The three phase injected current into the grid from STATCOM will cancel out the distortion caused by the nonlinear load and wind generator. Fig. 15(a) shows the source current waveform of the test system without STATCOM and the Fig. 15 (b) shows the corresponding FFT analysis waveform. From FFT analysis, it is observed that the Total Harmonic Distortion (THD) of the source current waveform of the test system without STATCOM is 27.21%.

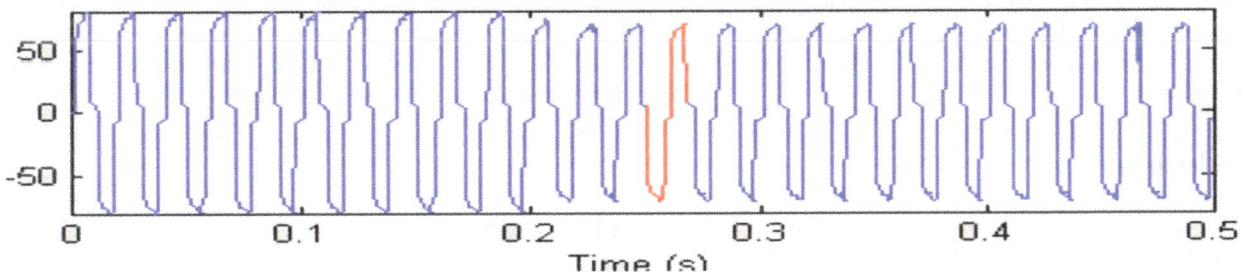

Fig. 15(a). Source current wave form with out STATCOM.

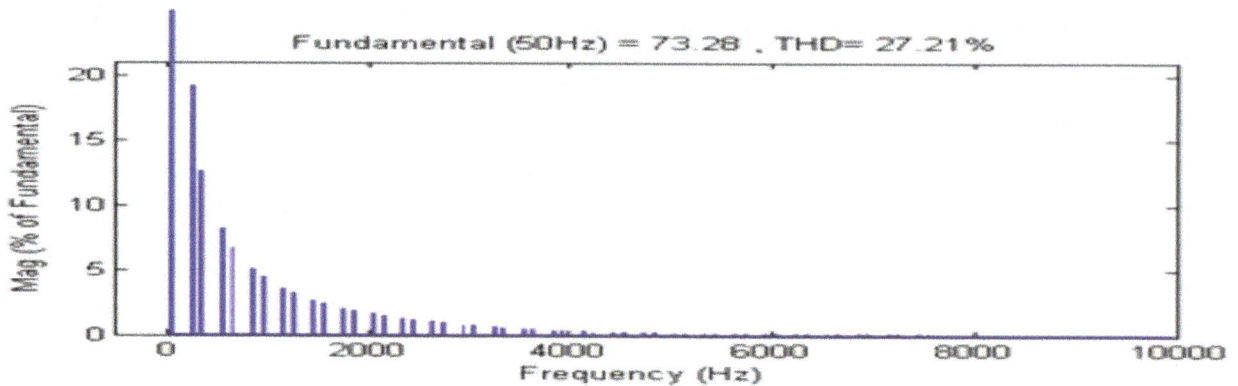

Fig. 15(b). *FFT analysis of source currnt wave form.*

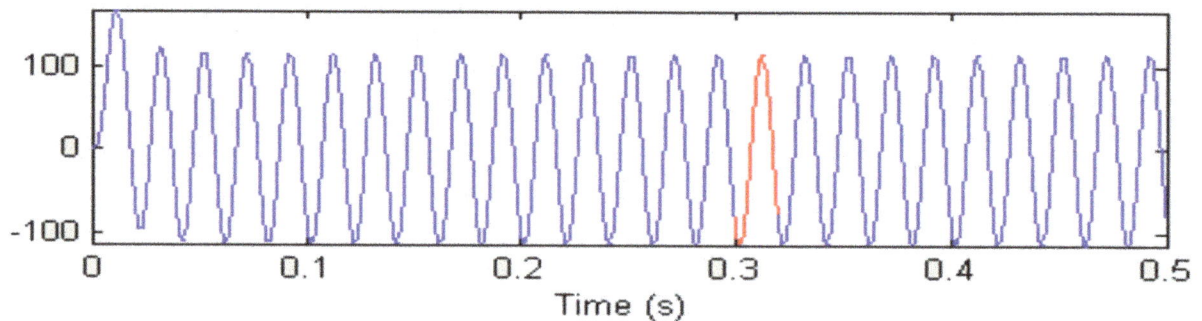

Fig. 16(a). *Source current wave form with STATCOM Compensation.*

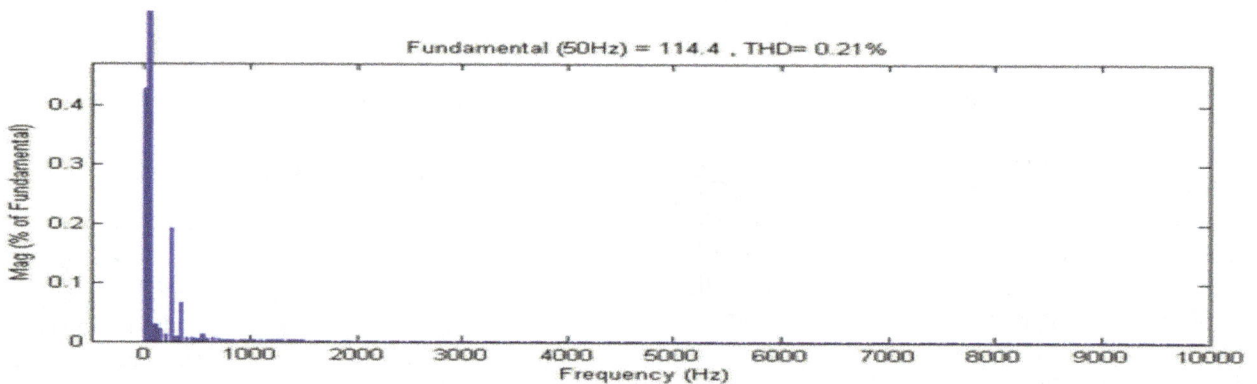

Fig. 16(b). *THD for the source current wave form with STATCOM.*

Fig. 16(a) shows the source current waveform of the test system with Hybrid F uzzy Logic Controller based STATCOM and the Fig. 16(b) shows the corresponding FFT analysis waveform. From FFT analysis, it is observed that the THD of the source current waveform of the test system with Hybrid FLC based STATCOM is 0.21 %. Thus, it is observed that there is a further reduction in the THD value of the source current waveform.

5. Conclusions

End user appliances are becoming more sensitive to the power quality condition. This Case presents a technical review of causes of Power quality Problems associated with renewable based distribution generation system (wind energy, solar energy). Simulation study has done on PV based grid connected system with four leg VSI to enhance the power quality. It has been shown that the grid interfacing inverter can be effectively utilized for power conditioning without affecting its normal operating of real power transfer. A hybrid fuzzy logic controller based STATCOM is presented for grid connected Wind Energy Generating System. The proposed Hybrid FLC based The proposed Hybrid FLC based STATCOM have improved the power quality of source current significantly by reducing the THD from 27.21% to 0.21%.

References

[1]   R. Mihalic, P. Zunko, and D. Povh"Improvement of transient stability using unified power flowcontroller" *IEEE Transactions on Power Delivery*, vol. 11, no. 1, pp. 485–492, Jan 1996.

[2] Mohamed A. Eltawila,b, Zhengming Zhao "Grid-connected photovoltaic power systems: Technical and potential problems—A review" journal homepage: www.elsevier.com/locate/rser

[3] T.R. Ayodele, A.A. Jimoh, J.L Munda, J.T Agee"Challenges of Grid Integration of Wind Power on Power System Grid Integrity: A Review" International journal of Renewable energy sources, Vol.2, No.4, 2012

[4] C. Kocatepe, A.Inan, O. Arikan, R. Yumurtaci, B.Kekezoglu, M. Baysal, A. Bozkurt, and Y. Akkaya, "Power Quality Assessment of Grid-Connected Wind Farms Considering Regulation in Turkey," Renewable and Sustainable Energy Reviews, vol. 13, pp. 2553-2561, 2009.

[5] R. Billinton and Bagen, "Reliability Consideration in Utilization of Wind Energy, Solar Energy and Energy Storage in Electric Power Systems," 9th International Conference on Probabilistic Methods Applied to Power Systems, KTH,Stockhlm, Sweeden, 2006, pp. 1-6.

[6] N. Kasa, T. Lida and G. Majumdar, "Robust control for maximum power point tracking in photovoltaic power system," *PCC-Osaka*, 2002, pp. 827-832.

[7] J. Matevosyan, T. Ackermann, S. Bolik, and L. Sder, "Comparison of international regulations for connection of wind turbines to the network," Nordic wind power conference, Goteborg, 2004.

[8] Edoardo Binda Zane, Robert Brückmann (eclareon), Dierk Bauknecht (Öko-Institut)"Integration of electricity from renewables to the electricity grid and to the electricity market – RES-INTEGRATION"

[9] Mukhtiar Singh, Vinod khadkikar, Ambrish Chandra, Rajiv Verma,"Grid Interconnection of Renewable Energy Sources at the Distribution level with Power-Quality Improvement Features", 0885-8977/2010 IEEE.

[10] Jingang Han, Tianhao Tang, Yao Xu, et at, "Design of storage system for a hybrid renewable power system", 2009 2nd Conference on Power Electronics and Intelligent Transportation System, Vol 2, pp. 67

[11] J. M. Carrasco,L. G. Franquelo, "Power electronicsystems for the grid integration of renewable energy sources: A survey,"IEEE 2002 pp. 827-832.

28

Improving Electricity Access in Ghana Challenges and the Way Forward

Aaron Yaw Ahali

Department of Economics and International Studies, the University of Buckingham, UK

Email address:

aofoee@gmail.com

Abstract: Growth in demographic requirements, increased urbanization and rural electrification coupled with an ever-increasing technological demand, and the aspiration to transform into a middle-income country have led to a fast growth in energy demand in the past two decades in Ghana. Yet there is a huge deficit in supply and this has become a major limitation to growth and quality of life. As Ghana has devoted itself to universal access to electricity by 2020, the real challenge is in reaching the capacity to meet this goal; and most importantly, ensuring that supply is adequate and reliable. With access to electricity in Ghana been low for some time now with no improvement in sight, there is the need for a study such as this. The paper sought to examine the challenges preventing the progress of accessible electricity in Ghana. The paper adopted a systematic review approach and used publications that focused on or related to the subject understudy. Key findings identified in the paper include; poor pricing, increasing demand and supply shortfalls coupled with irregularities, institutional restrictions, lack of credible off-taker and lack of policy and project continuity. To address these challenges, suggested recommendations include; exploring all means of getting power source, including LNG, Solar, landfill gas and nuclear power, establishing a vibrant and robust power ministry that can help reform and also help in revenue collection.

Keywords: Ghana, Electricity Access, Efficiency and Performance, Electricity Demand, Electricity Supply

1. Introduction

Globally, energy plays an important role in the socioeconomic development and economic growth [1]. Electricity powers modern society. It lights buildings and streets and runs computers. The extensive use of electric-powered machinery plays major roles in both industrial and household production. Without electricity, economic transformation through improved productivity in manufacturing and services, technological innovations, and promoting value-addition in resource-based economies would not be possible.

Against this backdrop, the availability of electricity cannot be overemphasised as it is a catalyst for sustainable economic development, its absence may perhaps have adverse consequences which can affect society negatively. Electricity cannot be eliminated in core areas of the economy such as agriculture, industries, transportation and service sector. Due to the rise in population, quality of life and rapid industrialization, the demand for energy is expected to grow. According to [2] insufficient supply of energy will limit socioeconomic activities, restrict economic growth and negatively impact living standards.

Globally, electricity is the form of energy which is extensively used and the preferred source of energy. [3] Observes that increase in population results in high energy demand. In most developing economies particularly African and some Southeast Asia countries, electricity supply is commonly known to be unreliable which causes huge interruptions with cost implications which thus affect the efficiency of production and competitiveness. Without a doubt Africa is endowed with many energy resources such as coal, natural gas, petroleum, solar, hydro, geothermal, nuclear etc. which can be used to power its electricity capacity, yet the region's power sector is acutely weak coupled with the fact that energy consumption on a whole and electricity consumption to be precise is very low.

It is worthy to mention that Ghana's energy sector has been through series of metamorphosis: from diesel generators and stand-alone electricity supply systems (owned by industrial mines and factories), to the hydro phase, and now to a thermal complement phase powered by gas and/or light crude oil. This phase included construction of the Takoradi and Tema Thermal Plants and the development of the West African Gas [4]. With increases in economic growth, prevalent penetration of technology each day and rising demographic needs, Ghana is always faced with the problem of not having enough electricity to meet its increasing developmental needs. The current energy crisis has taken most Ghanaians by surprise. This should not be the case, as this is a repetitive phenomenon in the history of the country, noticeably in 1983, 1994, 1997-98, and 2006-07, with increasing severity. The 2006-07 energy crises are estimated to have cost GDP growth by about 1.5% [5].The ongoing energy crisis threatens not only GDP growth, but also public safety and the prospect of transforming the economy.

To this end, the paper seeks to ascertain and investigate the various elements that obstruct access to reliable electricity in Ghana and also how Ghana's current power issues can be elevated to an economic suitable level. To be able to answer the aforementioned question, the current status of Ghana's electricity sector is analysed, this encompassed electricity production and consumption, fuel use, cost and energy security. The next segment examines the various challenges that impede Ghana's electricity access so as to propose steps that that can help improve access to electricity in Ghana.

The rest of the paper is set out as follows; Section 2 explicitly reviews the present electrical energy situation using Ghana as case study. The various challenges and issues pertaining to accessibility to electricity in Ghana are investigated. In the 3rd segment, some issues were investigated where section 4 provided recommendations and section 5 concludes the paper.

2. Study Area

Ghana is situated in the middle of the west coast of Africa and shares borders with three French-speaking countries. It is bordered to the north by Burkina Faso (formerly Upper Volta), to the west by Cote D'Ivoire, and to the east by Togo [6]. To the south of the country lies the Atlantic Ocean and Gulf of Guinea. The Greenwich Meridian which passes through London also traverses the country at Tema. Its total area is 238,540 square kilometres (91690 square miles). Important natural resources include manganese, bauxite, gold, timber and oil. Current exploration of oil in the country also suggests that Ghana has oil in commercial quantities [7]. The Volta River is formed at the centre of the country by the confluence of the Black Volta and the White Volta. The river has served as the source of the hydroelectric power for Ghana and its neighbouring countries for many years.

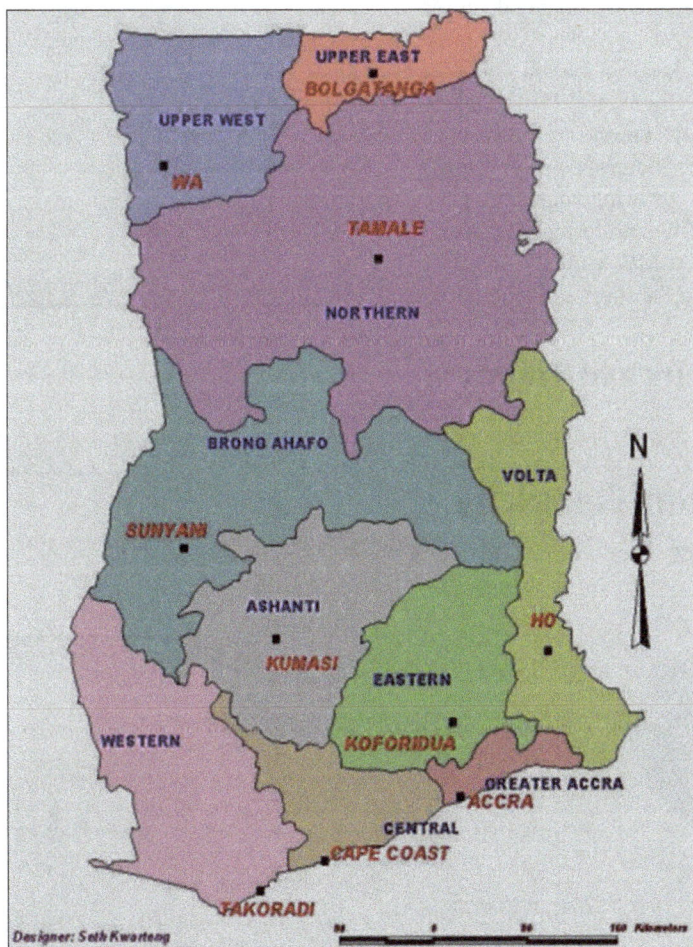

Source: http://www-pub.iaea.org/MTCD/Publications/PDF/CNPP2012_CD/countryprofiles/Ghana/Ghana.htm

Figure 1. A Map of Ghana.

Demand for accessible electricity in Ghana has been on the rise in modern times due to increase in economic growth, development and industrial activities. [8] Opined that in 2007, electricity accounted for about 9% of Ghana's 9.50Mtoe total final energy consumption. Biomass and petroleum fuels accounted for 64% and 27% respectively of final energy consumption. Approximately, 65% of electricity generated in Ghana is from a large hydro- power station, while the remaining 35% mostly emanate from various thermal power plants that operate on gas, diesel and light crude oil [9]

It is important to mention that, Ghana's electricity is plagued with several supply challenges. Presently, the power plants are unable to accomplish full generation capabilities due to limitations in fuel supply coupled with uncertainty of rainfall and water inflows into the hydroelectric power facilities. The high demand rate combined with intermittent hydrological shocks has caused the country to constantly rely on expensive oil and gas led methods of powering its plants. Again, Ghana's electricity prices are based on the cost of base-load hydropower priced at about $0.05/kWh, which means that the more the country uses oil-based forms of generating power into its plants, Volta River Authority (VRA) the mandated organisation in Ghana responsible for generating electricity will incur some financial loses annually.

Again, low tariff regime which does not permit full cost recovery has also reduced the ability to expand [10]. Lack of additional capacity has been noted to be one of the reasons why there are deficiencies in Ghana's current power supply which is not able to meet its growing demand [11]. Importantly, Ghana will need a different form of capacity addition which is approximately 200 MW annually in order to make up the increasing demand in the short and long term.

3. Energy Supply Capacity and Trends in Ghana

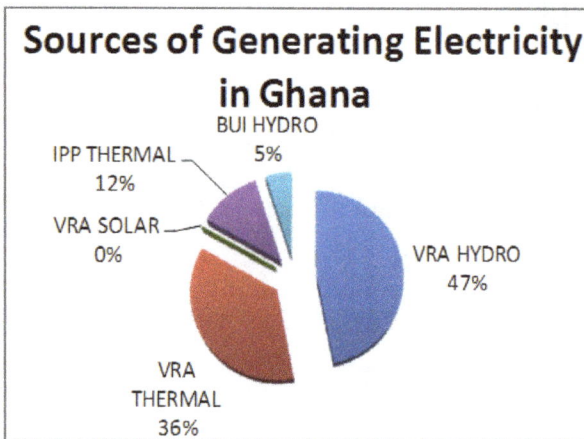

Source: Tuffour and Asamoah (2015)

Figure 2. Sources of Generating Electricity in Ghana.

The supply mix as of January 2015 shows hydro providing 52% of power needs and the rest from thermal, of which 20% can run on gas only and 80% on dual fuel generator (gas and light crude oil). Figure 2 below Ghana's installed sources of generating electricity as of January 2015.

Lack of rainfall, particularly throughout the past five years negatively exposes Ghana's historic dependence on hydro power to generate electricity via Akosombo, Kpong and Bui's installed capacity which is expected to be producing 1020, 160 and 400MW, respectively (See Table 1). The outcome is that, these plants do not perform as expected; hence they produce an output as low as percent.

Table 1. The result is that these plants operate under-capacity often as low as 75 percent.

Generating Station/Plant	Nameplate Capacity, MW	Dependable Capacity, MW
Hydro		
Akosombo	1,020	900
Kpong	160	140
Bui	400	342
Thermal		
TAPCO	378	300
TICO	252	200
TT1PP	126	110
TT2PP	49.5	45
MRP	85	80
T3	132	120
Sunon-Asogli	220	180
CENIT	126	110
EMBEDDED GENERATION		
Genser Power-IPP	5	2.1
RENEWABLE		
Solar	2.5	1.9
Total	2,956.0	2,531.0

Source: Tuffour and Asamoah (2015)

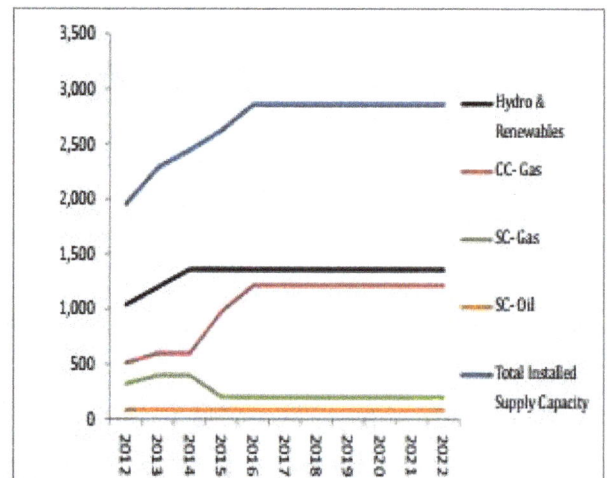

Source: Tuffour and Asamoah (2015)

Figure 3. Shows Installed Capacity Supply Capacity (MW); Expected 2012 and Projected 2013-2022 Legend – CC-Combined Cycle, SC-Single Cycle.

Also, with an installed capacity of 2,956MW (presently

estimated to be about 2,846MW); even regular maintenance shutdown will worsen and widen the gap between supply and demand. As shown in Table 1, the gap between supply and demand is much bigger if projected energy generation falls short of installed capacity. Due to inadequate reserve margin in capacity requirement, projections stipulated by the Volta River Authority (VRA) do not augur well in bringing into prominence the foreseeable deficit in electricity supply. The forecast of demand and available generation capacity with a 20% reserve margin (from 2012 to 2022) reveals significant deviation between forecast figures and projected installed capacity, calling for huge investments in generation mix (see figure 3 below).

4. Economic Cost of Failure of Electricity Supply

The absence of unreliable electricity has direct and indirect cost implications which have huge significance as it negatively affects consumers, utilities, and the economy at large. Stipulated estimates by the World Bank shows that, directly, the cost of power outages to most African nations is usually about 2% of GDP. In the context of Ghana, GDP growth averaged around 5% between 2000 and 2010, which implies that unreliable power supply in those years significantly affected potential economic growth. Furthermore, estimates put forward by the World Bank shows that Ghana's nominal GDP as of 2008 was approximately $16.1 billion, meaning that deficiency in power supply possibly cost the economy more than US$320 million annually. Figure 4 below shows Duration of Power Outages and Value Lost due to Power Outages:

This amount (US$320 million) is enough to support the activities of Ghana Grid Company Limited (GRIDCo), the national transmitter, from 2010 to 2016 and to ensure its effective performance to consumers. Supply disruption foists considerable cost on families because of damage to appliances and the waste of food, and compromises public safety in the delivery of health care services, often with tragic consequences and the greater the technology penetration into economic activities, the greater the economic losses. Direct costs to utilities, among other things, include cost of repairing damaged equipment, process restart costs; generation revenue losses, and reduced equipment's life span.

As shown in figure 4 the duration of power outages and the estimated value loss for selected countries in lower- (Kenya and Senegal), middle- (Morocco), and upper-income (Botswana and Malaysia) brackets. Ghana's average number of power outages during the 2006-07 crises was twice that of Botswana, four times of Morocco and 12 times more than Malaysia. Ghana's value loss of about 5.6% is only marginally exceeded by Kenya's 6.3%. The opportunity cost linked to loss of sales and revenue, the cost of doing business resulting from insecurity, as well as the cost of on-site power equipment such as generators and erratic power supplies have negative ramifications on households and businesses. They needlessly escalate cost, and reduce income and profitability with adverse effects on government revenue targets.

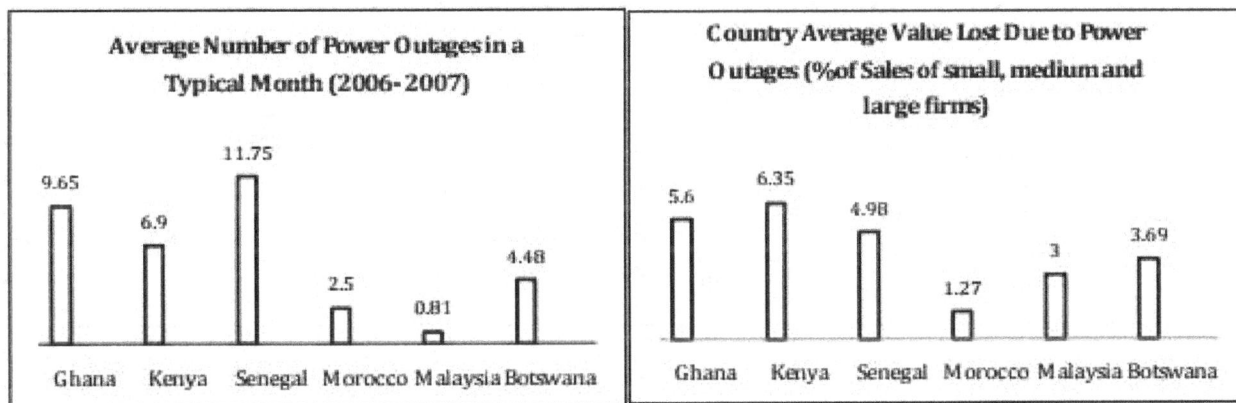

Source: Tuffour and Asamoah (2015)

Figure 4. Below shows Duration of Power Outages and Value Lost due to Power Outages.

5. Material and Methods

The study is qualitative in nature hence a systematic review technique which entails using explicit approach to search, appraise and synthesize available literature to satisfy the aim of the study was employed [12-13]. This method was adopted due to the nature of the paper which was broad based laying much emphasis on access to reliable electricity in Ghana. The vigorous and wide-ranging nature of systematic review helped in preventing any biasness. Also it allowed an array of data to be incorporated, thereby providing a more precise and consistent conclusions [12-13]. In broad terms, the study comprehensively dwelled on secondary sources such as books, journals, conference papers and reports that concerned themselves with the topic understudy. By employing various works on systematic review, rigorous steps were followed in order to retrieve data and provide the needed recommendations accordingly [14-15]. Finally, rigorous content analysis was done to clearly decipher the issues that concern themselves to the topic under study.

6. A Composition of Electricity Demand, Trends and Forecast in Ghana

It would be noted that there has been an immense increase between 2003 and 2013. Although demand in industrial sector spiked in 2000-2002 and again in 2005 and 2006, this declined but reasonably arose between 2008 and 2011, then fell off suddenly to 1.7% in 2013 (Table 2). From a small base in 2000, annual non-residential demand growth has augmented from the usual 9% in 2000-2010; this nearly doubled to 16.5% in 2010-2013, and doubled again to 33% in 2013. Demand in the residential sector has also increased in general by 6.2% annually during the last decade, but that rate more than doubled in 2013 alone. Figure 5: Historical Electrical Energy Demand, 2000- 2013

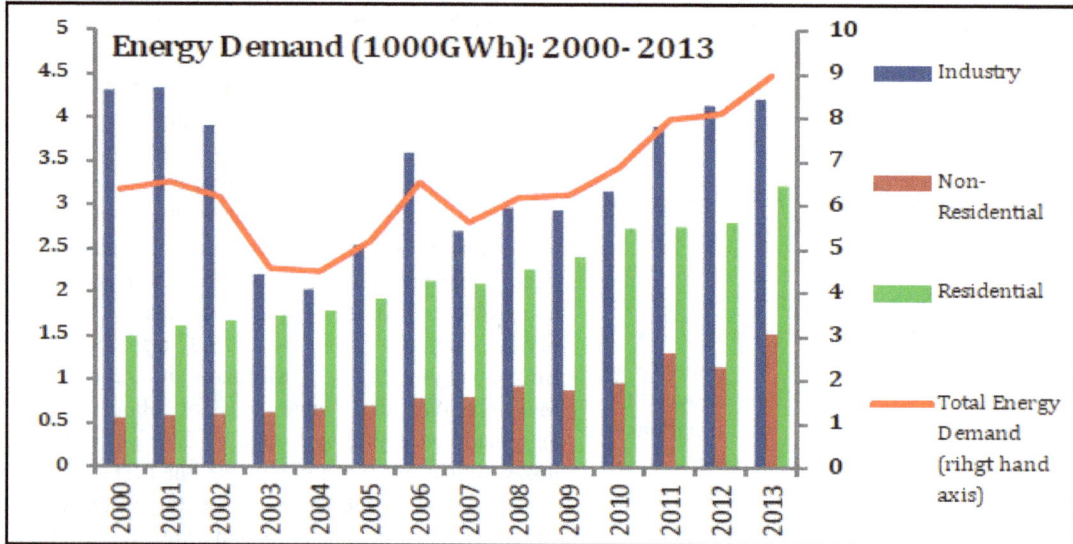

Source: Tuffour and Asamoah (2015)

Figure 5. Historical Electrical Energy Demand, 2000- 2013.

Following the trends emerging from Figure 5 and Table 2, it is worthy to mention that increases in the services sector showed that economic growth leaped from 29% of GDP in 2000 to 51% in 2010. This in part explains the growth in electricity demand since 2005. That growth emerged due to development in information and telecommunication, business services and innovation in the delivery of financial services, all dependent on the availability of energy. Also, a decline in the industrial sector is explained in part by the deteriorating share of industry and specifically manufacturing in GDP; industry declined from 25% in 2000 to 19% in 2010 and manufacturing from 9% in 2005 to 6.8% in 2010.

Table 2. Average Annual Percentage Growth in Demand.

	2000-2010	2010-2013	2013
Industry	2.2	9.7	1.7
Non-Residential	8.9	16.5	33
Residential	6.2	7.8	15.4

Source: Tuffour and Asamoah (2015)

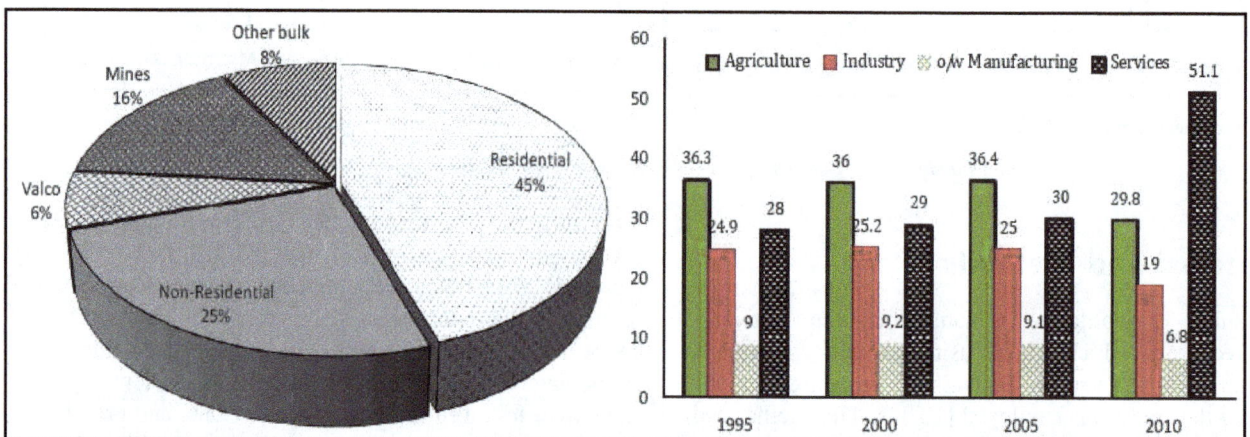

Source: Tuffour and Asamoah (2015)

Figure 6. Approximate Shares of various groups in electricity in sectoral contribution to GDP.

In a progressive manner, Ghana's retail electricity supply has a predominantly residential and non-residential customer base (nearly 70%). This has implications for pricing, revenue generation and the cash flow to make the industry financially viable. Figure 6 Below Provides approximate share of various groups in electricity use: 2013 and sectoral contribution to GDP.

An array of forecasts on consumption exists. For instance, in 2014, the Energy Commission (EC) indicated that peak load demand ranged between 1,900-2,200MW against the system peak load transmission of within 2,200-2,300MW. From the 2010 demand of 1,506MW, Ghana's requirement was expected to hit 2,764MW by mid-2015. Un-met demand in 2013 was projected to be within a 240-330MW thermal equivalent. With growth rates in demand ranging from 6-7%, it is anticipated that demand would reach between 3,598-3,898MW. A World Bank Report forecasts an additional 2,400MW of generation capacity by 2020, which would require more than doubling Ghana's generation capacity from 2012, subject to an average GDP growth of no more than 7% per annum.

7. Ghana's Energy in the Context of Global and Africa Energy Performance

Table 3 below shows Ghana's energy performance, access and security both in global and continental perspective. In the overall category of Energy Architecture Performance Index (EAPI) World [16], Ghana is lagging behind the world's top 10 performers. Although Ghana's average score of 0.45 is above the SSA average of 0.28, it falls behind considerably the global average of 0.84. Ghana's energy access and security ranked 105 out of 124 countries assessed in 2013. Remarkably, although Ghana has considerably advanced since 2000 in terms of increasing and reforming the economy, electricity production per capita has declined. Ghana's installed generation capacity of 132MW per million population in the mid-2000s fell short of 797MW per million among middle-income African countries [17].

Table 3 Ghana versus Global Top 10 in Energy Performance and Energy Access and Security: 2013 out of 125 countries.

Table 3. The Global Energy Architecture Performance Index Report 2014, World Economic Forum. Ranking out of 125 countries.

Country	Energy Architecture Performance Index (EAPI) 2014		Energy Access and Security Basket	
	Score	Rank	Score	Rank
Norway	0.75	1	0.96	1
New Zealand	0.73	2	0.85	5
France	0.72	3	0.81	18
Sweden	0.72	4	0.85	6
Switzerland	0.72	5	0.82	14
Denmark	0.71	6	0.88	3
Colombia	0.7	7	0.8	47

Country	Energy Architecture Performance Index (EAPI) 2014		Energy Access and Security Basket	
	Score	Rank	Score	Rank
Spain	0.67	8	0.78	30
Costa Rica	0.67	9	0.77	35
Latvia	0.66	10	0.77	36
Ghana	0.45	83	0.42	105

Source: Tuffour and Asamoah (2015)

8. Challenges in Ghana's Energy Sector

Apart from weak and sluggish policies, there are serious challenges confronting Ghana's energy sector. These major challenges are as a result of increasing demand and supply shortfalls coupled with irregularities. Major factors such as poor pricing coupled with a complex web of organization and institutional restrictions have all contributed to these challenges in the energy sector.

From a demand perspective, growth in demographical needs in key areas such as health, education, urbanization, rapid rural electrification, and technological advancement on daily basis have also played a role in generating these challenges. Then, in addition to these factors are issues of illegal connections, and problems of metering, billing and collection. While the preceding factors are obvious due to growth in population and high standards of living, it would be noted that factors in the second category work against the capability to generate revenue. Also, due to the exponential growth predominantly in the residential and non-residential customer base (70%) of the energy sector, as high energy sensitive production activities (manufacturing and mining) decline, makes economic pricing, strong billing, and the collection system critical to the financial viability of the industry.

Nonetheless, in recent years, the Ghanaian government have kept in check Electricity pricing by regulators. However, pricing for end users does not reflect the cost of inputs neither does it reveal the realism of the changing mix of the inputs away from a fairly cheaper hydro to gas and light crude oil. In addition to wrong pricing, the idea of subsidising some targeted consumers has damaged the financial capabilities of the Electricity Corporation of Ghana (ECG), a major supplier of Ghana's electricity at the retail level. Also, the absence of a good residential address system and a very weak billing of residential customers who consume about 45% of the electricity supplied suggest that commercial customers who only consume 25% are easily identified and are made to bear majority of burden in terms of paying bulk of ECG revenues.

In terms of supply, there are two main obstacles: insufficient supply and irregular supply. Factors such us ; poor investment which turns to obstruct generating capacity; failure to attract independent investors due to low incentives; poor governance and regulatory framework, and bureaucracy; declining hydro capacity; and failure to enhance the supply mix account for the inadequate supply. Also, input tariffs have merged with real cost coupled with poor pricing at the user end, this has intensified the financial incapabilities of both generating and distribution companies. Not only does this cause a decline in production per capita, but high transmission and distribution losses also causes a decline in supply significantly which results in short of the production capacity. Also, inaccessible supply of electricity originates from the lack of maintenance of generating, transmission, and distribution facilities, and the poor performance of ECG as the primary retailer.

Although there are independent power producers in Ghana (IPPs), there seems to be difficulty in catching the attention of new IPPs. Presently, issues pertaining to gas availability to power plants are a major blockade for prospective IPPs. In this light, it wouldn't be surprising if Ghana does not have access to sufficient supply of gas for power generation until 2015, or maybe even 2018. Similarly, poor governance and rigid regulatory framework is not attractive to IPPs. To start with, prospective IPPs do not have faith in the usual off taker ECG as the organization is in financial crises; and there are genuine concerns about its ability to pay power producers.

Again, there are concerns regarding matters of uncertainty about procedures and regulations. Finally, the IPP development process is awkward and takes a lot of time as presently Ghana does not have a single-window system for IPPs. Thus, it is important that the Government removes these barriers to IPPs in power generation. This process must be led by the sector Ministry (now Ministry of Power) by employing a full-time, high-level IPP manager to take on this duty, in association with the Ministry of Finance's (MoF's) Public Private Partnership unit. With VRA as a key actor on the supply side, Figure 7 above sums up the nature of the challenges on both the inputs side and the output side and, together with the challenges of ECG and the Northern Electricity Distribution Company (NEDCo), support each other in a vicious circle of a financially unsustainable system.

8.1. Lack of a Credible Off-taker

Presently, to be able to obtain licenses, procedures in the power sector makes provision for only "bulk customers," who are free to secure their power needs directly from wholesale suppliers via transmission services provided by an Independent System. Prospective off-takers of power are ECG, NEDCo, the mining companies, and other licensed bulk customers. VRA, which was Ghana's first Independent Power Provider (IPP) project, functioned as both owner and off-taker. Its dual functions were modernized as the rationale for the project was to complement hydro generation, and allow the country to optimize the yield from hydro and non-hydro sources. Nevertheless, VRA has since been reluctant to sign power purchase agreements (PPAs) with IPPs, because it regards IPPs as competing rather than complementary generation entities.

Apart from ECG, no other prospective buyer has signed a PPA to off-take power from any of the IPPs. Out of the four PPAs, only three underpinning IPP development in Ghana

has ECG as the off taker, while the other has VRA as the concurrent co-owner and off taker. The IPPs that have tried to penetrate the Ghanaian market have reported complexity in securing PPAs with other organizations.

8.2. ECG's Commercial Performance Needs Improvement

Due to low residential tariffs, most of ECG's revenues are generated from non-residential consumers who account for 56% of sales revenue, even though they account for only 12% of ECG's unit sales. Importantly, this cross-subsidy foists a major burden on commercial customers. Bearing in mind that PURC has failed to increase retail tariffs, it is reported that ECG incurred losses of US$44 million in 2012, and US$60 million in 2013. As indicated in a 2009 World Bank report (Box 1), ECG's losses are worsened by high technical losses; poor revenue collection, from both Government entities and private consumers; and rising dollar-denominated payment obligations. Tariff policies that provide subsidies to consumers have damaged the financial health of ECG and NEDCo.

8.3. Subsidies and Pricing

Going forward, for Ghana's power sector to be financially sustainable there is the need for the price of electricity to allow for full cost recovery across the entire value chain. Nonetheless, for a long time, prices paid by consumers of electricity have been below the cost of supply. The relatively low tariffs have made the sector unappealing to other stockholders. Since 2004, Government has virtually spent US$900 million on fuel subsidies for VRA. This is more than the cost of building the Bui Dam. Transferring this expenditure on infrastructure would have been of much more value in the long run.

8.4. Gas Pricing

Although the basic principles of Gas Pricing Policy was approved in mid-2012 by cabinet, based on a comprehensive study carried out by international consultants, Government has not been published its Gas Pricing Policy. Hence, the study advocates that gas pricing should indicate gas supply distributing priorities between different sectors in the market, such as power generation, petrochemicals, etc. Also, it established targets for maximum supply costs in the various sectors of the market and minimum gas prices for associated and non-associated gas.

8.5. Poor Revenue Collection

At the beginning of the year, the minster responsible for power established a Task Force to collect arrears of nearly GHc500 million owed ECG. As indicated a 2005 report, the relatively poor financial health of ECG and NEDCo stems from the poor payment culture of Government institutions as well as private consumers. To alleviate the poor payment culture, thus ECG must accelerate the fitting of prepaid meters at the power premises of consumers. Also, disciplinary measures must be put in place as a warning to the theft of electricity through bypassing of meters.

8.6. The Unproductive Clearinghouse Initiative

A cross-debt clearinghouse organization was set up to manage the inter-utility and Government debts, it encompasses VRA, ECG, NEDCo, and Ghana Water Company Ltd., with Government represented by Ministry of Finance (MoF). Quarterly, they must meet to reconcile the cross-indebtedness of the participants and net off such debts where suitable. However, this has not been effective, due to their inability to enforce payment expected from the net debtors. Since the MoF has not received any settlement payment from these institutions, and given the way that central banks make interbank clearing enforceable, the clearinghouse initiative is dependent on voluntary compliance. None-payment meant that VRA, which generates nearly 80% of its profits from these other clearinghouse members, is continually owed, with a build-up of receivables. In 2012 the MoF suspended the clearinghouse mechanism, but no other payment system has been instituted. The debt levels and the matrix of receivables of the utilities have negative repercussions for the utilities' operations and financial viability. While necessary, clearance of state arrears by the Government is not sufficient in the absence of better arrangements to prevent recurrence of the arrears.

9. Going Forward

9.1. The Future of Power Mix

Ghana gas: Ghana's power mix is going through metamorphoses from hydro to thermal, granting that the demand for power in Ghana is growing at a rate of 10% annually. It is wise to explore all potential power sources, including LNG, solar (for concentrated solar power), landfill gas, and nuclear. At best, Ghana Gas can supply gas that can generate 500 MW, which could raise Ghana's installed capacity from about 2,900 MW currently to about 3,400 MW, but still short of the expected installed capacity of 5,000 MW in 2016.

9.2. Future of WAGP and Challenges of Imported Gas

Guaranteeing acceptable and safe supply of natural gas is an essential determinant to improving the accessibility and affordable cost of power to both VRA and IPPs. Ghana has two different gas resources to turn to: domestic gas resources and Nigeria. Yet, sporadic supply disruptions on the West Africa Gas Pipeline (WAGP) have often led to serious near-term gas scarcity. The just agreed increase in the price of gas from WAGP could cause the supply of relatively high volumes of gas from Nigeria. This expectation, however, may only happen if other factors—such as militancy in the Niger Delta of Nigeria and efficient functioning of the value chain in the extraction, transmission and delivery of gas to Aboadze and Sunon-Asogli power plants - are stabilized.

9.3. The New Power Ministry

Clearly, the two key issues that are likely to influence the organizational dynamics of the power sector are the creation of a new Ministry of Power in December 2014, and the conclusion of agreements with the Millennium Challenge Corporation (MCA) for a US$500m investment in the energy sector. The new Ministry must rapidly establish urgent reforms, particularly to ECG in its operational, technical and management competence. Further to this, the Ministry must critically focus on the generation, supply, and efficiency of power in order to meet economic demands.

10. Conclusion

Access to reliable electricity is a major determinant which enhances rapid development in most economies in the world. Ghana's access to electricity has been below the economically acceptable level and has not improved in recent times due to some challenges. As highlighted in the paper, these challenges include; low efficiency and performance, security of fuel source for power generation, data inadequacy, regulatory barriers, lack of institutional arrangement, poor grid structure, dilapidated transmission and distribution network, low financial investment, lack of policy and project continuity. In this light a way forward has been presented in order to improve electricity access in Ghana. This much anticipated improvement may not be instantaneous but a gradual process.

References

[1] Emodi, V.N., Yusuf, S.D. and Boo, K.-J. (2014) The Necessity of the Development of Standards for Renewable Energy Technologies in Nigeria. *Smart Grid and Renewable Energy*, (5) 259-274.

[2] Oseni., M. (2012) Households's access to electricity and energy consumption pattern in Nigeria. *Renewable and Sustainable Energy Reviews*, 16 (1).

[3] Oyedepo, S., O. (2013) Energy in Perspective of Sustainable Development in Nigeria. *Sustainable Energy* (2) 14-25.

[4] Energy Commission (2014) *Energy (Supply and Demand) Outlook for Ghana*, April 2014. Foster, Vivien and Nataliya Pushak, *Ghana's Infrastructure: Continental Perspective*, Washington, DC: World Bank Policy Research Paper No. 5600, 2011.

[5] World Bank, (2011) Energizing Economic Growth in Ghana: Making the Power and Petroleum Sectors Rise to the Challenge, Energy Group Africa Region.

[6] Boateng, E. A. (1966) a Geography of Ghana 2nd Edition, Cambridge University Press, Cambridge.

[7] Government of Ghana (GOG) (2010) Budget statement for the year 2008 Ghana Official Portal [Online] accessed at http://Ghana.gov.gh/Ghana/budget_statements.jsp on 13th March 2015.

[8] Duku M, H. and Hagan E, B. (2011) A comprehensive review of biomass resources and bio fuels potential in Ghana. *Renewable Sustainable Energy Review* (15) 404–15.

[9]   Ghana Energy Commission (2011) Energy supply and demand outlook for Ghana: Ghana Energy Commission.

[10]   USAID (1999) an energy roadmap of Ghana: from crisis to the fuel for economic freedom. A report by the US Government Interagency Team.

[11]   Brew-Hammond A, Kemausour F (2007) Energy crisis in Ghana: drought, technology or policy. Kwame Nkrumah University of Science and Technology (KNUST), Ghana.

[12]   Victor, L., (2008) Systematic reviewing. *Social Science Update,* 58: 1-4.

[13]   Akobeng, A. K., (2005) Understanding systematic review and data analysis. *Achieves of Disease in Childhood,* 90(6), 845-848.

[14]   Victor, L., (2008) Systematic reviewing. *Social Science Update,* 58: 1-4.

[15]   Khan, K. S., Kunz, R., Kleijnen, J., and Antes, D., (2003) Five steps to conducting a systematic review. *Journal of the Royal Society of Medicine,* 96(3), 118-121.

[16]   World Economic Forum (2014) The Global Energy Architecture Performance Index Report.

[17]   USG-GoG Technical Team, August (2011) World Bank, Energizing Economic Growth in Ghana: Making the Power and Petroleum Sectors Rise to the Challenge, *Energy Group Africa Region.*

29

Optimal Location of Small Hydro Power Plants (SHPP$_S$) at Distribution System by Using Voltage Sensitivity Index

**Alaa Abdulwahhab Azeez Baker[1], Maamon Phadhil Yasen Al-Kababji[1],
Sameer Saadoon Al-Juboori[2, *]**

[1]Department of Electrical Power and Machines Engineering, Engineering College, Mosul, Iraq
[2]Department of Electronics & Control Engineering, Kirkuk Technical College, Kirkuk, Iraq

Email address:
alaaali.aam@gmail.com (A. A. A. Baker), al_kababjie@yahoo.com (M. P. Y. Al-Kababji), sameersaadoon@yahoo.com (S. S. Al-Juboori)

Abstract: This work presents a method to enhance the distribution network for both test and real systems by adding small hydro power plants (SHPPS). The voltage sensitivity index (VSI) was used to find the optimal locations to add small hydro power plants (SHPPS). The study has been applied to the system at unity and 0.9 lagging power factor. The test system is a standard IEEE 33-nodes radial distribution network. Maltab program was used to simulate the systems. The simulation results when connecting the (SHPPS) to the test system showed the improvement in voltage profile of the test system nodes in addition to power losses reduction. The reductions of the real and reactive power losses percentage reached (36%) and (14%) at unity power factor respectively, while at (0.9) lagging power factor, the reduction of the real and reactive power losses percentage were found (53%) and (56%) respectively.

Keywords: Distribution System, Loss Minimization, Voltage Profile, IEEE Bus System, Renewable Energy Sources (RESs), Small Hydropower Plant (SHP)

1. Introduction

The major considerations for any utility are to run at minimum cost, make maximum profit and to meet the customer demands all the time [1]. Nowadays, electrical utilities are undergoing rapid restructuring process and are planning to expand their electrical networks to meet the increasing load demand [2], but in fact traditional power plant expansion is a process that typically requires years for design, approval, installation and start-up. In a deregulated market environment, such processes are not the best alternative to follow deviations in the projected demand increase. Renewable energy resources have been considered as the best alternative to traditional fossil fuels [3]. The sizes of renewable energy based electricity generators would be very small as compared to large fossil fuel based power plant. Technically they are suitable for installation at low voltage distribution system, near loads centres [4]. Small hydropower system allows achieving self-sufficiency by using the best possible scarce natural resource that is the water, as a decentralized and low cost of energy production [5].

In the distribution systems, power losses reduction is one of the significant factors to improve the overall efficiency of the power delivery [6]. The term "distribution line losses" refers to the difference between the amount of energy delivered to the distribution system and the amount of energy customers billed. It is important to know the magnitude and causality factors for line losses because the cost of energy lost has recovered from customers [7]. Author in reference [8] represents techniques to minimize power losses in a distribution feeder by optimizing DG model in terms of size, location and operating point of DG. Sensitivity analysis for power losses in terms of DG size and DG operating point has been performed. The method in reference [9] has based on the branch current and power flow. The final algorithm arrives at opening of a branch in a loop carrying minimum resistive power flow to make the network radial causing minimum loss.

In reference [10] the researchers used a non-dominated sorting genetic algorithm (NSGA) for reconfiguring a Radial Distribution Corporation to minimize its operating costs considering real and reactive power costs while maximizing its operating reliability and satisfying the regular operating constraints. Reference [11] proposed a new method that presents an algorithm for reconfiguration associated with capacitor allocation to minimize energy losses on radial electrical networks considering different load levels. The proposed model has been solved using a mixed integer non-linear programming approach, in which a continuous function has been used to handle the discrete variables. In reference [12] the researchers used real wind, solar, load, and cost data and a model of a reconfigurable distribution grid to show that reconfiguration allows a grid operator to reduce operational losses as well as to accept more intermittent renewable generation than a static configuration can. In order to minimize the total active power losses and improve the voltage profile of the power system, several solutions have been proposed including the integration of Distributed Generation (DG) in the radial distribution network. The way that used in [13] proposed an Algorithm for Firefly (FA) which is a met heuristic algorithm inspired by the behaviour of fireflies flashing. The main objective of Firefly flash is to act as signalling system to attract other fireflies. In reference [1], another method has been used by the Power System Analysis toolbox in MATLAB to minimize the cost of electricity with optimal power flow for the southern grid of Kerala State Electricity Board. This work concentrates on the savings with the incorporation of a wind farm in the system.

The research in [2] works at minimizing the real power loss in radial distribution system by Optimal Allocation and Sizing of Distributed Generation using Artificial Bee Colony Optimization (ABC) technique. The Artificial Bee Colony algorithm is a population-based optimization technique that based on the intelligent foraging behaviour of the honeybee swarm. To reduce the losses in distribution system, bionic random search plant growth simulation algorithm (PGSA) has been proposed using optimal capacitor placement in [14]. The reference [15] presents a simple method for real power loss reduction, voltage profile improvement, by using Heuristic Search and PSO optimization methods and Imperialist Competitive Algorithm (ICA). Optimally determine distributed generation location and size are compared in a distribution network, the research also demonstrates that system losses may increase if DG units are connected at non-optimal locations or have non-optimal size.

The authors explain in [16] a comparison of novel, combined loss sensitivity index vector and voltage sensitivity index methods for optimal location and sizing of distributed generation(DG) in a distribution network, the power injections from renewable DG units located close to the load centres provide an opportunity for system voltage support, reduction in energy losses and reliability improvement. Reference [6] proposes a new long term scheduling for optimal allocation and sizing of different types of Distributed Generation (DG) units in the distribution networks in order to minimize power losses. The optimization process is implemented by continuously changing the load of the system in the planning time horizon. In order to make the analysis more practical, the loads changed linearly in small steps of 1% from 50% to 150% of the actual value.

This paper focuses to minimize operating losses considering real and reactive power losses and to improve the voltages profile of a radial distribution system; the results will be displayed on the IEEE 33-node radial distribution network, which proposed to be the test system. The results before and after the addition of small hydro plants (SHPPS) were compared using the voltage sensitivity index (VSI).

2. Optimal Location Based on Voltage Sensitivity Index (VSI)

The main objective of finding VSI is to find most sensitive node of the system from voltage sensitivity point of view [17]. It considered a numerical solution, which helps operator to monitor how close the system is to collapse, or to initiate automatic remedial action schemes to prevent voltage collapse. Nodes, having minimum voltage sensitivity index are selected and then, using equation (1) to calculate the voltage sensitivity index (VSI).

When connecting (SHPPS) at bus i, VSI is defined as [18]:

$$VSIi = \sqrt{\frac{\sum_{k=1}^{n}(1-V_k)^2}{n}} \qquad (1)$$

Where Vk is voltage at kth node and n is the number of nodes. The node with least VSI will be picked as the best location for the small hydro power plants (SHPPS) placement.

3. Results and Discussion

Results obtained for IEEE-33 test systems with VSI based method. The base MVA and base kV has been taken as: (MVA) Base=100MVA and (K.V) Base=33Kv.

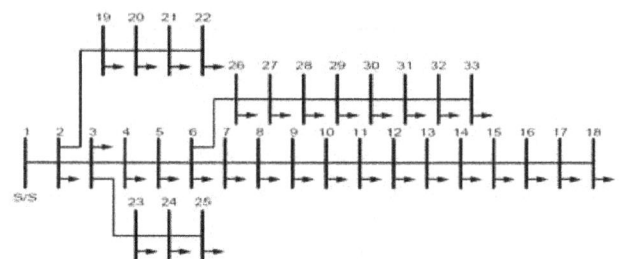

Figure (1). *Single Line Diagram of 33-Node Distribution System. [19]*

Table (1). *33-bus system with and without installation of small hydro power plants (SHPPS).*

	Without SHPPs	At Unity pf	At lagging pf
Real power from the substation MW	6.69034	4.84	4.705
Total real power loss MW	0.74634	0.350	0.241
Reactive power from the substation MVAR	3.96168	3.96168	2.911
Total reactive power loss MVAR	0.28168	0.242	0.09
Minimum bus voltage p.u	0.9184	0.9487	0.951
The source voltage p.u	1.001	1.003	1.004
Percentage of real losses %	11.15	7.2	5.14
Percentage of reactive losses %	7.11	6.11	3.09

Single Line Diagram of 33-Node Distribution System is shown in Figure (1) [19]. The results obtained at unity and 0.9 p.f (lag) for 33-bus system with and without installation of small hydro power plants (SHPPS) are shown in Table (1). The voltage profile for 33-bus system is shown in Figure (2). The size of (SHP) unit has been taken 2MW.

It is observed that, the losses are found lower with insulation of the small hydropower plants (SHPPS) at lagging power factor rather than (SHPPS) at unity power factor. This is due the reason of reactive power available locally for the loads and thereby decrease in the reactive power available from substation. The (VSI) profile for 33-bus system is shown in Figure (3).

The profile improves with (SHPPS) at unity and lagging power factor as shown in Figures (4) and (5) respectively. Thus, it is essential to consider the reactive power available from (SHPPS) and its impact on losses reduction and voltage profile improvement.

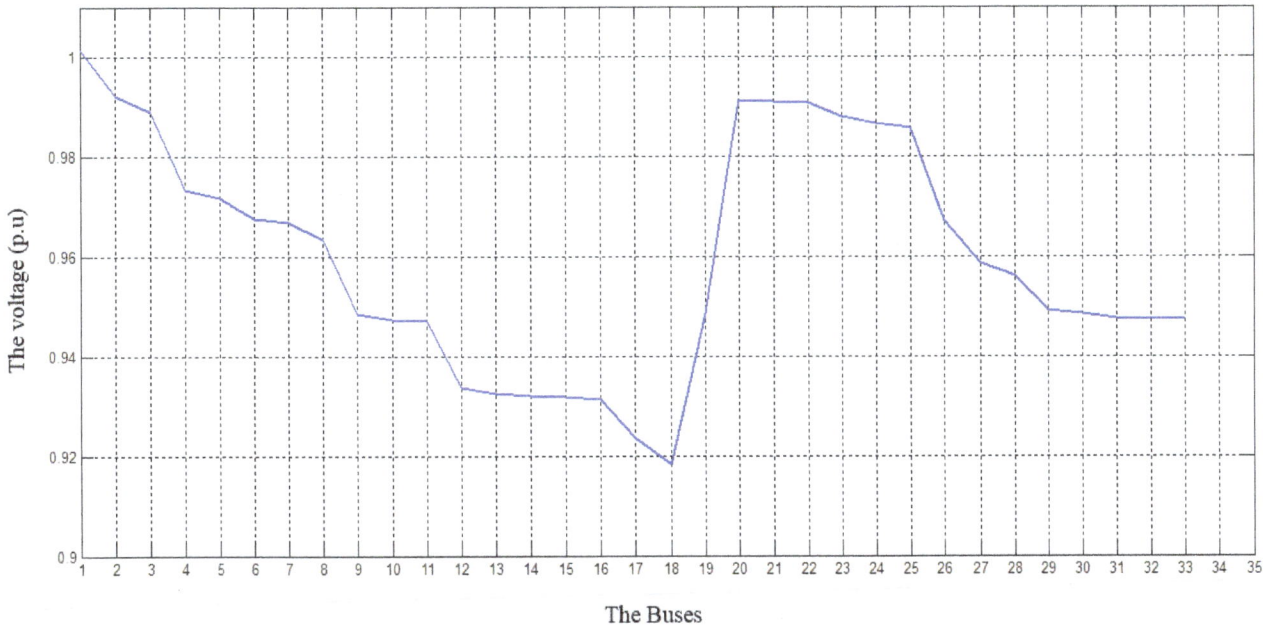

Figure (2). Voltage profile for 33-bus system.

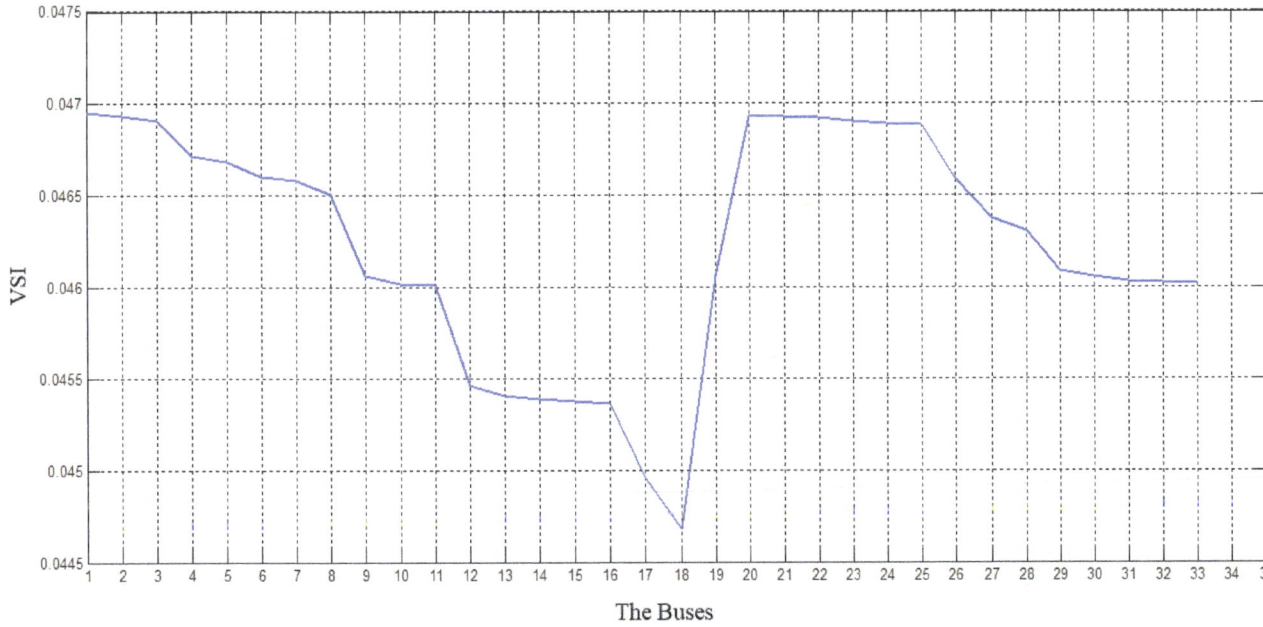

Figure (3). VSI profile for 33-bus system.

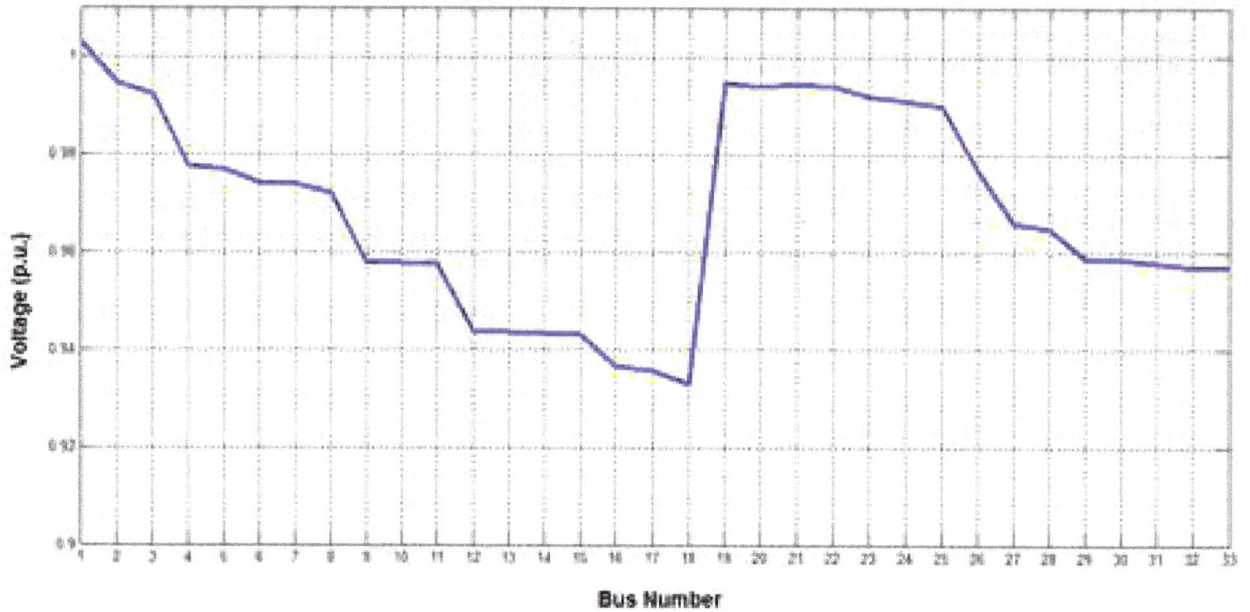

Figure (4). *Voltage profile for 33-bus system unity p.f.*

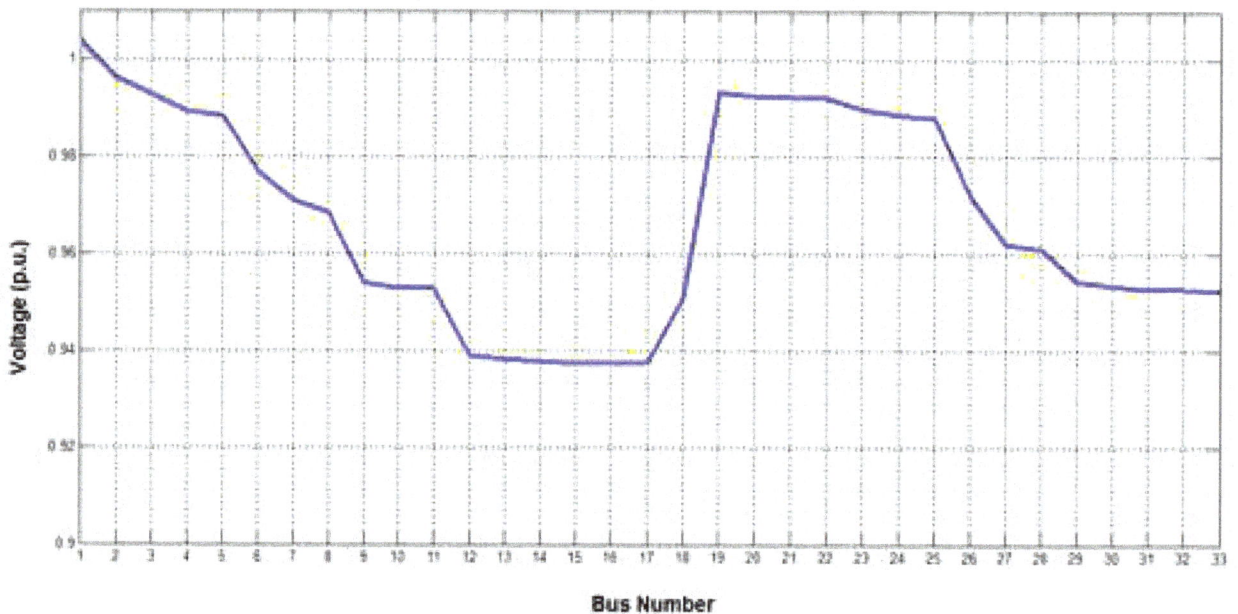

Figure (5). *Voltage profile for 33-bus system at 0.9 p.f lag.*

4. Conclusions

From the results obtained for real and reactive power losses, voltage profile; we can conclude that there is much reduction in real, reactive power losses and improvement in the voltage profile at 0.9 pf lag due to its reactive power supply to the system. Therefore, the proposed study at lagging power factor and supplying reactive power to the system is giving better results than unity power factor. From the results, it can be concluded that by introducing (SHPPS) in the system, voltage profile can be improved because (SHPPS) provide a portion of the real and reactive power to the load locally. The proper reactive power management and voltage profile with (SHPPS)

will help toward the cost reduction due to the lower kVA requirements from the substation.

References

[1] Sreerenjini K, Thomas P C, Anju G Pillai, V I Cherian, Tibin Joseph5 and Sasidharan Sreedharan. Optimal Power Flow Analysis of Kerala Grid System with Distributed Resources. IEEE. 2012 978-1-4673-2636-0/12.

[2] V. Selve, Jiji Johnny," Optimal Allocation of Distributed Generation to Minimize Loss in Distribution System", International Journal of Application or Innovation in Engineering & Management (IJAIEM), vol 4, no 3, pp 1-8, 2013.

[3]   Lettas, N.; Dagoumas, A.; Papagiannis, G.; Dokopoulos, P. A Case Study of the Impacts of Small Hydro Power Plants on the Power Distribution Network with the Combination of On Load Tap Changers, Power Tech, 2005 IEEE Russia, DOI: 10.1109/PTC.2005.4524627.

[4]   T. Niknam and B. Ba. Firouzi. A practical algorithm for distribution state estimation including renewable energy sources. Renewable Energy 34.2009. 2309–2316.S. Panda, "Multi-objective evolutionary algorithm for SSSC-based controller design", *Electr. Power Syst. Res.*, vol.79, no. 6, pp. 937-944, 2009.

[5]   Sa. Mishra, S. K. Singal and D. K. Khatod. Optimal installation of small hydropower plant—A review. Renewable and Sustainable Energy Reviews 15, 2011. 3862–3869.Mohammed Y. Suliman and S. M. Bashi," Instantaneous Active and Reactive Power Measuring in Three Phase Power System", 3rd International Scientific Conference of F.T.E,Najaf,Iraq,20-21 Feb 2013, Page(s): 926-936.

[6]   P. Karimyan, G. B. Gharehpetian, M. Abedi and A. Gavili. Long term scheduling for optimal allocation and sizing of DG unit considering load variations and DG type. Electrical Power and Energy Systems 54. 2014. 277–287

[7]   . Sameer S. Mustafa, Mohammed H. Yasen and Hussein H. Abdullah. Evaluation of Electric Energy Losses in Kirkuk Distribution Electric System Area. Iraq J. Electrical and Electronic Engineering. Vol.7 No.2, 201.

[8]   . M. A. Kashem, D. T. Le, M. Negnevitsky, and G. Ledwich, Distributed Generation for Minimization of Power Losses in Distribution Systems. IEEE. 2006.1-4244-0493.

[9]   S. P. Singh, G. S. Raju, G. K. Rao, M. Afsari. A heuristic method for feeder reconFigureuration and service restoration in distribution networks.Electrical Power and Energy Systems 31 (2009) 309–314

[10]   S. Chandramohan N. Atturulu, R. P. Kumudini Devi, B. Venkatesh. Operating cost minimization of a radial distribution system in a deregulated electricity market through recon

Figureuration using NSGA method. Electrical Power and Energy Systems 32 (2010) 126–132.

[11]   Leonardo W, de Oliveira, S. Carneiro Jr., Edimar J. de Oliveira, J. L. R. Pereira, Ivo C. Silva Jr., Jeferson S. Costa. Optimal recon Figureuration and capacitor allocation in radial distribution systems for energy losses minimization. Electrical Power and Energy Systems. 32, 2010. 840–848.

[12]   C. Lueken, Pedro M.S. Carvalho and Jay Apt. Distribution grid reconFigureuration reduces power losses and helps integrate renewables. Energy Policy 48, 2012, 260–273.

[13]   K. Nadhir, D. Chabane and T. Bouktir. Minimization of active power losses in radial distribution system by optimal location and size of distributed general using the firefy algorithm. Mediamira Science publisher.2013. Vol 54. No.1.40-45.

[14]   Tanuj M. and Y. S. Shishodia. Reduction in Power Losses in Distribution Lines using Bionic Random Search Plant Growth Simulation Algorithm International Journal of Recent Research and Review, Vol. III, September 2012, ISSN 2277 – 8322

[15]   V. V. S. N. Murthy and A. Kumar, Comparison of optimal DG allocation methods in radial distribution systems based on sensitivity approaches. Electrical Power and Energy Systems 53, 2013. 450–467.

[16]   Qian Kejun, Zhou Chengke, allan Malcolm, Y uan Yue. Effect of load models on assessment of energy losses in distribution generation planning. Electr Power Res 2011;2:1243–50

[17]   Qian Kejun, Zhou Chengke, allan Malcolm, Y uan Yue. Effect of load models onassessment of energy losses in distribution generation planning. Electr Power Res 2011; 2:1243–50.

[18]   Gopiya Naik S., D. K. Khatod and M. P. Sharma. Optimal Allocation of Distributed Generation in Distribution System for Loss Reduction. (2012) IPCSIT vol. 28.

[19]   Satish K., B. B. R. Sai, B. Tyagi, V. Kumar. Optimal placement of distributed generation in distribution networks. International Journal of Engineering, Science and Technology Vol. 3, No. 3, 2011, pp. 47-55

30

Modelling the Effects of Climate Change on Hydroelectric Power in Dokan, Iraq

Petter Pilesjo[1], Sameer Sadoon Al-Juboori[2, *]

[1]GIS Centre, College of Science, Lund University, Lund, Sweden
[2]Electronic and Control Engineering Dept., Kirkuk Technical College, Kirkuk, Iraq

Email address:
Petter.Pilesjo@gis.lu.se (P. Pilesjo), sameer.al-juboori@gis.lu.se (S. S. Al-Juboori)

Abstract: Due to shift in the average patterns of weather, climate change became one of the significant development challenges. Hydropower is currently being utilized in more than 150 countries, including 11,000 stations with 27,000 generating units. Increasing attention has been paid to hydropower generation in recent years, because it is renewable energy. Temperature and precipitation effects from global climate change could alter future hydrologic conditions in Iraq and, as a result, future hydropower generation. This is also valid for the Middle East and Iraq. The aim of this study (part1) is to evaluate potential climate change impacts on hydropower in Dokan region, and to recommend various options to maintain optimum required water level to ensure full capacity of electricity generation throughout the year. A simple approach assumes that hydropower systems will reduce generation if water supply reduces, and vice versa. The analysis of the approach was carried out to convert changes in water resource availability to changes in electric hydropower generation. By the year 2050 and based on 12GCMs, electric power generation in Dokan power plant will decrease by 20-40 MW. The other factors such as the site head, the turbine generating capacity and efficiency which were neglected, will be measured, calculated and discussed in part2 of the study.

Keywords: Climate Change, Hydroelectric, Dokan, Dam, Modeling

1. Introduction

One of the great challenges of the 21st century is climate change. Due to shift in the average patterns of weather, climate change and variability are now becoming one of the significant development challenges. A more variable climate is the expected outcome of increases in atmospheric concentrations of greenhouse gases resulting from human activities. Carbon dioxide (CO2) and many greenhouse gases occur naturally and keep the earth warm. Anthropogenic sources of CO2 since the industrial revolution, have added greatly to the atmospheric concentrations. Transportation and fossil fuels burning for electricity generation are the frequently cited as major sources. Increased levels of greenhouse gas concentrations over the next century are predicted to cause a significant rise in temperature which are greater than at any time in the past [1, 2].

The temperature of global average surface has increased by 0.74C°/ century. 1990s was the warmest decade in the 20th century, and1998 was observed as the warmest year. The emission of greenhouse gases is one of the major causes of global warming due to anthropogenic activities. As a result of global warming, climatic variables such as sea level, precipitation and atmospheric moisture, snow cover were observed [3].

2. Future Climate Prediction

The temperatures are assumed to rise by 3°C by the end of the next century, under present rates of economic and population increase. Simulation physical processes in the atmosphere and oceans predictions of future climate are based on the output of complex numerical Global Circulation Models (GCM's). Together with changes in climate patterns that might reduce rainfall in the Middle East, GCM simulations for the region, indicate higher future temperatures that will increase evapotranspiration. [4].

3. Global Distribution of Hydropower

In more than 150 countries around the world hydropower is currently being utilized, including 11,000 stations with 27,000 generating units. Europe has the highest installed capacity around 260GW, followed by Eastern Asia, led by China then South America, led by Brazil. In the past few years, China has commissioned significant hydropower capacity, and now exceeds the US as the country with the highest total installed capacity; see Figure 1 [5, 6].

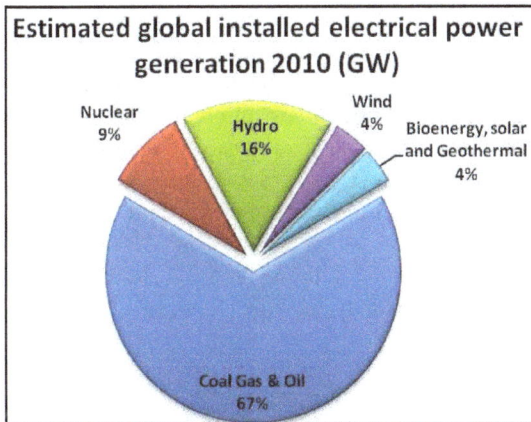

Figure 1. Global Installed Electrical power Capacity 2010 (GW) [6].

To analyze hydropower plants, types, capacities, generation, dams, global water resources, global runoff, rainfall, etc., GIS analysis has been utilized in different countries to understand and visualize regional scenarios of hydropower generation.

4. The Impacts of Climate Change on Water Resources Modeling

A range of sensitivities to climate change were displayed by river basins as observed by many studies. It can be seen that increased temperature results in non-linear variations in runoff due to changes in precipitation.

The runoff-rainfall processes non-linear nature will hit hydropower severe by changes in climate given. A reduction in rainfall by 10% gives a 25 to 50% loss in hydropower generation. As a result of temperature increase of a few degrees, a severe impact of higher evapotranspiration on hydropower might result in a substantial decrease in generated electricity. In general, increases in climate variability will lead to a lower energy security [7].

The main reason to apply climate models is the ability to explore different scenarios. Scenarios can study aspects that cannot directly be predicted, such climate change and population growth.

The potential for hydroelectric generation approximately follows runoff. A more accurate estimate of climate impacts on hydropower would involve assessment of hydro plants cost, the government's and international organizations policies and the economic development of the country.

A simple hydrological model would be able to derive suitable input-response river flow relationships if given reliable climate and river flow data. Using the suitable technical data and operational parameters, the hydrological model will convert input climate data into estimates of river flows. These results would be processed to estimate the electrical power generated [4, 8].

5. Iraq Electricity Sector Overview

In Iraq, electricity is supplied by 9.22%, 80.49% and 10.29% from hydro power, fossil fuel power plants and imported electricity respectively [9].

About 24,400 MW of new capacity will be added due to 2017, including 7, 000 MW of thermal power, 13,000 MW of gas-fired, and 400 MW of renewable energy. According to the country's energy master plan, a further 4,000 MW will be added by the conversion of simple-cycle power plants to combined-cycle technology [13].

Because the country is currently more interested in developing power plants in a short duration to get more energy quickly, oil- and gas-fired plants are currently the preferred choice. Although hydropower development is a part of the long term strategy, there is also more focus on developing large dams for flood protection and irrigation. Wind and solar fields are also desired. The full hydropower potential may be estimated as high as 80,000 GWh annually. Table1 shows the installed capacity of hydro power plants in Iraq [9, 10, and 11].

Table 1. The hydro power plants installed capacity in Iraq, 2012 [9].

No.	Project name	Installed capacity (MW)
1	Dokan Dam (our case study)	400
2	Darbandikhan Dam	240
3	Mosul Main Dam	750
4	Mosul Dam pump storage plant	200
5	Mosul Regulating Dam	60
6	Haditha Dam	660
7	Samaraa Barrage	80
8	Hemrin Dam	50
9	Adhaim Dam	40
10	Al-Hindiyah Barrage	15
11	Shatt Al-Kuffa Regulator	6

6. Dokan Dam in Iraq

The location of the Dokan Dam is shown in Figure2 and 3 below. It is located in the north of Iraq on the Lesser Zab River approximately 295km north of Baghdad and 65km southeast of Sulaimaniyah city. The Dokan Dam reservoir had a total design of 6,870 Million Cubic Meters [12, 13].

Figure 2. *The Dokan Dam is located on the Lesser Zab River [Google Earth].*

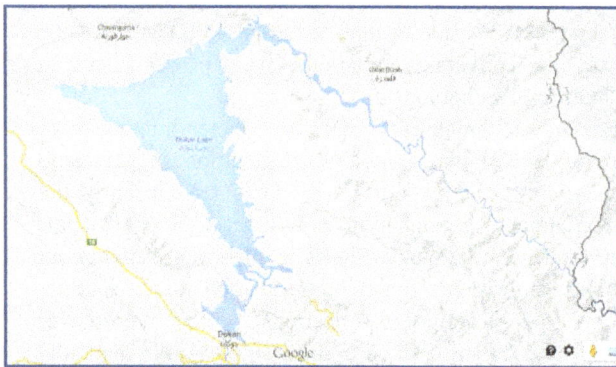

Figure 3. *Dokan Lake [Google Earth].*

7. Assessing Climate Change Impacts on Hydropower Generation Systems

To assess climate change impacts on hydropower generation systems, a simple approach is based on the assumption that the current hydropower generation system may only be limited by water availability and that if water supply reduces, the hydropower systems will reduce generation and vice versa. Hydropower generation with this approach changes in annual mean flows. To simulate observed regional patterns of the twentieth century, multi-decadal changes in stream flow, an ensemble of 12 climate models was used with statistically and significant skill. The hydropower generations by countries as shown in Figure 4, were mapped in a GIS database system [5, 14].

Database management expedites the analysis on various tables that make up the GIS database. Computed future (2050) changes in runoff are based on results from 12 GCMs. The future (year 2050) generation based on the current generation levels is produced by mapping the changes as shown in Figure 5.

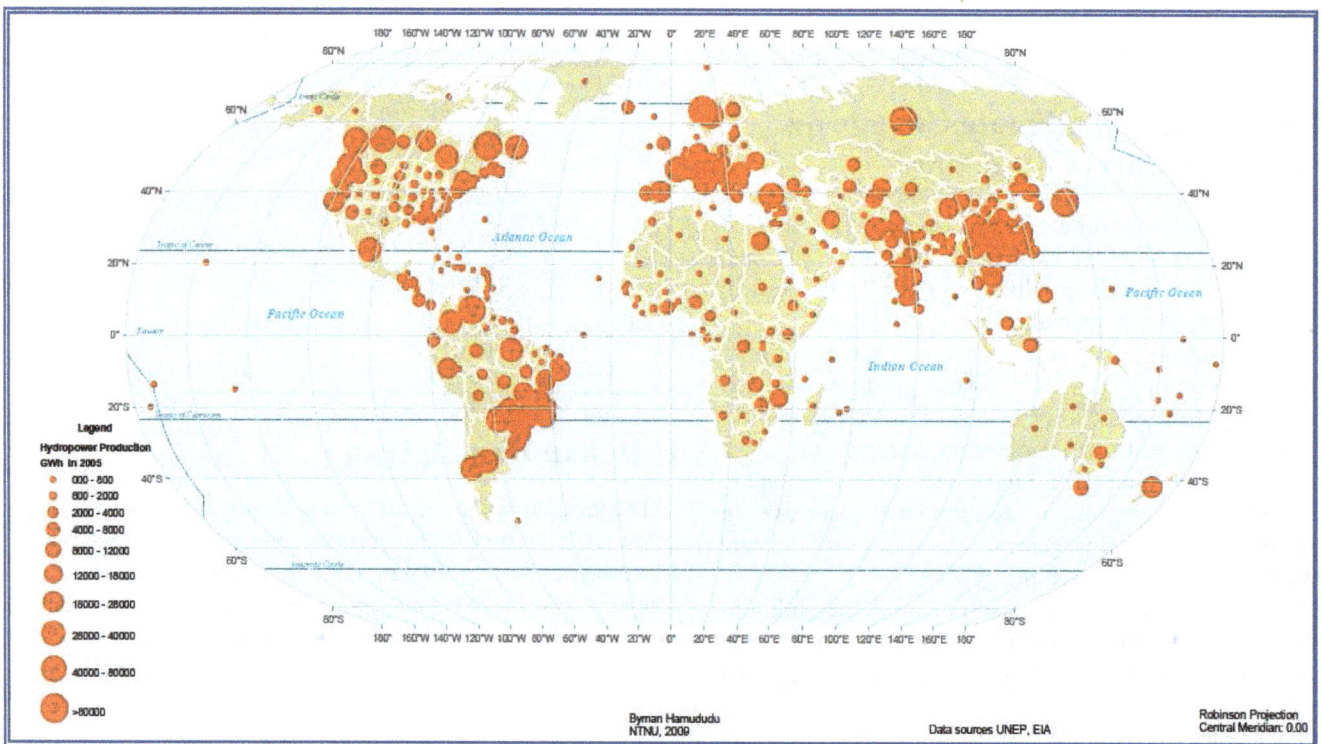

Figure 4. *Hydropower generation (GWh) in 2005[14].*

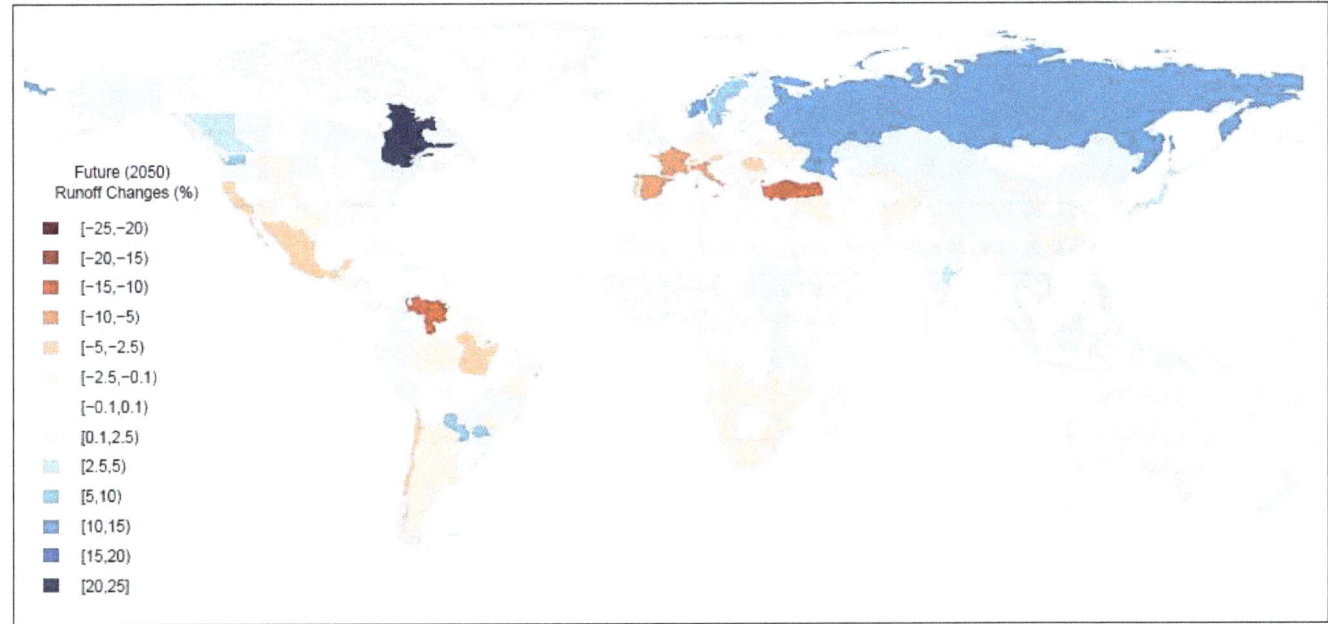

Figure 5. Runoff changes (%) based on 12 GCMs under A1B scenario for future (2050) [5].

Water resource availability changes are converted to changes in hydropower generation. The runoff was assumed to be the main determinant of limitation to hydropower generation. The analysis methodology is based on the fact that hydropower generation is a function of flow (Q, in m^3s-1), Head (H, in m) and efficiencies. Assuming that the changes in water resources will impact hydropower produced in the future, the most varying factor used in the procedure for the water resources for each country is the flow (Q [5].

On average, runoff can be thought of as the difference between the precipitation and evaporation over long periods of time, and this makes it the available water to be used for hydropower, domestic consumption, irrigation etc. as shown in Figure 6.

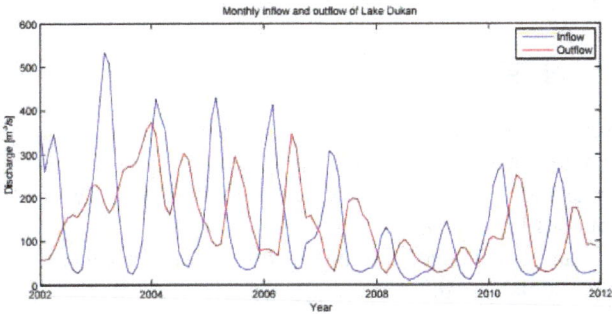

Figure 6. A comparison between Inflow and Outflow of Dokan Lake during (2002 – 2012) [14].

Compared with control period, Figure 7 shows the average annual rainfall for the north of Iraq for A2 and B2 scenarios. [14]

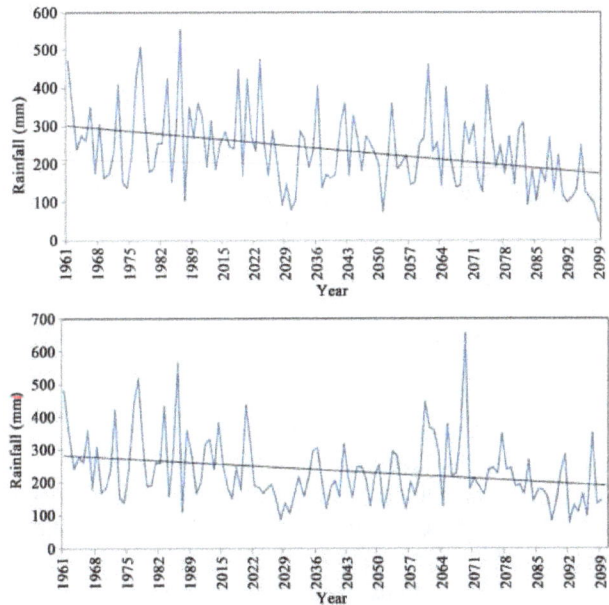

Figure 7. Average annual rainfall in the north of Iraq for A2 scenario (upper) and B2 scenario (lower) compared with control period.

Estimated changes in runoff which are used as predictors in projecting hydropower generation are the bases for country values, GCM estimates. Dokan Dam and Zab River catchment water balance recorded data are shown in Tables2, 3, and4 [13].

Table 2. Rainfall and evaporation water balance in North of Iraq for the period 1970-2011[13].

	Jan.	Feb.	Mar.	Apr.	May	June	Jul.	Aug.	Sep.	Oct.	Nov.	Dec.
Rain, mm.	127.3	125.3	135.9	93.1	26.9					27.67	92.6	134.8
Energy Evaporation transpiration, mm	26.95	31.76	48.60	77.0	111.65	111.65	158.33	153.04	136.19	103.47	64.97	37.53
Surplus of water, mm.	100.3	93.54	87.3	16.1							27.63	97.27
Water deficit, mm.					84.75	111.65	158.33	153.04	136.19	75.8		-

Table 3. Characteristic for the annual discharge of the Little Zab River in Dokan Dam [13].

The year	Properties of the year	Medium Discharge m³/sec	Annual revenue Billion m³	The average height of the water basis mm/y
1958-2001	General	193.29	6.09	812.52
1988	Wet	446	14.06	1896.66
2000	Dried	58	1.82	245.5
1996	medium	201	6.33	853.90

Table 4. Entering and release water rate in the lake (1953-2011). [13].

Month	Monthly rate of import m³/sec.	Monthly released discharge rate m³/sec.
October	49.38	200
November	110.4	175
December	166.9	170
January	200.8	163
February	327.8	176
March	429	209
April	424.3	188
May	260.1	169
June	114.6	162
July	59.54	223
August	40.65	273
September	39.19	245
The annual average (m³/sec.)	184.7	196
The annual average (billion m³)	5.82	6.19

8. Results and Discussion

The overall objective of this study was to evaluate the impacts of climate change on hydropower to present a global picture of impacts.

According to recorded data in Figures 7 and 8, a comparison shows a decrease in inflow and outflow of Dokan Lake during (2002 – 2012) and the average annual rainfall has been reduces in the north of Iraq for both A2 and B2 scenarios.

A lot of simplifications were made. Only the mean discharge flow has been used as a factor to hydropower generation, while the other factors such as the site head and the turbine generating capacity and efficiency were currently neglected and will be measured, calculated, analyzed and discussed in (part2), a new model will be build.

From the analysis, based on 2005 global hydropower generation presented in Figure 5, it can be seen that by year 2050, the hydropower generation have totally decreased by 5-10%. Runoff changes in Iraq based on 12 GCMs under A1B scenario, Figure6 decreases by 5 to 10 % which will have negative effects on energy sector and the future production of electricity in Iraq.

Accordingly hydropower generation reduces by 5 – 10% and the electric power generation in Dokan power plant decreases by 20-40 MW.

To mitigate the effects of climate change on hydropower plants, improvements in the present hydro-power resources for water recycling and/or development of micro-dams for storage of excess water need exploration.

Finally, construction of new plants with better technology (e.g., high efficiencies) in the hydropower sector could help to reduce the gap that may be created by the effects of climate change on electric power generation.

9. Conclusions

Based on this study part1 we conclude that:

1. Some countries will experience decreases in climate potential while others increases, but with a great degree of risks in both cases.
2. It can be concluded that one of the most environmental and social sensitive power generation technologies is hydropower technology.
3. In some countries, careful planning and design are required due to the occurrences of extreme weather events caused by climate change, so as to come up with sustainable hydropower projects.
4. A need for national adaptation strategies to water supply shortages is so important.

References

[1] IPCC: Special Report on Renewable Energy Sources and Climate Change Mitigation, Technical Report, Intergovernmental Panel on Climate Change: Geneva, Belgium, 2011.

[2] Subimal, G.; Chaitali M.: Assessing Hydrological Impacts of Climate Change: Modeling Techniques, and Challenges. The Open Hydrology Journal, 4, 115-121 (2010).

[3] Atsushi, I.: Estimating Global Climate Change Impacts On Hydropower Projects: Applications in India, Srilanka And Vietnam, The World Bank Sustainable Development Network Finance, Economics, and Urban Development Department September 2007.

[4] Harrison, G.P.; Whittington, H.W.; Gundry, S.W.: Climate Change Impacts on Hydroelectric Power. In: Proceedings of 33rd University Power Engineering Conference, Edinburgh, 391-394 (1998).

[5] Byman, H.; Aanund, K.: Assessing Climate Change Impacts on Global Hydropower. Energies. 5, 305-322 (2012).

[6] Manjeet, D.: Climate Change Impacts on Reservoir based Hydropower Generation in Nepal: A case study of Kulekhani Hydropower Plant. M.Sc. thesis submitted to School of Environmental Management and Sustainable Development (SchEMS) Baneshwor, Kathmandu September, 2011.

[7] Advait, G.: Climate change impacts on hydropower and the electricity market: A case study for Switzerland Master's Thesis Faculty of Science University of Bern, 2014.

[8] Peter, D.: Climate Change and Hydropower, Impact and Adaptation Costs: Case Study Kenya. Report available at www.futurewater.nl. 2009.

[9] Indexmundi: Electricity Import to Iraq in 2011. Available at www.indexmundi.com/g/g.aspx?v=83&c=iz&l=en. 2012.

[10] Milly, P.C.D.; Dunne, K.A.; Vecchia, A.V.: Global pattern of trends in stream flow and water availability in a changing climate. Nature. 438, 347–350 (2005).

[11] Ratcliffe, V.: Power Generation a Top Priority in Iraq. Available at www.meed.com/supplements/2012/iraq-projects/power-genera tion-a-top-priority-in-iraq/3129589.article. (2012).

[12] The World Bank: Dokan and Darbandikhan Emergency Project. Inspection Report 1537 (2006).

[13] Iraq Ministry of water resources, Center for the study of water resources projects for the northern region: Evaluation of some irrigation projects in Dukan watershed as controlling and conservation of water resources. Preparing by: Amwag Abbas Talab, Assistant general manager of irrigation and drainage, 2013.

[14] Nadhir, Al-Ansari; Ammar A.; Ali, S.: Present Conditions and Future Challenges of Water Resources Problems in Iraq. J. Water Resour. Prot. 6, 1066-1098 (2014).

31

Assessing the Impact of Load and Renewable Energies' Uncertainty on a Hybrid System

Amin Shokri Gazafroudi

EE Department, Imam Khomeini International University, Qazvin, Iran

Email address:

shokri_amin@ymail.com

Abstract: As increasing of fossil fuels and the trend of expiring, using of alternative fuels has been on the agenda of most countries particularly in the past two decades. In the meantime, the using of wind energy and solar radiation are extremely popular as sources of green energy and high-efficiency. Hence, the prediction of wind and solar power is important. The power output of these power plants depends on wind speed, temperature and radiation. In this paper, the uncertainty of wind and solar power generation, and load forecasting are considered based on correlation analysis on wind power, solar radiation, and ambient temperature time series. Predicted values are given to the hybrid system (wind–fuel cell–photovoltaic) to provide electrical load for the 24-hours. Finally, the proposed model is applied to demonstrate its effectiveness based on actual examples information of load, wind, radiation, temperature of wind farm and solar power plants.

Keywords: Renewable Energies, Neural Network, Correlation Analysis, Forecasting Method, Uncertainty

1. Introduction

Increasing human population has direct effect on electrical energy demand which has increased the electrical power generation [1]. So that in 2007, 26 % of pollution and greenhouse gas emissions are resulted from the production of electricity and heat [2]. Therefore in recent years, electricity generation based on renewable energies, has been highly regarded in different countries [3]. Electrical power generation based on wind energy, has been fastest growing among the renewable energy sources [4]. This is due to its environmental benefits. It is estimated that in 2020, 12% of the world's electricity will supply from wind power [6].So, wind power generation will play an important role in electricity supply. Inability of wind power to follow the load pattern is important point and required to design a hybrid system in are as separated from the main system. In this paper, a hybrid system (wind-fuel cell– photovoltaic) is designed to provide the load in a day. So, the radiation and temperature forecasts which are solar system inputs. They are as important as wind power forecasting.

An accurate forecasting of the electrical load, wind speed, solar radiation and temperature can have positive impacts on the production scheduling. In this paper, short-term prediction of load, wind speed, solar radiation and ambient temperature are applied by feed forward neural network according to their historical time series. Finally, the forecasted data are given in to a hybrid system. The proposed model is shown in Fig.1.

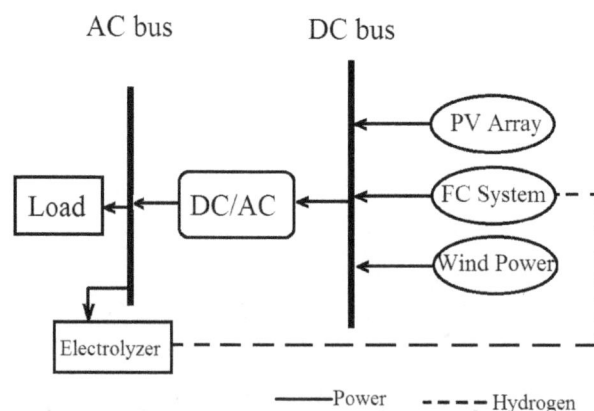

Fig. 1. System Schema.

Different sections of this paper are divided as follows. In Section 2, load model is presented. In Section 3, radiation, temperature and wind speed are forecasted. In section 4, designing and simulating the Solar panels, fuel cell and wind power plants are presented. Finally, the conclusion drawn from proposed hybrid system is provided in section 6.

2. Load Model

The 24 hour load curve is shown in Fig. 2. According to he load curve, load has two peaks. One of them is in the middle of day and related to add cooling device. The second one is at early evening and related to add lighting. The actual oad curve in comparison with predicted load curve in 24-hour and a week period are shown in Fig. 2 and Fig. 3.

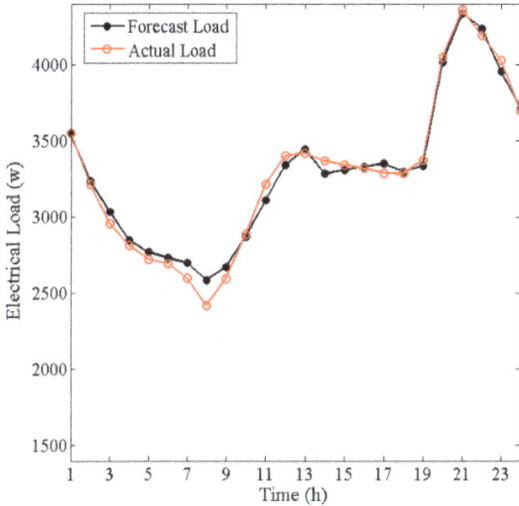

Fig. 2. *24-hour load forecast.*

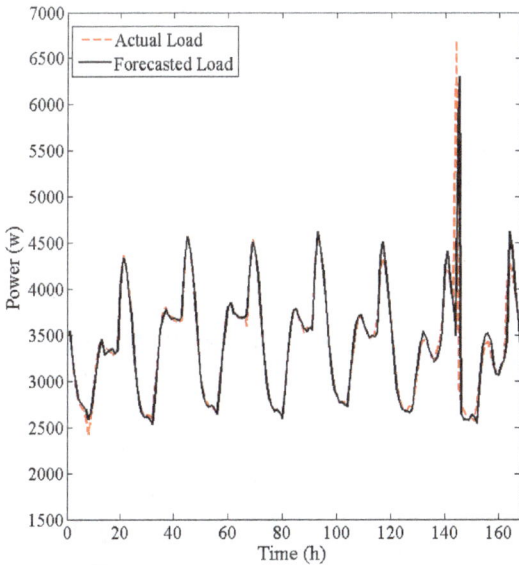

Fig. 3. *A week load forecast.*

3. Wind Speed, Radiation and Temperature Forecast

In this section, artificial neural networks and correlation analysis are presented as methods for cognitive behavior of wind speed, radiation and temperature data, and the predicted data are resulted by a forecast engine based on artificial neural network. Wind speed, radiation and temperature time series are shown respectively in Fig. 4, Fig. 5 and Fig. 6.

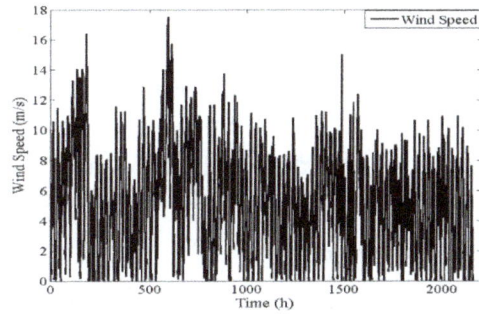

Fig. 4. *Wind speed time series.*

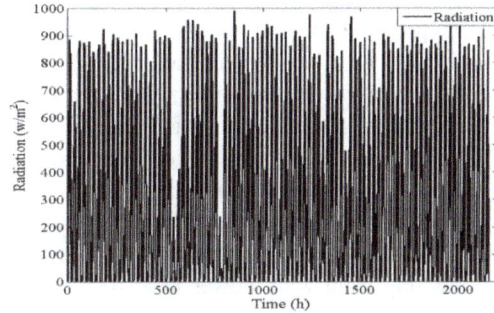

Fig. 5. *Radiation time series.*

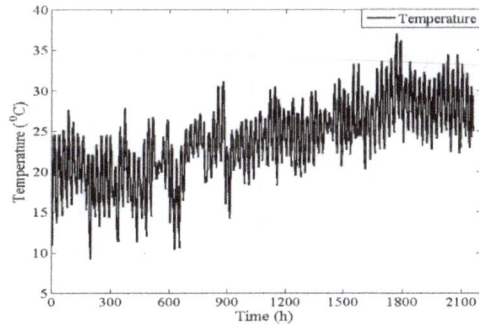

Fig. 6. *Temperature time series.*

3.1. Forecasting Method

Neural network is the nonlinear technique that can be applied to detect patterns and find the complex behavior of a system through its input and output data. A neural network (NN) needs learning to find the complex behaviors of data. Despite the NN is simple, but it is a powerful and flexible tool to forecast. NN have an ideal output by choosing appropriate data number for training, inputs, hidden layers and neurons in each layer. NN can approximate any nonlinear function [6]. Thus, depending on how much the system inputs are appropriate; we will expect a better response of system outputs.

Before modeling a system, it is necessary to find the relation between the measured parameters. The correlation coefficient matrix is a normalized value that expresses the strength of linear relationship between two variables. Correlation coefficient $r_{X,Y}$ between two random variables X and Y, with mean values μ_X and μ_Y and standard deviation σ_X and σ_Y, expresses covariance between normalized variables.

The correlation coefficient obtained from the following equation:

$$r_{X,Y} = \frac{cov(X,Y)}{\sigma_X \sigma_Y} = \frac{E((X-\mu_X)(Y-\mu_Y))}{\sigma_X \sigma_Y} \qquad (1)$$

The appropriate inputs of the neural network are selected by taking the minimal value of the correlation coefficient.

3.2. Wind Speed Forecasting

Wind power is proportional to air density, wind speed and environmental obstacles [7][8]. Air density is an exponential function of air temperature and pressure. So wind power is dependent on weather conditions such as wind speed, wind direction and temperature. Also, the wind power has great relationship with its past values in any moment [4][5].Wind power is a non-linear function of wind parameters such as wind speed, air density and turbulent kinetic energy [8]. Wind power forecasting methods can be divided into physical models and time series or statistical models [13][14].

In physical model, wind speed is estimated in are going under the physical laws governing atmosphere, its behavior and the corresponding from wind power in a specified level of wind farm [15]. These models consider the physical characteristics of the environment, orography, atmospheric conditions (such as pressure and temperature) to estimate the future wind speed and wind power [16].

In statistical model tries to predict the future wind's speed and power by finding the relationship between the parameters of the historical data[15]. In this paper, the statistical model is applied to forecast the short-term wind speed. Wind speed autocorrelation curve is shown in Fig. 7 and the correlation between the wind speed and radiation is shown in Fig. 8.

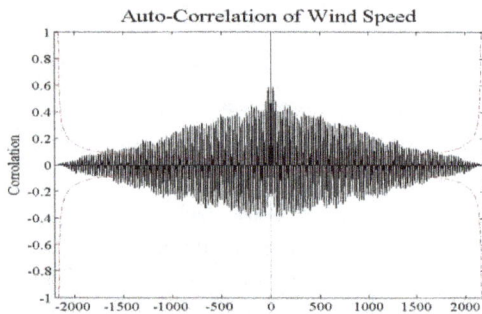

Fig. 7. Wind speed autocorrelation curve.

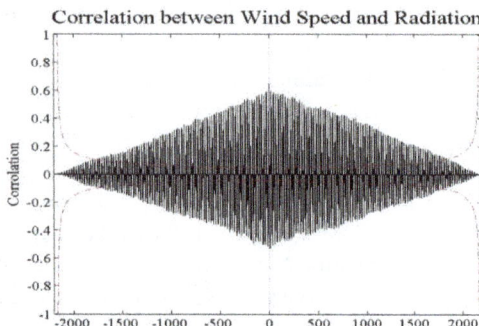

Fig. 8. Correlation between the wind speed and radiation.

The selected appropriate inputs related to wind speed depending on the amount of their cross-correlation which presented in Tables 1.

Table 1. *Selected inputs for* $W(t)$.

Rank	Selected Inputs	Cross-Correlation	Rank	Selected Inputs	Cross-Correlation
1	$W(t-1)$	0.847	3	$S(t-1)$	0.637
2	$W(t-2)$	0.718	4	$S(t-2)$	0.615

In Table 1, $W(t)$ represents the wind speed and $S(t)$ represents the solar radiation. As indicated in Table 1, wind speed at any moment has the highest correlation with wind speed in an hour ago, wind speed in two hours ago, solar radiation in an hour ago and solar radiation in two hours ago. Despite the correlation analysis and find the correlation between wind speed and temperature, the temperature isn't found as an input in Table 1. This is due to the highest correlation between wind speed and temperature is 0.336, which is not the useful input for wind speed forecast. The actual wind speed curve in comparison with predicted wind speed curve in 24-hour, and a week period are shown in Fig.9 and Fig. 10.

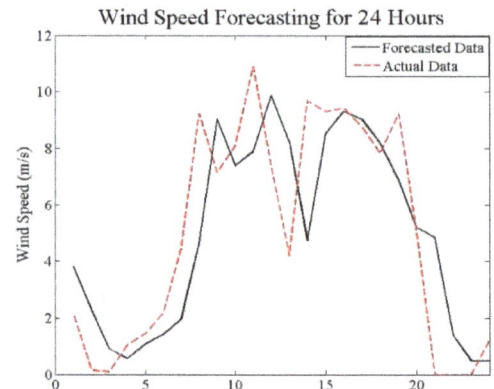

Fig. 9. 24-hour wind speed forecast.

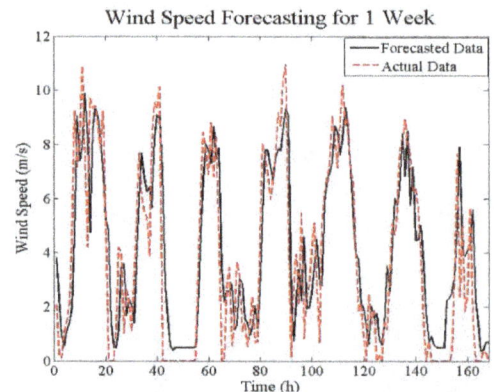

Fig. 10. A week wind speed forecast.

3.3. Solar Radiation Forecasting

The appropriate inputs for solar radiation forecasting are selected by correlation analysis as wind speed prediction. Those data are selected as inputs for solar radiation forecasting engine which their cross-correlation are more than 0.85.The selected appropriate inputs related to solar radiation depending on the amount of their cross-correlation are presented in Tables 2.

Table 2. Selected inputs for $S(t)$.

Rank	Selected Inputs	Cross-Correlation	Rank	Selected Inputs	Cross-Correlation
1	S(t-1)	0.909	3	S(t-48)	0.864
2	S(t-24)	0.882	4	S(t-72)	0.852

As indicated in Table 2, solar radiation at any moment has the highest correlation with solar radiation in an hour ago, solar radiation in 24 hours ago, solar radiation in 48 hours ago and solar radiation in 72 hours ago. Periodic behavior of radiation is clear from the Fig. 12.

Despite the correlation analysis and find the correlation between solar radiation and temperature, the temperature isn't found as an input in the Table 2. This is due to the highest correlation between solar radiation and temperature is 0.426, which is not the useful input for solar radiation forecast. The actual solar radiation curve in comparison with predicted solar radiation curve in 24-hour and a week period are shown in Fig. 11 and Fig. 12.

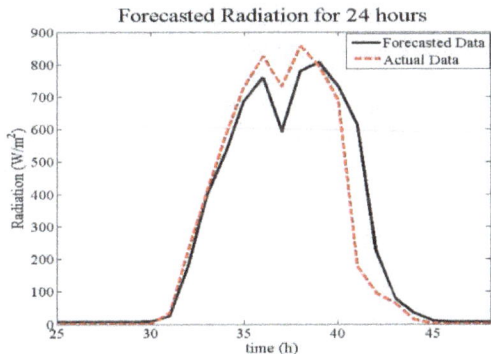

Fig. 11. 24-hour solar radiation forecast.

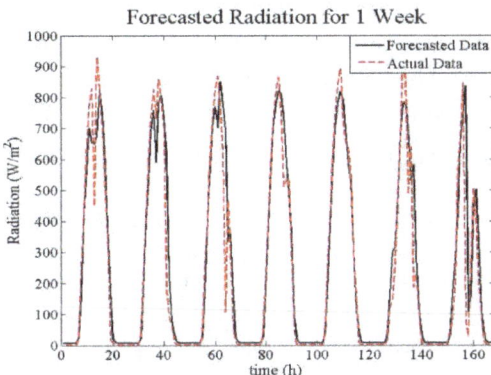

Fig. 12. A week solar radiation forecast.

3.4. Temperature Forecasting

The appropriate inputs for temperature forecasting are selected by correlation analysis. Those data are selected as inputs for temperature forecasting engine which their cross-correlation are more than 0.85. The selected appropriate inputs related to temperature depending on the amount of

their cross-correlation are presented in Tables 3.

Table 3. Selected inputs for $T(t)$.

Rank	Selected Inputs	Cross-Correlation	Rank	Selected Inputs	Cross-Correlation
1	T(t-1)	0.972	3	T(t-23)	0.858
2	T(t-2)	0.914	4	T(t-24)	0.867

In Table 3, $T(t)$ represents the temperature. As indicated in Table3, temperature at any moment has the highest correlation with temperature in an hour ago, temperature in 2 hours ago, temperature in 23 hours ago and temperature in 24 hours ago. This correlation indicates that the temperature has a periodic behavior such as solar radiation. The actual temperature curve in comparison with predicted temperature curve in 24-hour and a week period are shown in Fig. 13 and Fig. 14.

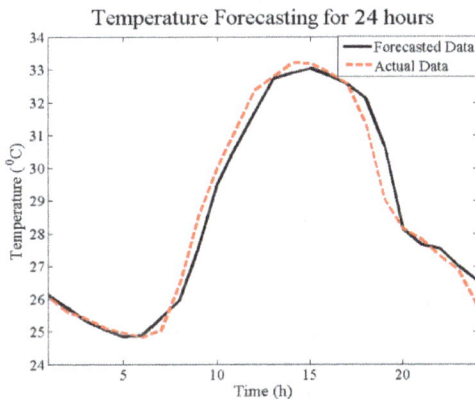

Fig. 13. 24-hour temperature forecast.

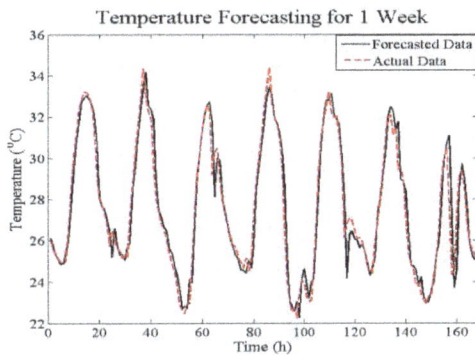

Fig. 14. A week temperature forecast.

4. Designing and Simulating the Hybrid System

4.1. Wind Power Plants Simulation

Analysis of electric energy and the reliability benefits needed to accurately simulation of winds peed for each specific site. Wind speed time series can be modeled with many distributions such as Weibull and normal distributions. Assuming a normal distribution is used to describe the wind speed data particular site, the normal distribution density function is:

$$f(s) = \frac{1}{\sqrt{2\pi}\sigma} \exp\left[-\frac{(s-\mu)^2}{2\sigma^2}\right] (-\infty \le s \le +\infty) \qquad (2)$$

Wind (m/s)	P (kW)
0	0.00
1	0.00
2	0.00
3	0.14
4	0.34
5	0.67
6	1.16
7	1.81
8	2.71
9	3.82
10	5.00
11	5.70
12	6.00
13	6.00
14	6.00
15	6.00
16	6.00
18	6.00
20	6.00

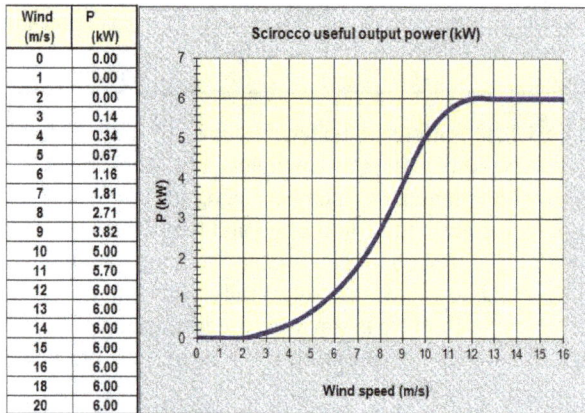

Fig. 15. Power curve of a wind turbine.

Wind speed can be approximated by the following equation [17][19]:

$$p(s) = \begin{cases} p_r\left(A + B\times S + C\times S^2\right) & v_{ci} \le s \le v_{cr} \\ p_r v_{cr} & \le s \le v_{co} \\ 0 \, otherwise \end{cases} \qquad (2)$$

Where p_r is rated power capacity, v_{ci} is cut-in speed and v_{co} is cut-out speed. The parameters A, B and C are given by Eq. (4):

$$A = \frac{1}{(v_{ci}-v_r)}\left[v_{ci}(v_{ci}+v_r) - 4(v_{ci}\times v_r)\left(\frac{v_{ci}+v_r}{2v_r}\right)^3\right]$$

$$B = \frac{1}{(v_{ci}-v_r)^2}\left[4(v_{ci}+v_r)\left(\frac{v_{ci}+v_r}{2v_r}\right)^3 - 3(v_{ci}\times v_r)\right] \qquad (3)$$

$$C = \frac{1}{(v_{ci}-v_r)^2}\left[2 - 4\left(\frac{v_{ci}+v_r}{2v_r}\right)^3\right]$$

In section 3-2, we forecast wind speed for a 24-hour period. According to Eq. (3) and parameters A, B and C, forecasted wind power and actual wind power are shown in Fig. 16.

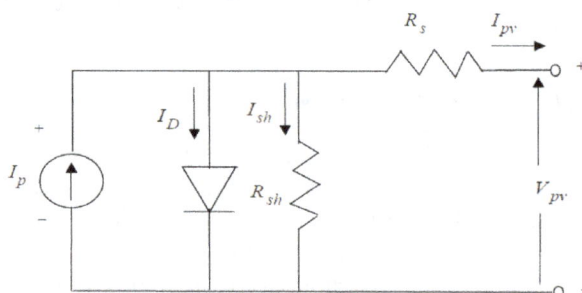

Fig. 16. Electrical model of the PV array.

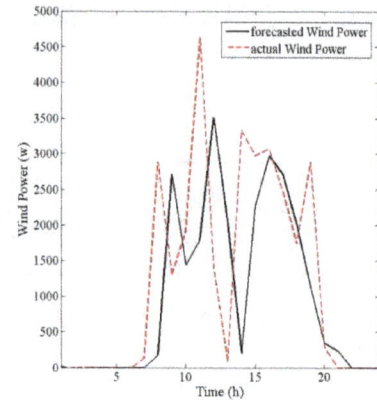

Fig. 17. 24-hour power output of wind power plant.

As shown in Fig. 16, when the wind speed is lower than v_{ci} and upper than v_{co}, output of wind power plant is zero.

4.2. Solar Arrays Simulation

PV arrays can be simulated by an electrical circuit (Fig.17), which includes a current source and a diode. Current power supply circuit is representative of solar radiation in this circuit [19].

Current is fixed to a value of voltage array, after that with voltage increasing, the diode opened and the output current is reduced drastically [20][21]. Output voltage and current are obtained from the reference [22] equations. Simulation is performed by considering the values which mentioned in [22] and [23], the MA 36/45 solar arrays results is mentioned finally according to [23].

Table 4. Electrical information for a MA36/45 PV array [22].

Maximum Power	45 W
No Load Voltage	205 volt
Voltage in Maximum Power	16.7 volt
Short Circuit Current	2.96 volt
Current in Maximum Power	2.74 volt

According to the system load, series and parallel models of solar panels are shown in Fig. 18. Forecasted solar power and actual solar power are shown in Fig. 19.

4.3. Fuel Cells Simulation

As the clean energy conversion technology, fuel cell is the promising alternative to electrical power generation [24]. In this paper, PEM fuel cell is simulated because of the quick Startup and high efficiency and power-voltage characteristic has been evaluated based on [24] simulations.

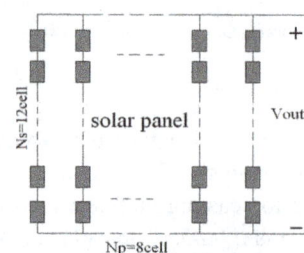

Fig. 18. Series and parallel models of solar panels.

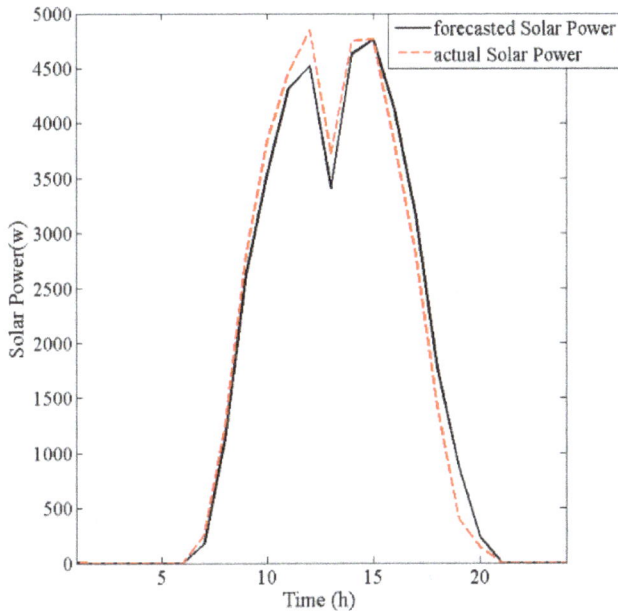

Fig. 19. 24-hour power output of solar power plant.

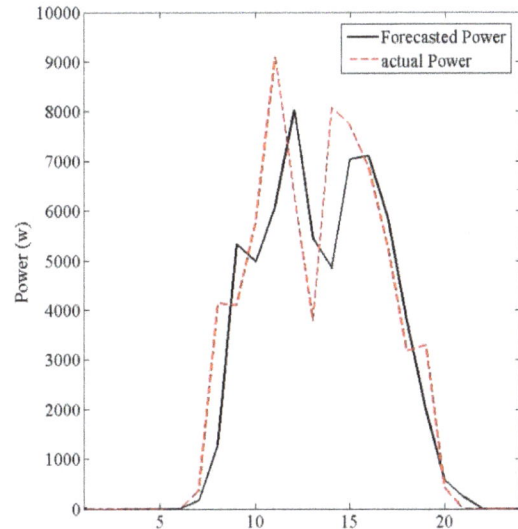

Fig. 20. 24-hour total power generation by wind and solar.

4.4. Simulation Results

In this section, the impact of uncertainty of load forecasting, wind, radiation and temperature are investigated on system. Hence, the system will be discussed in the four cases:

A. Load, wind speed, radiation and temperature are actuality.

B. Load, wind speed, radiation and temperature are forecasted.

C. Wind, radiation and temperature are actuality and load is forecasted.

D. Wind, radiation and temperature are forecasted and load is actuality.

As shown in Fig. 21, power used in the fuel cell is not significantly different between the four cases and it is because of accurate load forecasting. Note that the wind and radiation don't work in this situation. So the uncertainty in load forecast has little impact on the system.

Electrolyzer power curves are shown in Fig. 22. Due to suitable gust of wind and high solar radiation during the day, system power generation exceeds the load and the extra Power generation spend to product hydrogen and stores it in the hydrogen tank. As shown in Fig. 22, uncertainties have a large impact on the electrolyzer power. Since, electrolyzer power affected by wind power and wind forecast has high uncertainty. Therefore, cases B and D are very different. Cases A and C are so similar and it's also due to involved the uncertainties of wind speed and radiation forecasts in system.

The amount of hydrogen stored in a tank is shown in Fig. 23. Case C and A are similar. Cases B and D are very different, due to wind uncertainty. Whatever the wind uncertainty is greater, the incorrect scheduling and consequently more fuel consumption and costs will suffer.

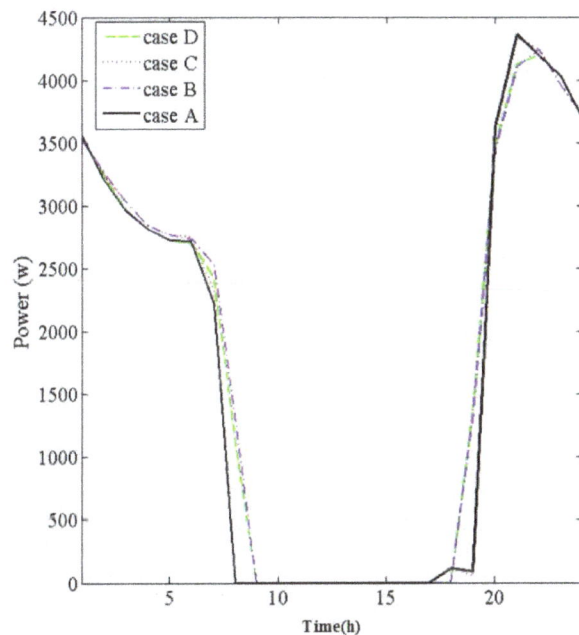

Fig. 21. 24-hour fuel cell output power.

5. Conclusion

With hybrid industrial development in the field of renewable energies, using hybrid energies (wind-fuel cell–photovoltaic) in remote or off-grid areas is deemed important and profitable. Hence, the accurate predictions of wind speed, radiation and temperature are important. We can do the accurate scheduling of generation, fuel and cost by considering the exact short-term and long-term of load, wind, radiation and temperature forecasting.

In this paper, a hybrid system (Wind-FC-PV) is simulated to supply the variable load, which wind speed, radiation and load are system inputs. According to the forecasting results and hybrid systems outputs, wind speed uncertainty is more than the radiation and temperature uncertainties. Thus, wind forecast plays an important role in hybrid systems.

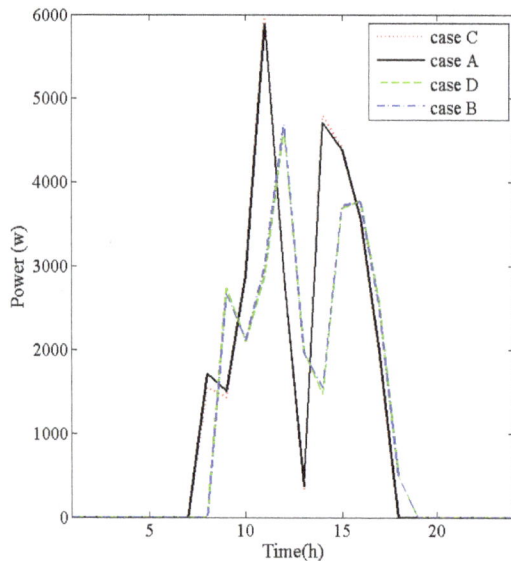

Fig. 22. *24-hour electrolyzer power.*

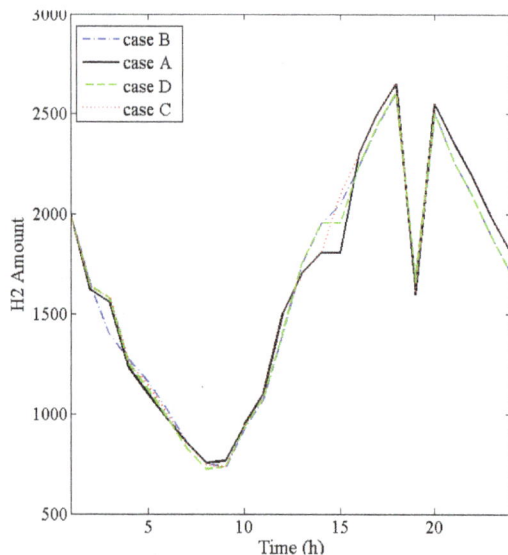

Fig. 23. *24-hour hydrogen tank.*

References

[1] Johnson PL, Negnevitsky M, Muttaqi KM. "Short term wind power forecasting using adaptive neuro-fuzzy inference systems", Australasian Universities Power Engineering Conference, AUPEC 2007, Dec. 2007, pp. 1–6.

[2] A report prepared by European Environment Agency (EEA). Greenhouse gas emission trends and projections in Europe 2009 tracking progress towards Kyoto targets, Report No. 9/2009. Available: [online] www.eea.europa.eu

[3] Ortega-Vazquez MA, Kirschen DS. "Estimating the spinning reserve requirements in systems with significant wind power generation penetration". IEEE Transactions on Power Systems 2009; 24(1):114–124.

[4] N. Amjady, F. Keynia, and H. R. Zareipour. "A new hybrid iterative method for short-term wind speed forecasting". Euro. Trans. Elect. Power, DOI: 10.1002/etep.463.

[5] N. Bigdeli, K. Afshar, Amin Shokri Gazafroudi, M. Yousefi Ramandi, "A comparative study of optimal hybrid methods for wind power prediction in wind farm of Alberta, Canada", Renewable and Sustainable Energy Reviews, Volume 27, November 2013, Pages 20-29.

[6] S. Fan, J. R. Liao, R. Yokoyama, L. Chen, and W.-J. Lee, "Forecasting the wind generation using a two-stage network based on meteorological information", IEEE Trans. Energy Convers., vol. 24, no. 2, pp. 474482, Jun. 2009.

[7] Y. K. Wu and J. S. Hong, "A literature review of wind forecasting technology in the world", in Proc. IEEE Power Tech. Conf., Jul. 2007, pp. 504509.

[8] F. Ö. Thordarson, H. Madsen, H. A. Nielsen, and P. Pinson," Conditional weighted combination of wind power forecasts", Wind Energy, vol. 13, no. 8, pp. 751763, Nov. 2010.

[9] G. Sideratos and N. Hatziargyriou, "Using radial basis neural networks to estimate wind power production", in IEEE Power Engineering Society General Meeting, 2007, DOI: 10.1109/PES.2007.385812.

[10] G. Giebel, The State-Of-The-Art in Short-Term Prediction of Wind Power a Literature Overview 2003 [Online]. Available: http://anemos.cma.fr/download/ANEMOS_D1.1_StateOfThe Art_v1.1.pdf

[11] I. J. Ramirez-Rosado, L. A. Fernandez-Jimenez, C. Monteiro, J. Sousa, and R. Bessa, "Comparison of two new short-term wind-power forecasting systems", Renewable Energy, vol. 34, no. 7, pp. 18481854, Jul. 2009.

[12] M. Lei, L. Shiyan, J. Chuanwen, L. Hongling, and Z. Yan, "A review on the forecasting of wind speed and generated power", Renewable Sustain. Energy Rev., vol. 13, no. 4, pp. 915920, May 2009.

[13] L. Ma, S. Y. Luan, C. W. Jiang, H. L. Liu, and Y. Zhang, "A review on the forecasting of wind speed and generated power," Renew. Sust. Energy Rev., vol. 13, no. 4, pp. 915–920, May 2009.

[14] R. J. Bessa, V. Miranda, and J. Gama, "Entropy and correntropy against minimum square error in offline and online three-day ahead wind power forecasting", IEEE Trans. Power Syst., vol. 24, no. 4, pp. 16571666, Nov. 2009.

[15] Methaprayoon K, Yingvivatanapong C, Lee WJ, Liao JR. "An integration of ANN wind power estimation into unit commitment considering the forecasting uncertainty". IEEE Transactions on Industry Applications 2007; 43(6):1441–1448.

[16] www.solacity.com/Scirocco.htm.

[17] Billinton R, Chowdhury AA. "Incorporation of wind energy conversion systems in conventional generating capacity adequacy assessment" IEE Proc. Gener. Transm. Distrib. 1992; 139:47e56.

[18] Giorsetto P, Utsurogi KF. "Development of a new procedure for reliability modeling of wind turbine generators" IEEE Trans. Power App. Syst.1983; 102:134.

[19] M. Farhat and L. Sbita, "Advanced fuzzy MPPT control algorithm for photovoltaic systems", science Academy Transactions on Renewable Energy Systems Engineering and Technology , vol. 1, no. 1, pp. 29-36, March 2011.

[20] N. Hidouri, L.sbita, "Water Photovoltaic Pumping System Based on DTC SPMSM Drives," Journal of Electric Engineering: Theory and Application, Vol.1, no. 2, pp. 111-119, 2010.

[21] M.S. Benghanem, Saleh N. Alamri, "Modeling of photovoltaic module and experimental determination of serial resistance," Journal of Taibah University for Science (JTUSCI), Vol. 1, pp. 94-105, 2009.

[22] PhD thesis, YUN TIAM TAN FEBRUARY 2004. "IMPACT ON THE POWER SYSTEM WITH A LARGE PENETRATION OF PHOTOVOLTAIC GENERATION". PhD Thesis, Department of Electrical Engineering and Electronics UMIST.

[23] Hassan Moghbelli, Robert Vartanian "Implementation of the Movable Photovoltaic Array to Increase Output Power of the Solar Cells".

[24] Mohammad Sarvi, Mohammad Mehdi Kazeminasab, Masoud Safari Shal "A New Fuzzy Control Method for Maximum Power Point Tracking of PEMFC's System " The 2nd Conference on Hydrogen and Fuel Cell K. N. Toosi University of Technology, 2012.

32

Rotating Machines Based Islanding Detection Using Fuzzy Logic Method Analysis

Lucas Ongondo Mogaka[1], D. K. Murage[2], Michael Juma Saulo[1]

[1]Electrical and Electronics Department, Technical University of Mombasa, Mombasa, Kenya
[2]Electrical and Electronics Department, JKUAT, Nairobi, Kenya

Email address:

mogaka.Lucas@gmail.com (L. O. Mogaka), dkmurage25@yahoo.com (D. K. Murage), michaelsaulo@yahoo.com (M. J. Saulo)

Abstract: An electric rotating machine can be defined as any form of apparatus which has a rotating member and generates, converts, transforms, or modifies electric power, such as a motor, generator, or synchronous generator. Although there are many variations, the two basic rotating machine types are synchronous and induction machines. The recent increasing use of rotating machines among other distributed generators is due to a number of advantages including peak shaving, improvement of the quality of power and reliability, power efficiency, environmental friendliness among others. Despite the above mentioned benefits of distributed power generation in the power grid, they have one major drawback, unintentional islanding. If this islanding condition is not detected in time or goes undetected, the distributed generator loses synchronism with the rest of the utility supply. This may lead to out of phase reconnection of the two systems and thus destroying the distributed generators and even lead to a total blackout in the power system. Again, upon the occurrence of an island, rotating machine based generators have another possible consequence of self-excitation. There is therefore need of fast detection of islanding condition especially when rotating machine based generators are integrated into the main power grid. There are many islanding detection methods and each has its merits and demerits. Their usage depends on certain factors including type of distributed generation in consideration and cost of implementation. Furthermore, the rotating machine based generators have the capability of sustaining an island. This makes the islanding detection and protection of these generators a bit challenging when compared with inverter based generators. This paper presents a passive islanding detection method, fuzzy logic algorithm, particularly on rotating machine based generators and its results analyzed under different conditions. After this analysis, it is concluded that the proposed method for islanding detection for rotating machine based generators is robust and accurate when implemented in the distribution network. This is because fuzzy logic control helps to improve the interpretability of knowledge-based classifiers through its semantics that provide insight in the classifier structure and decision-making process over crisp classifiers.

Keywords: Fuzzy Logic Control, Rotating Machines, Islanding Detection

1. Introduction

The increase of distributed resources in the electric utility systems is being witnessed currently due to recent and ongoing technological, social, economical and environmental aspects [1]. This trend has significantly brought many psitive and negative effects to both the power system infrastructure and the electric power market at large.

The Distributed Generation (DG) technologies can be broadly and generally categorized into inverter-based or rotating machine-based generation, that is both synchronous and induction machine based generators [2]. There is a variety of rotating machine-based DG technologies in use currently.

These include and not limited to; Small hydraulic units, with and without governor, driving synchronous generators with automatic voltage regulators, diesel units with governors and voltage regulators, wind turbines connected to the system through directly coupled induction generators, and Wind turbines connected to the system through doubly-fed induction generators [3].

The integretion of the distributed generation to the power system has a number of advantages which favours its use. These include peak saving when most loads are connected into the system, saving on the costs of upgrading the transmission and distribution part of the grid, environmental friendliness of most DGs among other merits.

However, despite these merits and the continued use of DGs in the power grid, it has some drawbacks. These include; posing of health and safety hazards to the maintainance personnel, it creates and compromises power quality and associated problems for customers load, out of phase re-connection of the recloser switches and most importantly unintentional islanding. Islanding condition occurs when the DG continues to power a part of the grid system even after the connection to the rest of the system has been lost, either intentionally or unintentionally. Thus, this calls for accurate and efficient methods of detecting the islanding condition for the said DG to operate within required conditions.

As per IEEE standard 1547-2003, the distributed generators must sense any unplanned power grid formed and trip it within two seconds, failure to which may lead to several problems in terms of power quality, safety and operational problems [11]. To curb this, there are a number of islanding detection methods being used currently and each has its advantages and disadvantages. Hence, there are some parameters oftenly used to check the suitability of any given islanding detection method. These include; reliability of the method used, its impact on the power grid, operation time and cost effectiveness to both the DG owner and the power utility companies [2]

There are a number of islanding detection methods being used currently and each has its advantages and disadvantages. Hence, there are some parameters oftenly used to chech the suitability of any given islanding detection method. These include; reliability of the method used, its impact on the power grid, operation time and cost effectiveness to both the DG owner and the power utility companies [2].

2. Islanding Detection Methods Classification

Islanding detection methods can be broadly classified into local and remote islanding detection methods. The remote methods include power line communication (PLC) and supervisory control and data acquisition (SCADA) methods. These methods do not have Non-Detection Zones (NDZ) and are more reliable in their usage compared with the local methods. However, the major weakness of these methods is that they are more expensive to implement thus making them uneconomical for small systems. The NDZs are common problems in the islanding detection in power systems and can be defined as a loading condition for which an islanding detection method would fail to operate in a timely manner [4].

The local islanding detection methods are further categorized into active and passive islanding detection methods. Passive techniques are based on measurement of the information at the local site and comparing it with the preset value in determining the occurrence of an island, such as under or over frequency, under or over voltage, voltage phase jump, voltage unbalanced, total harmonic distortion, rate of change of frequency, vector surge, phase displacement monitoring, rate of change of generator power output, comparison of rate of change of frequency [1] among others.

These methods include; the rate-of-change of power signal, the rate-of-change of voltage and change in power factor, the vector surge technique, the rate-of-change of frequency, the phase-shift method, the harmonic impedance estimation technique [5]. Some of the merits of these islanding detection methods include; they are cheap and easy to implement and have no effect on the system when used. It also works pretty well when the mismatch between the generated power and the size of the load is very large.

The parameters typically used to detect islanding conditions in the passive islanding detection method are frequency and voltage. The first discussed passive method is the Over/under voltage and over/ under frequency (OUV/OUF), which is one of the oldest used passive anti-islanding detection technique. These techniques basically monitor the systems voltage and frequency in order to decide whether or not an islanding has taken place [10]. The under/over voltage (UVP/OVP)and under/overfrequency (UFP/OFP) is the oldest technique adapted to protection the distribution system. The protection relays for this technique are placed on a distribution feeder to determine the various types of abnormal conditions. UVP/OVP and UFP/OFP are used to monitor of the grid voltage/frequency exist the limits im posed by the relevant standards [4] Thresholds for UOV and UOF.

On the other hand, active islanding detection methods have very small non-detection zones compared to passive methods. However, they introduce external signals into the power system thus lowering the power quality. According to these methods, some signals are added into the system which aid in detecting islanding condition. These signals have no effect when the distributed generator is in parallel operation with the mains; but they are quickily detected in case of loss of the main grid. Some of active methods in use currently include; positive feedback for active and reactive power loops in governor and excitation system of synchronous DGs, injection of a negative sequence of current through the interface Voltage-Sourced Converter (VSC) [6], Sandia frequency and voltage shift methods and harmonic amplification factor, which is based on the voltage change at the Point of Common Coupling (PCC) [7].

Generally, in islanding detection, two major steps are involved. This include the extraction of features from the signal and then classification of the extracted features for islanding dection. This is well explained in the figure below.

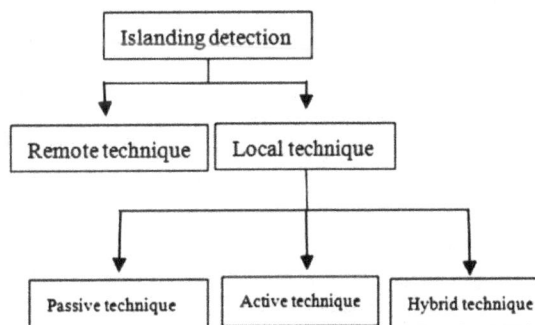

Figure 1. General islanding detection steps.

The rest of this paper is organized as follows; section 3 discusses the theories of fuzzy logic algorithm used in islanding detection, section 4 explains the methodology that is used in this study, the results of the study are discussed in section 5 and finally the conclusions are drawn in section 6.

3. Fuzzy Logic Controller

A fuzzy logic controller is a control algorithm based on several linguistic control rules and it is used to analyze continuous signals. The fuzzy rule base has the ability to handle more uncertainties in the signal being analyzed that fall along the slope of the fuzzy trapezoidal membership function unlike the crisp classifiers like decision tree which have sharp boundaries, and large data base. Thus, the superior approximation capabilities of the fuzzy systems over crisp classifiers help to develop algorithms that meet the real time application with wide range of uncertainties. Hence the fuzzy logic controller can easily and accurately be used in islanding detection for rotating machine based generators. Fuzzy logic system has also an ability to express non-linear input and output parameters into a set of qualitative IF – THEN rules.

Some of the recent applications of fuzzy logic in islanding detection are as follows; in [8], fuzzy logic was introduced from the transformation of decision tree, where the combination of fuzzy membership functions (MFs) and the rule base were used to develop the fuzzy rule base. This technique was easy to implement for online islanding detection and could handle uncertainties such as noise. In [9], however, the band pass filter was used to replace the function of discrete wavelet transform and still worked out pretty well.

Basically a Fuzzy inference system (FIS) is composed of five functional blocks as shown in Figure 2 below. These include fuzzification interface, defuzzification interface, decision making unit, data base and rule base blocks.

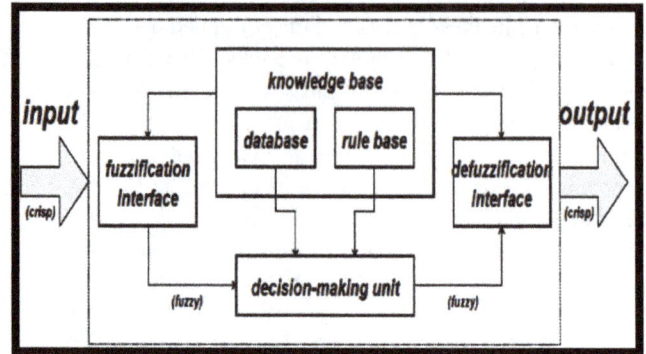

Figure 2. Fuzzy inference system [1].

4. Methodology

The figure 3 below was used for this analysis.
Load data
Nominal ph-ph voltage: 400 vrms, Nominal frequency: 50 Hz, Active power: 500kw, Configuration: Y (grounded)
Syncronous machine (Generator) data
Rotor type: salient pole, Nominal power: 2MVA, Voltage: 400vrms, Frequency: 50hz, Field current: 100A
Grid (3-phase source) data
Phase to phase rms voltage: 25kv, Frequency: 50hz, Internal connection: Yg, 3-phase short circuit level at base voltage: 100mva, X/R ratio: 7

In this system, the following three main features are chosen and measured to be used in the algorithm. These include frequency deviation, rate of change of power (ROCOP) and rate of change of frequency (ROCOF). However, under normal circumstances, there are eleven features that can be measured and used in islanding detection. These include; the frequency deviation, the voltage deviation, rate of change of frequency, rate of change of voltage, rate of change of power, rate of change of frequency over power, the total harmonic distortion of the current, total harmonic distortion of the voltage, the power factor deviation under, the absolute value of the phase voltage times power factor and the gradient of the voltage times power factor. Among all these features, only the three mentioned above are sufficient in islanding detection.

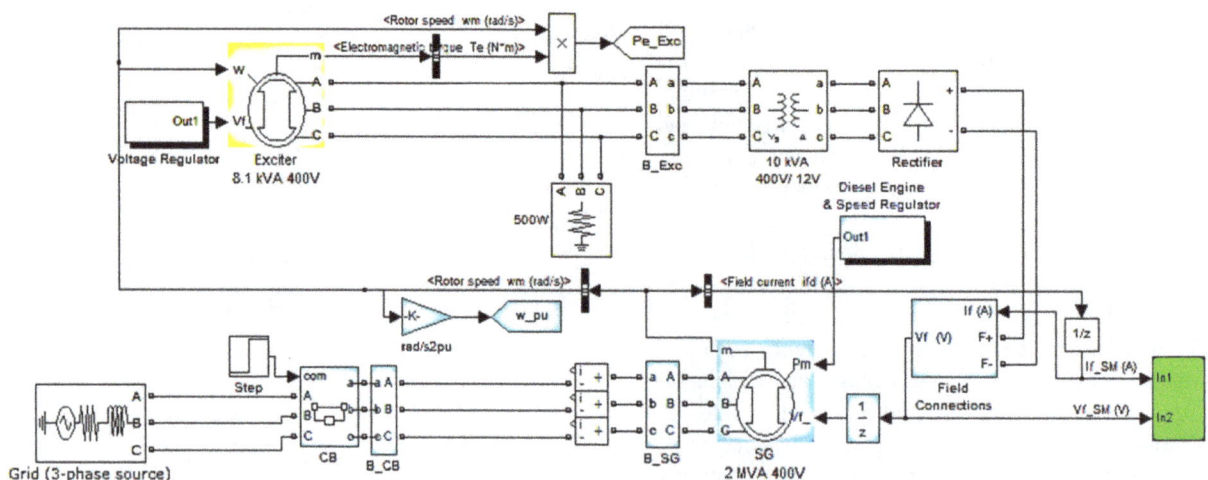

Figure 3. System used for the analysis.

_hom = 0.2507 V
from rotor terminals)

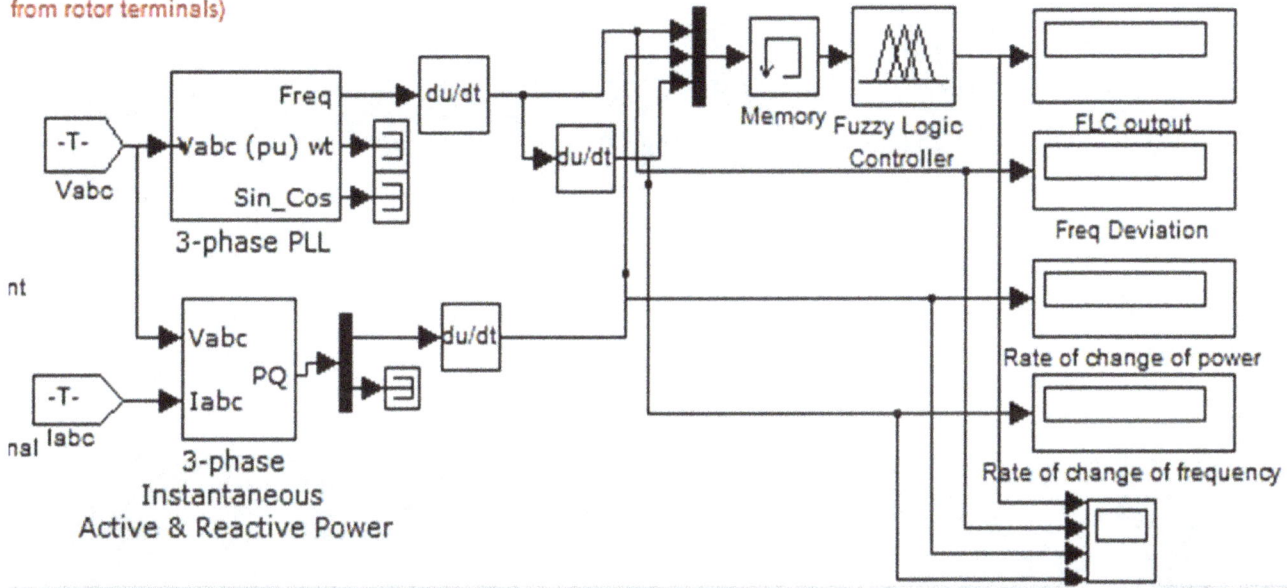

Figure 4. *Fuzzy Inference system for islanding detection.*

The above mentioned features are extracted under different islanding and non-islanding conditions of the network. That is, when the circuit breaker (CB) is both closed and also tripping it to create the condition of islanding of the distributed generator while supplying the bus loads at the PCC. The extracted features are then subjected to fuzzy logic algorithm as shown in the figure below to determine the islanding and no islanding condition.

In this study, Mamdani fuzzification model is used in implementing the rules and centroid method is used in defuzzification of the output as shown in figure 5 below.

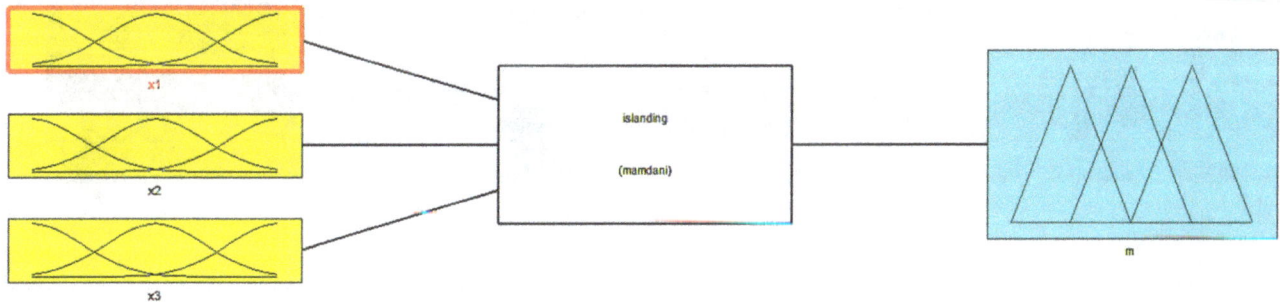

Figure 5. *Fuzzy inference system showing mamdani fuzzification.*

5. Results and Discussion

After setting up the model as shown above and simulating, the results of the analysis are as follows. First, the algorithm was able to distinguish islanding condition from non-islanding condition as shown in figure 6 below.

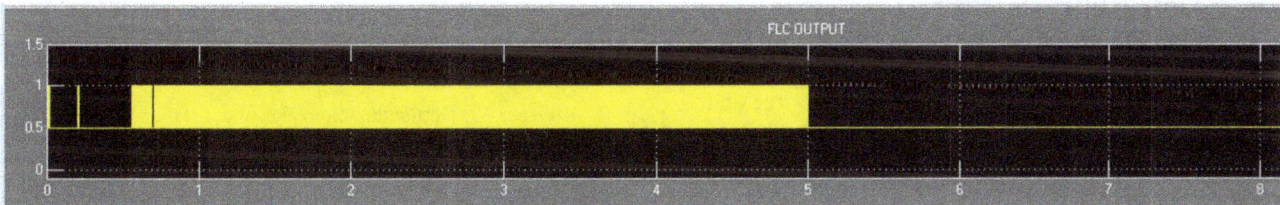

Figure 6. *FLC output.*

Then the frequency measurement was done and the result is as shown in figure below. As it can be seen, the frequency varies from the normal on the 5th second as soon as the CB is opened.

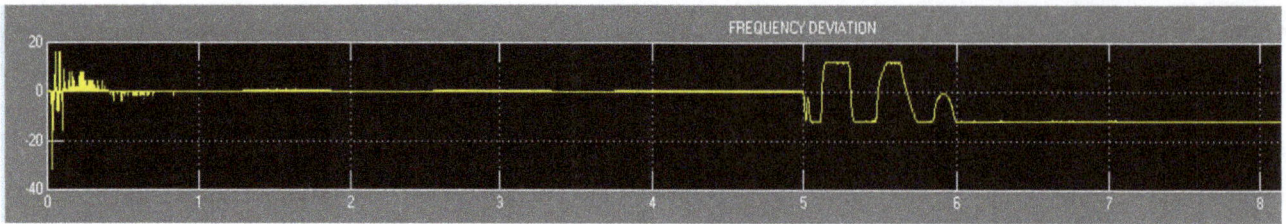

Figure 7. Frequency deviation measurement.

The rate of change of power (ROCOP) was also measured and its result is as shown in figure 8 below. It is noted that there is a sharp change on the 5th second when the CB opens to create an island. Then later return to normal position. This is due to the voltage regulator which restores the system back immediately. ROCOP is normally used to measure the impact the active power variations have on frequency and the voltage of the system. When the DG is operating with mains, the system the impact is usually small. However, this is not the case when the DG is operating in an island.

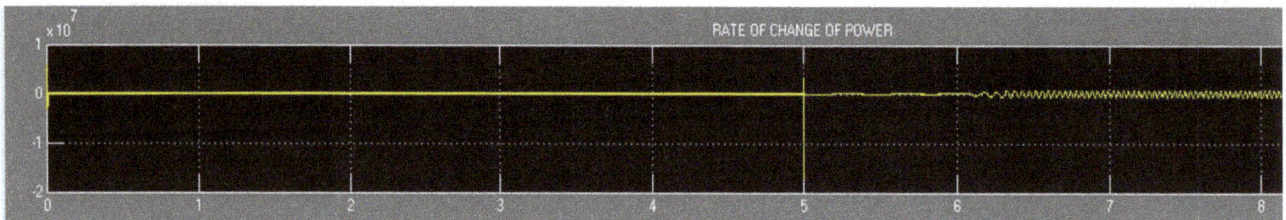

Figure 8. Rate of change of power.

Lastly, the rate of change of frequency was measured and the result was as shown in figure 9 below. Unlike frequency deviation, there is no significant change when the CB is opened to create an island.

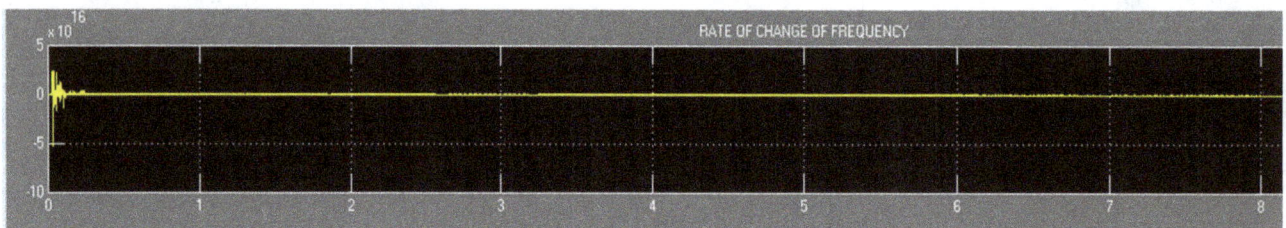

Figure 9. Rate of change of frequency.

6. Conclusion

This proposed method for islanding detection for rotating machine based generators is robust and accurate when implemented in the distribution network. This is because fuzzy logic control helps to improve the interpretability of knowledge-based classifiers through its semantics that provide insight in the classifier structure and decision-making process over crisp classifiers [5]. Another advantage of fuzzy logic is its ability to handle more noise falling along its trapezoidal membership functions and it does not have sharp boundaries like other crisp classifiyers like decision tree and others. This method is therefore suitable in developing relays for real time islanding and non-islanding detection in a large power grids.

Acknowledgement

The authors would like to express the greatest gratitude to the Technical University of Mombasa for the continued support from time to time when required.

References

[1]   S. Noradin and G. Behrooz, "Adaptive neuro-fuzzy inference system (ANFIS) islanding detection based on wind turbine simulator," *International Journal of Physical sciences,* vol. 8, no. 27, pp. 1425-1436, 23 July 2013.

[2]   A. Etxegarai, I. Zamora, P. Eguia and L. Valverde, "Islanding detection of synchronous distributed generators," in *International Conference on Renewable Energies and Power Quality,* spain, 28th to 30th March, 2012.

[3]   B. A. Ozeer, G. Hernandez-Gonzalez and T. H. and EL-Fouly, "GuilleDistributed Generation Analysis Case Study 5 – Investigation of Passive Anti-Islanding Detection for Rotating Machine-based DG Technologies," Varennes Research Centre, Canada, 2012.

[4]   H. Vahedi, R. Noroozian, A. Jalilvad and G. B. Gharehpetian, "A new method for islanding detection of inverter based distributed generation using DC-link voltage control," *IEEE Transaction on power delivery,* vol. 26, no. 2, pp. 1176-1186, 2011.

[5] K. M. Shareef, K. H. Reddy and K. S. Kumar, "Islanding detection in Distributed Generation by using Fuzzy Rule based approach," *Journal of Electrical and Electronics Engineering,* vol. 2, no. 2, pp. 30-36, July-Aug 2012.

[6] B. Bahrani, H. Karimi and R. Iravani, "Non-Detection zone assessment of an active islanding detection method and its experimental evaluation," *IEEE transaction on power delivery,* vol. 26, no. 2, pp. 517-525, 2011.

[7] M. Bakhshi, R. Noroozian and G. B. Gharehpetian, "Islanding detection of synchronous machine based DGs using average frequency based index," *Iranian journal of electrical and eletronic engineering,* vol. 9, no. 2, pp. 94-106, June 2013.

[8] S. Samanta, k. El-arroudi, G. Joós and k. I., "A fuzzy rule-based approach for islanding detection in distributed generation," *IEEE Transaction on Power Delivery,* pp. 1427-33, 2010.

[9] J. Pham, N. Denboer, N. Lidula, G. Member, N. Perera and A. Rajapakse, "Hardware implementation of an islanding detection approach based on current and voltage transients.," *Electrical power and energy conference (EPEC),* pp. 152-157, 2011.

[10] H. Zeineldin, E. El-Saadany, M. Salama, "AImpact of DG Interface Control on Islanding Detection and Nondetective Zones," IEEE Trans.on Power Del, vol. 21, no. 3, pp. 1515 – 1523, July 2006.

[11] Pukar Mahat, Zhe Chen, and Birgitte Bak-Jensen,, "Review on islanding operation of distribution system with distributed generation," International Conference on Electric Utility Deregulation and Restructuring and Power Technologies, Nanjing, China, p. 2743 2748, April 2008.

Permissions

All chapters in this book were first published in IJEPE, by Science Publishing Group; hereby published with permission under the Creative Commons Attribution License or equivalent. Every chapter published in this book has been scrutinized by our experts. Their significance has been extensively debated. The topics covered herein carry significant findings which will fuel the growth of the discipline. They may even be implemented as practical applications or may be referred to as a beginning point for another development.

The contributors of this book come from diverse backgrounds, making this book a truly international effort. This book will bring forth new frontiers with its revolutionizing research information and detailed analysis of the nascent developments around the world.

We would like to thank all the contributing authors for lending their expertise to make the book truly unique. They have played a crucial role in the development of this book. Without their invaluable contributions this book wouldn't have been possible. They have made vital efforts to compile up to date information on the varied aspects of this subject to make this book a valuable addition to the collection of many professionals and students.

This book was conceptualized with the vision of imparting up-to-date information and advanced data in this field. To ensure the same, a matchless editorial board was set up. Every individual on the board went through rigorous rounds of assessment to prove their worth. After which they invested a large part of their time researching and compiling the most relevant data for our readers.

The editorial board has been involved in producing this book since its inception. They have spent rigorous hours researching and exploring the diverse topics which have resulted in the successful publishing of this book. They have passed on their knowledge of decades through this book. To expedite this challenging task, the publisher supported the team at every step. A small team of assistant editors was also appointed to further simplify the editing procedure and attain best results for the readers.

Apart from the editorial board, the designing team has also invested a significant amount of their time in understanding the subject and creating the most relevant covers. They scrutinized every image to scout for the most suitable representation of the subject and create an appropriate cover for the book.

The publishing team has been an ardent support to the editorial, designing and production team. Their endless efforts to recruit the best for this project, has resulted in the accomplishment of this book. They are a veteran in the field of academics and their pool of knowledge is as vast as their experience in printing. Their expertise and guidance has proved useful at every step. Their uncompromising quality standards have made this book an exceptional effort. Their encouragement from time to time has been an inspiration for everyone.

The publisher and the editorial board hope that this book will prove to be a valuable piece of knowledge for researchers, students, practitioners and scholars across the globe.

List of Contributors

Lucas Ongondo Mogaka and Michael Juma Saulo
Electrical and Electronics Department, Technical University of Mombasa, Mombasa, Kenya

D. K. Murage
Electrical and Electronics Department, JKUAT, Nairobi, Kenya
hazmy and Badr Habeebullah
Mechanical Engineering, King Abdulaziz University, Jeddah, Kingdom of Saudi Arabia
Rahim Jassim
Saudi Electric Services Polytechnic (SESP), Baish, Jazan Province, Kingdom of Saudi Arabia

Huiru Zhao, Yaowen Fan and Nana Li
School of Economics and Management, North China Electric Power University, Beijing, China

Fuqiang Li and Yuou Hu
North China Grid Company Limited, Beijing, China

Miguel Meque Uamusse
Department of Water Resource, Lund University, Lund, Sweden
Faculdade de Engenharia, Universidade Eduardo Mondlane, Maputo, Mozambique

Kenneth Person
Department of Water Resource, Lund University, Lund, Sweden

Petro Ndalila
Department of Mechanical Engineering, Mbeya University of Science and Technology, Mbea, Tanzania

Alberto JúlioTsamba
Faculdade de Engenharia, Universidade Eduardo Mondlane, Maputo, Mozambique

Frede de Oliveira Carvalho
Departamento de Engenharia Química, Universidade Federal de Alagoas, Brazil

Dexin Xie and Dongyang Wu
School of Electrical Engineering, Shenyang University of Technology, Shenyang, China

Jian Wang and Zhanxin Zhu
TBEA Shenyang Transformer Co., Ltd., Shenyang, China

Mahasidha R. Birajdar and Sandip A. Kale
Mechanical Engineering Department, Trinity College of Engineering and Research, Pune, India

P. Sivakumar and M. Gunasekaran
Department of Physics, Periyar E.V.R. College, Tiruchirappalli, India

Sashank Srinivasan
Department of Mechanical Engineering, Birla Institute of Technology and Science-Pilani, Hyderabad Campus, India

Vikranth Kumar Surasani
Department of Chemical Engineering, Birla Institute of Technology and Science-Pilani, Hyderabad Campus, India

Zhiguang Cheng, Yana Fan and Tao Liu
Institute of Power Transmission and Transformation Technology, Baobian Electric Co., Ltd, Baoding, China

Behzad Forghani
Infolytica Corporation, Place du Parc, Montreal, Canada

Zhigang Zhao
School of Electrical Engineering, Hebei University of Technology, Tianjin, China

Sheetal Agrahari and Suresh Balpande
Department of Electronics Engineering, Shri Ramdeobaba College of Engineering and Management Nagpur, Maharashtra, India

Samomssa Inna and Jiokap Nono Yvette
University Institute of Technology (IUT) of the University of Ngaoundere, Department of Chemical Engineering and Environment, Ngaoundere, Cameroon

Kamga Richard
National Advanced School of Agro-Industrial Sciences (ENSAI) of the University of Ngaoundere, Department of Applied Chemistry, Ngaoundere, Cameroon

Mohammed Y. Suliman
Department of Electrical Engineering, Technical College, Mosul, Iraq

Sameer Sadoon Al-Juboori
Department of Electronic and Control Engineering, Technical College, Kirkuk, Iraq

Tao Cui and Weiying Zheng
National Center for Mathematics and Interdisciplinary Sciences, State Key Laboratory of Scientific and Engineering Computing, Institute of Computational Mathematics and Scientific/Engineering Computing, Academy of Mathematics and Systems Science, Chinese Academy of Sciences, Beijing, China

Xue Jiang
Department of Mathematics, Beijing University of Posts and Telecommunications, Beijing, China

Lin Li, Keke Liu, Xiaoying Zhang and Ning Zhang
Department of Electrical and Electronics Engineering, North China Electric Power University, Beijing, China

Yanli Zhang, Yuandi Wang, Dianhai Zhang, Ziyan Ren and Dexin Xie
School of Electrical Engineering, Shenyang University of Technology, Shenyang, China

Ali Razmjoo
Department of Energy Systems Engineering, Faculty of Engineering, Islamic Azad University-South Tehran Branch, Iran

Mohammad Ghadim
Department of Mechanical Engineering, Islamic Azad University, roudehen branch Tehran, Iran

Mehrzad Shams
Mechanical Engineering Faculty, Energy Conversion Group, K. N Toosi University of technology, Tehran, Iran

Hoseyn Shirmohammadi
Department of industrial Engineering, West Tehran Branch, Islamic Azad University, Tehran, Iran

Li Wan-peng, Sun Ya-qiao, Yao Meng and Dou Lin
College of Environmental Science and Engineering, Chang'an University, Xi'an, Shanxi, China

Halidini Sarakikya and Iddi Ibrahim
Department of Electrical Engineering, Arusha Technical College, Arusha, Tanzania

Jeremiah Kiplagat
Department of Energy Engineering, Kenyatta University, Nairobi, Kenya

Naga Sarada Somanchi, Anjaneya Prasad B, Ravi Gugulothu, Ravi Kumar Nagula and Sai Phanindra Dinesh K
Department of Mechanical Engineering, JNTUH College of Engineering, Kukatpally, Hyderabad, Telangana State, India

Zaiyi Liao
Dept of Architectural Science, Ryerson University, Toronto, Canada

Wei Xuan
Dept of Architecture, Hefei University of Technology, Hefei, China

Yang Liu
China State Grid Smart Grid Research Institute, Beijing, China
Institute of Power Transmission and Transformation Technology, Baobian Electric Co., Ltd, Baoding, China

Guang Ma
China State Grid Smart Grid Research Institute, Beijing, China

Tao Liu, Lanrong Liu and Chongyou Jing
Institute of Power Transmission and Transformation Technology, Baobian Electric Co., Ltd, Baoding, China

Xiaojun Zhao, Dawei Guan, Fanhui Meng and Yuting Zhong
Department of Electrical Engineering, North China Electric Power University, Baoding, China

Shubhangi S. Nigade and Abhimanyu K. Chandgude
Department of Mechanical Engineering, KJEI's Trinity college of Engineering and Research Pune, Maharashtra, India

Sangram D. Jadhav
Department of Mechanical Engineering, Government of Maharashtra Dr. B. A. Technological University Mangaon, Maharashtra, India

Nikhil C. Raikar and Sandip A. Kale
Mechanical Engineering Department, Trinity College of Engineering and Research, Pune, India

Yong Du, Wanqun Zheng and Jingjun Zhang
Department of Electrical Engineering, Hebei University of Engineering, Handan, Hebei Province, China

Yuting Zhong, Dawei Guan and Fanhui Meng
Department of Electrical Engineering, North China Electric Power University, Baoding, China

S. Yasmeena and G. Tulasiram Das
EEE Department Vishnu Institute of Technology, Bhimvaram, A. P., India

Aaron Yaw Ahali
Department of Economics and International Studies, the University of Buckingham, UK

Alaa Abdulwahhab Azeez Baker and Maamon Phadhil Yasen Al-Kababji
Department of Electrical Power and Machines Engineering, Engineering College, Mosul, Iraq

Sameer Saadoon Al-Juboori
Department of Electronics & Control Engineering, Kirkuk Technical College, Kirkuk, Iraq

Petter Pilesjo
GIS Centre, College of Science, Lund University, Lund, Sweden

Amin Shokri Gazafroudi
EE Department, Imam Khomeini International University, Qazvin, Iran

Index

www.ingramcontent.com/pod-product-compliance
Lightning Source LLC
Chambersburg PA
CBHW080515200326
41458CB00012B/4219